Design Solutions for nZEB Retrofit Buildings

Elzbieta Rynska
Warsaw University of Technology, Poland

Urszula Kozminska
Warsaw University of Technology, Poland

Kinga Zinowiec-Cieplik
Warsaw University of Technology, Poland

Joanna Rucinska
Warsaw University of Technology, Poland

Barbara Szybinska-Matusiak
Norwegian University of Science and Technology, Norway

A volume in the Advances in Civil and Industrial Engineering (ACIE) Book Series

IGI Global
DISSEMINATOR OF KNOWLEDGE

Published in the United States of America by
IGI Global
Engineering Science Reference (an imprint of IGI Global)
701 E. Chocolate Avenue
Hershey PA, USA 17033
Tel: 717-533-8845
Fax: 717-533-8661
E-mail: cust@igi-global.com
Web site: http://www.igi-global.com

Library of Congress Cataloging-in-Publication Data

Names: Rynska, Elzbieta, 1962- editor.
Title: Design solutions for nZEB retrofit buildings / Elzbieta Rynska, Urszula Kozminska, Kinga Zinowiec-Cieplik, Joanna Rucinska, and Barbara Szybinska-Matusiak, editors.
Description: Hershey PA : Engineering Science Reference, [2018] | Includes bibliographical references and index.
Identifiers: LCCN 2017028148| ISBN 9781522541059 (hardcover) | ISBN 9781522541066 (ebook)
Subjects: LCSH: Architecture and energy conservation. | Buildings--Energy conservation. | Buildings--Retrofitting.
Classification: LCC NA2542.3 .D47 2018 | DDC 720/.472--dc23 LC record available at https://lccn.loc.gov/2017028148

This book is published in the IGI Global book series Advances in Civil and Industrial Engineering (ACIE) (ISSN: 2326-6139; eISSN: 2326-6155)

British Cataloguing in Publication Data
A Cataloguing in Publication record for this book is available from the British Library.

Advances in Civil and Industrial Engineering (ACIE) Book Series

Ioan Constantin Dima
University Valahia of Târgovişte, Romania

ISSN:2326-6139
EISSN:2326-6155

Mission

Private and public sector infrastructures begin to age, or require change in the face of developing technologies, the fields of civil and industrial engineering have become increasingly important as a method to mitigate and manage these changes. As governments and the public at large begin to grapple with climate change and growing populations, civil engineering has become more interdisciplinary and the need for publications that discuss the rapid changes and advancements in the field have become more in-demand. Additionally, private corporations and companies are facing similar changes and challenges, with the pressure for new and innovative methods being placed on those involved in industrial engineering.

The **Advances in Civil and Industrial Engineering (ACIE) Book Series** aims to present research and methodology that will provide solutions and discussions to meet such needs. The latest methodologies, applications, tools, and analysis will be published through the books included in **ACIE** in order to keep the available research in civil and industrial engineering as current and timely as possible.

Coverage

- Productivity
- Operations research
- Hydraulic Engineering
- Transportation Engineering
- Optimization Techniques
- Urban Engineering
- Quality Engineering
- Production Planning and Control
- Ergonomics
- Structural Engineering

IGI Global is currently accepting manuscripts for publication within this series. To submit a proposal for a volume in this series, please contact our Acquisition Editors at Acquisitions@igi-global.com or visit: http://www.igi-global.com/publish/.

The Advances in Civil and Industrial Engineering (ACIE) Book Series (ISSN 2326-6139) is published by IGI Global, 701 E. Chocolate Avenue, Hershey, PA 17033-1240, USA, www.igi-global.com. This series is composed of titles available for purchase individually; each title is edited to be contextually exclusive from any other title within the series. For pricing and ordering information please visit http://www.igi-global.com/book-series/advances-civil-industrial-engineering/73673. Postmaster: Send all address changes to above address.

Titles in this Series

For a list of additional titles in this series, please visit: www.igi-global.com/book-series

Dynamic Stability of Hydraulic Gates and Engineering for Flood Prevention
Noriaki Ishii (Osaka Electro-Communication University, Japan) Keiko Anami (Osaka Electro-Communication University, Japan) and Charles W. Knisely (Bucknell University, USA)
Engineering Science Reference • copyright 2018 • 660pp • H/C (ISBN: 9781522530794) • US $245.00 (our price)

Handbook of Research on Trends and Digital Advances in Engineering Geology
Nurcihan Ceryan (Balikesir University, Turkey)
Engineering Science Reference • copyright 2018 • 765pp • H/C (ISBN: 9781522527091) • US $345.00 (our price)

Intelligent Vehicles and Materials Transportation in the Manufacturing Sector Emerging Research and Opportunities
Susmita Bandyopadhyay (University of Burdwan, India)
Engineering Science Reference • copyright 2018 • 230pp • H/C (ISBN: 9781522530640) • US $155.00 (our price)

Recent Advances in Applied Thermal Imaging for Industrial Applications
V. Santhi (VIT University, India)
Engineering Science Reference • copyright 2017 • 306pp • H/C (ISBN: 9781522524236) • US $205.00 (our price)

Performance-Based Seismic Design of Concrete Structures and Infrastructures
Vagelis Plevris (Oslo and Akershus University College of Applied Sciences, Norway) Georgia Kremmyda (University of Warwick, UK) and Yasin Fahjan (Gebze Technical University, Turkey)
Engineering Science Reference • copyright 2017 • 320pp • H/C (ISBN: 9781522520894) • US $205.00 (our price)

Engineering Tools and Solutions for Sustainable Transportation Planning
Hermann Knoflacher (Vienna University of Technology, Austria) and Ebru V. Ocalir-Akunal (Gazi University, Turkey)
Engineering Science Reference • copyright 2017 • 374pp • H/C (ISBN: 9781522521167) • US $210.00 (our price)

Design Solutions and Innovations in Temporary Structures
Robert Beale (Independent Researcher, UK) and João André (Portuguese National Laboratory for Civil Engineering, Portugal)
Engineering Science Reference • copyright 2017 • 501pp • H/C (ISBN: 9781522521990) • US $200.00 (our price)

Modeling and Simulation Techniques in Structural Engineering
Pijush Samui (National Institute of Technology Patna, India) Subrata Chakraborty (Indian Institute of Engineering Science and Technology (IIEST), Shibpur, India) and Dookie Kim (Kunsan National University, South Korea)
Engineering Science Reference • copyright 2017 • 524pp • H/C (ISBN: 9781522505884) • US $220.00 (our price)

IGI Global
DISSEMINATOR OF KNOWLEDGE

701 East Chocolate Avenue, Hershey, PA 17033, USA
Tel: 717-533-8845 x100 • Fax: 717-533-8661
E-Mail: cust@igi-global.com • www.igi-global.com

Editorial Advisory Board

List of Reviewers

Table of Contents

Detailed Table of Contents

The aim of the chapter is to present existing definitions of building renovation to nearly zero energy building (nZEB). The EU buildings stock has low energy efficiency and is responsible for the biggest energy consumption. This chapter describes first of all the legal background in EU and general definition of nZEB renovation. In the next part, country-specific definitions are cited. Most of them are setting requirements for primary energy demand. The example of Poland is used to show the possible process of defining the nZEB renovation standard.

Typical methods of improving energy efficiency of buildings to the nZEB standard, such as better external envelope insulation, using better windows, or introducing a proper ventilation system, may prove inadequate. Often, it is necessary to apply unconventional systems and renewable energy sources. Each case has to be analyzed considering the location, building technology, usage of the building, and even conservation officer's requirements. To verify the effectiveness of standard activities in various climates, the following locations in Europe were chosen: Athens, Stockholm, and Warsaw. For evaluation, two methods were applied: the monthly balance method and the simplified hourly method 5R1C.

This chapter describes a general issue of selecting renewable energy sources (RES) and technical systems. To achieve the nearly zero-energy building (nZEB) standard, application of an RES (e.g., solar, wind, geothermal, hydropower, and biomass energy) is necessary. Each type of RES has specific characteristics and can be used to produce electricity and/or heat in certain systems. A short review of various systems using renewable energy sources is presented. To find the required and satisfactory solution that guaranties meeting the nZEB standard, an analysis must be carried out considering a number of aspects: local availability, structure and time-dependence of energy demand, building construction, economic conditions, legal regulations, and specific requirements. Finally, two examples of modernisation towards the nZEB standard are included.

For designers proceeding with work in accordance with nZEB, management strategy means that participants of different disciplines have to accept that surrounding environment parameters are in constant interaction with design. This further points out the necessity to understand the correlations taking place between local and existing environments, when buildings should be fitted out with systems working in interaction between the local and general biosphere parameters. Hence, within the construction business, such development means integrated proceedings in all design phases and construction sphere itself, integration of the artificial systems allowing for the buildings function and parameters characteristic to the local surroundings. This condition also applies to nZEB buildings, except that the parameters are more limited to passive and active solutions used for achieving effective energy choices, while maintaining required user comfort parameters and environmental balance. This chapter is dedicated to management procedures for nZEB investments from a Polish perspective.

This chapter describes the concept of an expert system supporting improvements in a building's energy efficiency. An expert system is a computer program or group of programs that facilitates using knowledge and decision making. The main elements of an expert system are the knowledgebase and a conclusion system. The knowledgebase collects information about a particular area, which was beforehand written in using particular rules. The conclusion system uses the knowledgebase and user input facts to generate conclusions or prove the user's hypothesis. The proposed expert system contains information on building technology, modernization and installation activities, and the values for financial, environmental, and technical indicators characterising these technologies. The user of an expert system defines the problem to be solved using questions. Using the knowledgebase, the system will present the optimal solution or information that the technologies in the existing knowledgebase will be applicable in the case defined by the user.

In the integrated building modelling, all aspects have the same importance. It means that the architect, construction, and HVAC engineers create a building with interactions. BIM guarantees permanent information about important indicators generated parallel during the whole design process. Sophisticated simulation gives opportunity to observe important factors, for example, thermal comfort, ventilation

human population, our world is undergoing rapid urban development. This state overlaps with climate changes and growing scarcity of resources, which has a high impact on the ongoing transformation of our built environment. Many of those issues are mirrored by European legislation, especially in Energy Performance of Buildings Directive, which makes nearly zero-energy buildings a standard by 2020. Many of the technologies are already available. nZEB renovation process will be a challenge for parties involved in the design and construction process.

This chapter is dedicated to the modernisation process of existing buildings aiming to achieve the nearly zero-energy standard. The process is described from the designer's perspective. Related issues, requirements, constraints, design options, and local determinants are analysed, and optimal architectural solutions for selected case studies are also presented. The analysis is based on the KodnZeb project, which included the modernisation of two existing buildings, located in Warsaw (the Faculty of Building Services, Hydro- and Environmental Engineering and Student Housing Muszelka), which differ in architectural features function, location, and needs. Thus, two diverse approaches to the modernisation process are examined. The chapter concludes with general guidelines and recommendations for similar architectural projects.

This chapter is a case study for the energy retrofit of an existing single-family residential building. The main assumption of the project was creating a model example for an energy retrofit with the aim of achieving the nZEB standard in existing residential building. The discussed building was built between the 1960s and the 1970s. The building was built using mixed technologies. The flooring on the ground floor was replaced; the foundation, external walls, and roof were thermally insulated. The windows and doors were replaced with higher parameter ones. Moreover, a modern biomass boiler was installed in the building along with the installation of a mechanical bidirectional ventilation unit with a heat recovery. Before the renovation, the building used about 133.4 GJ final energy for heating annually. After the renovation, the building uses about 8.89 GJ annually. The author describes all the stages of the renovation, the technical solutions, the calculations of economic and environmental benefits of the conducted renovations.

One of the most interesting ways of returning nature to the city is green architecture, which embraces hybrid buildings at the base of the coexistence of the natural and building worlds. By definition, green architecture does not have to be green at all. Nevertheless, it takes on a number of hybrid forms integrated with vegetation. Built and natural form in the composition is inseparably "tangled" – one follows other to form a cohesive whole. Green architecture becomes part of a larger green infrastructure system. The idea of green infrastructure leads to passage from passive protection, to the active-present in every aspect of human life. It is a development tool respecting the laws of nature. Architecture enters the world of

Foreword

It is important to take immediate action to mitigate climate change. We currently consume Earth's resources far beyond any sustainable levels and emit a vast amount of CO_2 every year. In the light of COP21, a global commitment to limit global warming well below 2 degrees, aiming at 1.5 degrees was agreed upon. The EU issued ambitious climate and energy targets for 2020 and 2030. Given that, at least two-thirds of today's buildings will still be standing in 2050 and considering their vast energy consumption, a longer term vision is necessary to align with the challenges ahead. The current energy performance of buildings has a strong impact on the achievable energy savings and potential cost of building renovation. Thus, the higher the current efficiency of the building stock, the more expensive further improvement and the stronger the political incentives will have to be.

Since sustainability of the European society and economy should be based on renewable energy and high resource efficiency. For the building sector, this implies the large scale deployment of nearly zero-energy buildings (nZEB). European legislation (recast Energy Performance of Buildings Directive) makes nZEBs a standard for new buildings by 2020. As concrete numeric thresholds or ranges are not defined in the EPBD, these requirements allow room for interpretation and thus give Member States the liberty to define their nZEB in a very flexible way taking into account their country specific climate conditions, primary energy factors, ambition levels, calculation methodologies and building traditions. This is also the main reason why existing nZEB definitions differ significantly from a country to country. It is thus a challenging task to find a common denominator for defining an nZEB on a European scale.

Article 7 of the EPBD (n.d.). states that:

Member States shall take the necessary measures to ensure that when buildings undergo major renovation, the energy performance of the building or the renovated part thereof is upgraded in order to meet minimum energy performance requirements set in accordance with Article 4 so far as this is technically, functionally and economically feasible.

Major renovations, as defined in Article 2 of the EPBD recast, include the renovation of a building where:

1. The total cost of the renovation relating to the building envelope or the technical building systems is higher than 25% of the value of the building, excluding the value of the land upon which the building is situated; or
2. More than 25% of the surface of the building envelope undergoes renovation.

Awareness among investors and citizens about the multiple benefits and feasibility of nZEBs (e.g., cost savings over the life-time, on-site electricity generation, increase of internal comfort) is another nearly insurmountable barrier to nZEB market penetration. All EU Member States must raise awareness about the benefits of investing in nZEB. Effective communication is a key to increase interest in nZEB and transform the market. Most Member States must become better at guiding investors and the construction sector through the legislative and financial maze of regulations, schemes and subsidies. In many cases, the required legislation is in place but its complexity can be a costly barrier, hampering the nZEB transition.

As one of the Partners in Nearly Zero-Energy Building Strategy 2020 (ZEBRA2020), where eight Partners monitored the market uptake of nZEBs across Europe (creating an observatory) and thereby generated data and evidence for policy evaluation and optimization, I am a keen observer of any other projects and developments which mirror and further develop the ZEBRA2020 findings and recommendations. This is important as the EPBD specifically requires Member States to define what constitutes a nearly Zero-Energy Building, whereas in Poland, this definition still has to be approved.

This book contains several values. First of all it contains interdisciplinary international points of view which at the end promote demonstration projects and energy efficiency procedures, exemplifying the benefits and viability of highly performing buildings. The nZEB brand is foreseen as part of a positive sustainability narrative presented in such a way as to attract representatives and students of different disciplines including Architects, HVAC and Environmental Engineers, Landscape Architects and many other specialists who hope that part of their individual research can be used for the benefit of communities and a more sustainable environment. Secondly it can be used as a positive knowledge benchmark proving that similar attitudes and scientific research is conducted in various countries. Thirdly – this particular book can play the role of a dissemination knowledge source to be used by local communities.

Andrzej Wiszniewski
National Energy Conservation Agency (NAPE), Poland

Andrzej Wiszniewski *is the Chairman of the National Energy Conservation Agency Management Board, Associate Professor at the Faculty of Building Services, Hydro and Environmental Engineering, Warsaw University of Technology, Poland. Co-founder and Member of the Association of Energy Auditors Management Board and the Warsaw Section of the Association of Energy Engineers. Leader of various scientific research teams, expert and organizer of post diploma studies and education programs dealing with effective use of energy sources and sustainable energy supplying systems.*

REFERENCES

EPBD. (n.d.). Retrieved from https://issuu.com/glassforeurope/docs/the_energy_performance_of_buildings/14

Preface

A lot has been said about building investment processes and design strategies, as well as environmentally friendly buildings and effective energy strategies. Many countries are far in the front line of advanced development, others are still struggling with the challenges which they are facing (Kurnitski et al., 2013, 2014; Lindkvist et al., 2014; Jagemar et al., 2011).

When considering our urbanized space, it should be noted that most of structures which have been built in the past 30 years under certain climatic conditions and according to the then – contemporary technical knowledge, will be present in our surroundings for at least next 30 years and will have to withstand adverse changing climatic impacts as well stricter effective energy expectations (Aelenei et al., 2015). This implies that most of our efforts should be placed on the modernization of the building substance constructed in various techniques and standards. Due to geographic distribution and diverse user expectations, as well as alternative approaches, it is evident that countries should develop individual strategies even if based on the transformed knowledge and know-how of more developed countries.

Even brief research run on the possible alternative solutions used for modernization of the building substance proves that nZEB might be one of the major paths which should be followed. Hence, the most important issue is that when designing according to nZEB strategies, all designers participating in the process have to understand the conditions of each particular development as well as use surrounding environmental parameters as complementary interactive influence on design solutions provided (Dalla Mora et al., 2015), (Kantola et al., 2016). On the nZEB level the parameters are usually discussed from the energy effective point of view and user comfort needs. In case of retrofit buildings, existing structures are even more diverse and any transfer procedure of existing solutions between countries is likely to be initiated with the achievement of new solutions before high standard buildings will appear (Hamdy et al., 2013; O'Riain et al., 2016; Beccali et al., 2013; Dalla Mora et al., 2015).

Hence, the target audience should be found within all disciplines connected with the design, construction and management process of buildings and their surroundings. Its contents should be of interest not only to the representatives of scientific disciples, but also contractors, students, users and all of those people who wish to know more in order to join in with the sustainable development on an everyday basis. The other added value of this book is that at the end, due to the interest of many people it contains international research covering a wide scope of scientific issues.

THE CHALLENGES

This publication contains mainly the outcomes and achievements of a Project "Thermo-modernization of two chosen Public Buildings according to nZEB Standards", financed by the Program Operator PL04 "Energy efficiency and promotion of alternative energy sources" with funds drawn from the Bilateral Cooperation Financial Mechanism of European Economic Area 2009-2014.

Nevertheless, the editors of this book thought that introduction of different disciplines and therefore approaches - which included impact of green areas, use of wind for ventilation purposes in urbanized complexes, as well as management issues and various nZEB or energy efficient approaches, originating from different countries would allow producing a book which can be used as a knowledge benchmark, with a far wider reading impact. This in fact proved to be the major challenge, as a Landscape Architect perceives the build area from a different perspective than a Designer. At the same time Environmental Engineer creates emphasis on different solutions than an Architect. It proved that in end – regardless of the approach – we all would like to prove a synergic value to our surroundings. As editors of this book we hope we have managed to do point out that interdisciplinary approach is the only possible solution when dealing with such complex issue as nZEB buildings. Possible solutions of those challenges appear to be in accordance with findings in various countries not only across Europe (Zeiler, 2010; Kantola et al., 2016; Christoforidis et al., 2016); Karlessi et al., 2017).

ORGANIZATION OF THE BOOK

The book is organized into 14 chapters, starting with a chapter presenting existing definitions of building renovation to Nearly Zero Energy Building (nZEB). The EU buildings stock has low energy efficiency and is responsible for the biggest energy consumption. This chapter describes first of all the legal background in the EU and a general definition of nZEB renovation. Then it continues to the Polish country specific definitions. Most of them are setting requirements for primary energy demand. The example of Poland was used to show the possible process of the defining the nZEB renovation standard.

The second chapter delves on the selection of design methods in the modernization process of buildings to the nZEB Standard. The author points out that the typical methods of improving energy efficiency of buildings to the nZEB standard, such as better external envelope insulation, using better windows or introducing a proper ventilation system, may prove to be inadequate. Often, it is necessary to apply unconventional systems and renewable energy sources. Each case has to be analyzed individually, considering the location, building technology, age and technical standard of the building, and sometimes include historic conservation requirements. The author chose three characteristic European locations verify the effectiveness of standard activities in various climates based on two alternative methods - the monthly balance method and the simplified hourly method 5R1C.

Chapter 3 describes a general issue of renewable energy sources (RES) and technical systems selection. To achieve the nearly zero-energy building (nZEB) standard, application of an RES (e.g. solar, wind, geothermal, hydropower, and biomass energy) is necessary. Each type of RES has specific characteristics and can be used to produce electricity and/or heat in certain systems. A short review of various systems using renewable energy sources is presented within the text followed by a choice analysis of required and satisfactory solutions guaranteeing the nZEB standard. This analysis must consider a number of aspects: local availability, type of structure and time-dependence of energy demand, building construction process, local economic conditions, legal regulations, and other specific requirements. Two examples of modernization to the nZEB standard conclude the chapter contents.

The fourth chapter considers management issues within the whole complex building investment process set on the canvas of sustainable development and its conditions. It evolves around the interdisciplinary design process, pointing out that the main issue to be covered is proceeding with work in accordance with nZEB, management strategy. This means that participants of different disciplines have to accept that surrounding environment parameters are in constant interaction with design. This further emphases the necessity to understand the correlations taking place between local and existing environments, when buildings should be fitted out with systems working in interaction between the local and general biosphere parameters. Hence, within the construction business, such development means integrated proceedings in all design phases and construction sphere itself, integration of the artificial systems allowing for the buildings function and parameters characteristic to the local surroundings. This condition also applies to nZEB buildings, except that the parameters are more limited to passive and active solutions used for achieving effective energy choices, while maintaining required user comfort parameters and environmental balance. This Chapter will be dedicated towards management procedures for nZEB investments from the Polish perspective.

The next chapter describes the concept of an expert system supporting improvements in a building's Energy efficiency. An expert system is a computer program or group of programs, which facilitates using knowledge and decision-making. The main elements of an expert system are the knowledgebase and a conclusion system. The knowledgebase collects information about a particular area, which was written beforehand and contains the requirement to use particular rules. The system uses the knowledgebase and user input facts to generate conclusions or prove the user's hypothesis. The proposed expert system contains information on building technology, modernization and installation activities, and the values for financial, environmental, and technical indicators characterizing these technologies. The user of the expert system defines each problem which will be solved through the use of questions. Using the knowledgebase, the system will present the best case solution or information that none of technologies in the existing knowledgebase can be applied in the user defined case.

The following chapter uses energy simulations as a tool in the integrated modeling design process where all aspects have the same importance. This means that the Architect, Structure and HVAC engineers create a building through interactive design. BIM is one of the choices that guarantees permanent information about important indicators generated parallel during the whole design process. Sophisticated simulation gives opportunity to observe important factors for example: thermal comfort, ventilation quality, building's energy performance, operational and investment costs. For an effective design process, careful selection of the appropriate people to form a team is part of providing a high standard process. It is particularly important to have a similar approach to the modernization process of buildings. Without careful analysis of the initial Concept solutions and interdisciplinary cooperation at various design and construction stages, it is not possible to make a complex retrofit of a building. The aim is not only reducing of the energy use, but also maintaining the desired user thermal comfort.

Chapter 7 contains practical guidelines for improved process for design and retrofitting of energy efficient buildings, with an aim to integrate buildings better with the neighborhood energy grid, also through energy matching process. This chapter describes the role of energy simulations in integrated building retrofitting process, process of selecting adequate technologies for the retrofitting when aiming at nearly zero energy building standard. The author points out, that the feasibility of performing holistic analyses of retrofitting options can be increased through the integration of BIM, well equipped and linked databases and a multi-criteria decision making approach. Multiple Criteria Decision Making methods aid to take into account a number of building energy performance and user preference related criteria and the trade-offs between the different criteria for each retrofitting option. The real life viewpoints and benefits of utilizing the developed methods and processes are discussed, especially from the Eastern European point of view.

Improving the energy quality and indoor environmental quality in retrofit buildings is the theme of Chapter 8. The author points out that energy requirement levels are differently specified in various countries. In many cases different technologies are applied. For this reason, improving the energy quality and indoor environmental quality in retrofit buildings requires a detailed study case. Modification of the building structure, technical systems only, is not sufficient. Therefore it is also necessary to change the user behavior and create uniform zones in terms of usage. Preparation of the modernization of building requires elaboration of a detailed multidisciplinary concept. It is a challenge to take all aspects into account that is why in modernized buildings, the concept design phase often requires more work, time and provision of nonstandard solutions.

Chapter 9 provides chosen case studies of nZEB retrofit buildings. It starts with the enquiry as to why contemporary buildings are named intelligent and the cities are called smart. Whether this name arises from the requirement to achieve effective use of natural resources while maintaining at least current standard of living when faced with global climatic changes and growing scarcity of resources or is there a different reason. The author points out that presently more than often, energy efficient and environmentally friendly urban solutions tend to concentrate on the possibilities of upgrade of already existing buildings forming majority of contemporary urbanized landscape. Due to the increasing number of human population our world is undergoing rapid urban development. This state overlaps with climate changes and growing scarcity of resources which has high impact on the ongoing transformation of our built environment. Many of those issues are mirrored by European legislation, especially in Energy Performance of Buildings Directive which makes nearly Zero-Energy Buildings a standard by 2020. Many of the technologies are already available. Especially nZEB renovation process will be a challenge for parties involved in the design and construction process. This chapter contains both undergoing design and already finalized Norwegian building investments with the main aim as how to make them more energy efficient.

Chapter 10 was prepared by two architects involved in the complex design process being the outcome of the retrofitting of existing buildings to nZEB standard as well as one of the products of an EU grant mentioned in the earlier pages of the Preface. The process is described from the designer's perspective. Related issues, requirements, constraints, design options and local determinants are analyzed and optimal architectural solutions for selected case studies are also presented. The analysis is based on the KodnZeb project, which included the modernization of two existing buildings, located in Warsaw (the Faculty of Building Services, Hydro- and Environmental Engineering and Student Housing Muszelka), which differ in architectural features function, location and needs. Thus, two diverse approaches to the modernization process are examined. Chapter concludes with general guidelines and recommendations for similar architectural projects.

Chapter 11 is a case study for the energy retrofit of an existing single-family residential building located in Poland. A very complex subject especially when lack of general education on energy efficiency issues is a typical state of information in Poland. The main assumption of the project was to create a model energy retrofit example with the aim of achieving the nZEB standard in existing residential building. Discussed building was built between the 60's and the 70's of the former Century, using mixed technologies and does not present a very high workmanship standard. During the modernization process the flooring layers on the ground floor were replaced; the foundations, external walls and roof were insulated. The windows and doors were replaced with higher parameter elements. Moreover, a modern biomass boiler was installed in the building along with the installation of a mechanical bidirectional ventilation unit with heat recovery. Prior to renovation process, the building used approximately 133.4 GJ of final energy for annual heating purposes. After the renovation the building uses approximately 8.89 GJ annually. The author described all the stages of renovation, proposed technical solutions, calculation analysis of economic and environmental benefits provided by the conducted upgrade works.

The last chapters in this book are dedicated to the use of alternative solutions supporting the energy efficiency standard in modernized buildings and small urban areas. Chapter 12 considers the issue of integration of green elements with Architectural Forms. For the author – a Landscape Architect - one of the most interesting ways of return of the Nature to the city is green architecture, especially the type, which embraces hybrid buildings at the base of the coexistence of the natural and building worlds. By definition, green architecture does not have to be green at all - literally. Nevertheless, it takes on a number of hybrid forms integrated with vegetation. Built and natural forms are inseparably "tangled" in the 3D composition - one follows the other to form a cohesive whole. Green architecture becomes part of a larger green infrastructural system. This idea leads to the transition from passive protection, to active - present in every aspect of a human life. It is a development tool respecting the laws of Nature. Architecture enters the world of Nature to an unprecedented level as never before, clings to it through the cooperation of designers with specialist consultants representing environmental sciences representing knowledge from cellular microbiology to macro scale processes in ecosystems. This requires designers to be particularly sensitive to the natural world, understanding and accepting existing rules.

Chapter 13 follows the same idea with the issue of climbing plants impact on buildings and their environment. Benefits and risks of plant growing on the walls are discussed with respect to their durability. Economic benefits from the presence of vines are shown including energy savings for heating and cooling systems. Additionally, it is pointed out that the phytoremediation (cleaning up the environment by plants) properties of vines can form part of the strategy to lower the smog zone impacts. It should be stated that climbing plants can contribute to damage only in places where facades are damaged, plaster cracked, or where plants are incorrectly planted.

The last of the chapter proves that it is possible to shape our urban environment in a way which enhances the use of wind directions. The authors state that the possibilities for energy-efficient, natural ventilation of buildings in an urban environment depend on the airflow around their forms. This Chapter deals with the issue of dependence of air exchange in urban spaces on the building forms used in them and on relative position of these buildings. The authors focused on the problem of air stagnation in dense urban development. This phenomenon increases the energy demand of buildings and if properly solved could be used to enhance the nZEB solutions. The purpose of the study is to present this problem and identify it on the selected example of existing Warsaw settlements. The existing situation was compared with a revised spatial layout. The conclusions relate to the volume features of building arrangements that are exposed to the problem of insufficient ventilation.

The outcomes of this book impacts the field of the nZEB design and contributes to the subjects also covered by other scientific researchers in view of energy and economic targets (Becchio et al., 2015), as well as identification and management of risks involved in the transition to nZEB (Kantola et al., 2016). It also dwells on the possible structure of the definition of nearly zero-energy buildings (Kurnitski et al., 2011). A lot of attention is focused on the interactions between architectural design and other disciplines, the choice and context awareness of such investments (Uribe et al., 2015). Part of the chapters are dedicated to the issue of decision making and support tools which differs between countries (Kang, 2015).

Elżbieta Ryńska
Warsaw University of Technology, Poland

Urszula Koźmińska
Warsaw University of Technology, Poland

Kinga Zinowiec-Cieplik
Warsaw University of Technology, Poland

Joanna Rucińska
Warsaw University of Technology, Poland

Barbara Szybińska-Matusiak
Norwegian University of Science and Technology, Norway

REFERENCES

Aelenei, L., Petran, H., Tarrés, J., Riva, G., Ferreira, A., Camelo, S., & Gonçalves, H. et al. (2015). New Challenge of the Public Buildings: nZEB Findings from IEE RePublic_ZEB Project. *Energy Procedia*, *78*, 2016–2021. doi:10.1016/j.egypro.2015.11.195

Beccali, M., Galatioto, A., Leone, G., & Longo, S. (2013). Is the NZEB benchmarking approach suitable for assessing energy retrofit design? *Applied Mechanics and Materials*, *361-363*, 402–407. doi:10.4028/www.scientific.net/AMM.361-363.402

Becchio, C., Bottero, M., Corgnati, S., & Ghiglione, C. (2015). nZEB design: Challenging between energy and economic targets. *Energy Procedia*, *78*, 2070–2075. doi:10.1016/j.egypro.2015.11.226

Christoforidis, G., Melandri, D., Branciforti, V., Nousdilis, A., Papagiannis, G., & Penalvo, E. (2016). Meeting of energy skills: An educational framework for energy professionals in NZEB. In *EEEIC 2016 - International Conference on Environment and Electrical Engineering*. Institute of Electrical and Electronics Engineers Inc.

Dalla Mora, T., Cappelletti, F., Peron, F., Romagnoni, P., & Bauman, F. (2015). Retrofit of an historical building toward NZEB. *Energy Procedia*, *78*, 1359–1364. doi:10.1016/j.egypro.2015.11.154

Dalla Mora, T., Cappelletti, F., Peron, F., Romagnoni, P., & Bauman, F. (2015). Towards nZEB buildings: A historical building case study. *REHVA Journal*, *52*, 16–20.

Hamdy, M., Hasan, A., & Siren, K. (2013). A multi-stage optimization method for cost-optimal and nearly-zero-energy building solutions in line with the EPBD-recast 2010. *Energy and Building*, *56*, 189–203. doi:10.1016/j.enbuild.2012.08.023

Jagemar, L., Schmidt, M., Allard, F., Heiselberg, P., & Kurnitski, J. (2011). Towards nZEB - Some Examples of National Requirements and Roadmaps. *REHVA Journal*, *48*(3), 14–17.

Kang, H. (2015). Development of a decision support tool for nZEB (Nearly zero emission building) at the early design stage. *Journal of Asian Architecture and Building Engineering*, *14*(2), 475–481. doi:10.3130/jaabe.14.475

Kantola, M., & Saari, A. (2016). Identifying and managing risks involved in the transition to the EU nZEB decree. *Facilities*, *34*(5/6), 339–349. doi:10.1108/F-03-2014-0032

Kantola, M., & Saari, A. (2016). Project delivery systems for nZEB projects. *Facilities*, *34*(1-2), 85–100. doi:10.1108/F-03-2014-0025

Karlessi, T., Kampelis, N., Kolokotsa, D., Santamouris, M., Standardi, L., Isidori, D., & Cristalli, C. (2017). The Concept of Smart and NZEB Buildings and the Integrated Design Approach. *Procedia Engineering*, *180*, 1316–1325. doi:10.1016/j.proeng.2017.04.294

Kurnitski, J., Allard, F., Braham, D., Goeders, G., Heiselberg, P., Jagemar, L., … Railio, J. (2012). How to define nearly net zero energy buildings nZEB. *REHVA Journal*, 6-12.

Kurnitski, J., Buso, T., Corgnati, S. P., Derjanecz, A., & Litiu, A. (2014). nZEB definitions in Europe. *The REHVA European HVAC Journal, 51*(2), 6–9. Retrieved from http://www.rehva.eu/fileadmin/REHVA_Journal/REHVA_Journal_2014/RJ_issue_2/P.06/06_09_nZEB_definitions_in_Europe_RJ1402_WEB.pdf%0Ahttp://www.rehva.eu/publications-and-resources/rehva-journal/2014/022014/nzeb-definitions-in-europe.html

Kurnitski, J., Feldmann, C., Heiselberg, P., Massarella, L., Sartori, I., Voss, K., & Wahlström, Å. (2013). *Present Energy Performance Requirements and nZEB Targets in Some Selected Countries. In Cost Optimal and Nearly Zero-Energy Buildings (nZEB)* (pp. 31–46). Definitions, Calculation Principles and Case Studies. doi:10.1007/978-1-4471-5610-9_3

Lindkvist, C., Karlsson, A., Sørnes, K., & Wyckmans, A. (2014). Barriers and Challenges in nZEB Projects in Sweden and Norway. *Energy Procedia*, *58*, 199–206. doi:10.1016/j.egypro.2014.10.429

Ó'Riain, M., & Harrison, , J. (2016). Cost-optimal passive versus active nZEB. How cost-optimal calculations for retrofit may change nZEB best practice in Ireland. *Architectural Science Review*, *59*(5), 358–369. doi:10.1080/00038628.2016.1184130

Singh, B., Roy, P., Spiess, T., & Venkatesh, B. (2015). *Sustainable Integrated Urban & Energy Planning, the Evolving Electrical Grid and Urban Energy Transition. The Centre for Urban Energy*. Ryerson University.

Uribe, O., Santos, M., Garcia-Alegre, M., & Guinea, D. (2015). A context-awareness architecture for managing thermal energy in an nZEB building. *2015 IEEE 1st International Smart Cities Conference*. Institute of Electrical and Electronics Engineers Inc.

Zeiler, W. (2010). Integral design of NZEB a perquisite for 0 impact buildings. *Proceedings SB10 Euregion: Towards 0-impact buildings and environments*. Academic Press.

Acknowledgment

The editors would like to acknowledge the help of all the people involved in the project and, more specifically, to the authors and reviewers that took part in the review process. Especially the Polish-Norwegian team from Warsaw University of Technology and Norwegian University of Science and Technology, involved in the Project titled "Thermo-modernization of two chosen Public Buildings according to nZEB Standards".

The main theme of the book presents the outcome of this Project, financed by the Program Operator PL04 "Energy efficiency and promotion of alternative energy sources" with funds drawn from Bilateral Cooperation Financial Mechanism of European Economic Area 2009-2014.

Project coordinator: National Fund for Environmental Protection and Water Management, Poland

Project Leader: Warsaw University of Technology, Poland

Without their support, this book would not have become a reality.

First, the editors would like to thank each one of the authors for their contributions. Our sincere gratitude goes to the chapter's authors who contributed their time and expertise to this book.

Second, the editors wish to acknowledge the valuable contributions of the reviewers mentioned earlier in this book regarding the improvement of quality, coherence, and content presentation of chapters. The authors Jacek Borowski, Szymon Firlag, Mari Hukkalainen, Hanna Jedrzejuk, Joanna Klimowicz, Krzysztof Klobut, Michał Pierzchalski, Arkadiusz Weglarz, and Katarzyna Zielonko-Jung also served as referees. We highly appreciate their double task.

Elzbieta D. Rynska
Warsaw University of Technology, Poland

Urszula Kozminska
Warsaw University of Technology, Poland

Kinga Zinowiec-Cieplik
Warsaw University of Technology, Poland

Joanna Rucinska
Warsaw University of Technology, Poland

Barbara Matusiak
Norwegian University of Science and Technology, Norway

Chapter 1
Definition of nZEB Renovation Standard

Szymon Firląg
Warsaw University of Technology, Poland

ABSTRACT

The aim of the chapter is to present existing definitions of building renovation to nearly zero energy building (nZEB). The EU buildings stock has low energy efficiency and is responsible for the biggest energy consumption. This chapter describes first of all the legal background in EU and general definition of nZEB renovation. In the next part, country-specific definitions are cited. Most of them are setting requirements for primary energy demand. The example of Poland is used to show the possible process of defining the nZEB renovation standard.

INTRODUCTION

The existing buildings are responsible for as much as 40% of the EU's final energy requirements, and over one third of carbon dioxide emissions (BPIE, 2014). About 75% of buildings are energy inefficient and, depending on the Member State, only 0.4-1.2% of the stock is renovated each year (EPBD, 2016). Furthermore, Europeans spending 90% of their time indoors, it is therefore important that energy renovation supports healthy indoor climate of buildings (BPIE, 2014).

Acceleration of the cost-effective renovation of existing building can improve energy efficiency of whole the EU. It is at the same time the easiest and fastest way of gaining energy savings. The main objective of this chapter is to present possible requirements for cost-effective renovation standard. It is also shown on the example of Poland what is the renovation market and how the scale of renovation can be improved.

Legal requirements for renovated building are described in the Energy Performance of Buildings Directive (EPBD, 2010) and the Energy Efficiency Directive (EED, 2012). According to the EPBD (Article 9), Member States should also develop policies in order to encourage the renovation of buildings to Nearly Zero Energy Building (nZEB) levels. So far the nZEB requirements established by the European

DOI: 10.4018/978-1-5225-4105-9.ch001

Commission only address directly new buildings to be constructed from 2020 onwards. For the purpose of the EPBD, the following definition of nZEB applies: *'nearly zero-energy building' means a building that has a very high energy performance. The nearly zero or very low amount of energy required should be covered to a very significant extent by energy from renewable sources, including energy from renewable sources produced on-site or nearby.* No mandatory requirements have been introduced for nZEB renovations. According to the Energy Efficiency Directive (EED, 2012), Article 4 Member States should establish long-term strategies for mobilizing investment in the renovation of national buildings stocks. The Article 5 EED sets a 3% annual renovation target for buildings owned and occupied by central government. Also the package "Clean Energy for All Europeans", published by the European Commission in the November 2016, recognizes the important role buildings renovation as a way of fulfilling the "Energy efficiency first" role.

GENERAL nZEB DEFINITION FOR BUILDINGS THAT UNDERGO MAJOR RENOVATIONS

The definition of major renovation process in case of building is described in the EPBD Directive (EPBD,2010): *'major renovation' means the renovation of a building where:*

1. The total cost of the renovation relating to the building envelope or the technical building systems is higher than 25% of the value of the building, excluding the value of the land upon which the building is situated; or
2. More than 25% of the surface of the building envelope undergoes renovation.According to COHERENO project report (COHERENO, 2013) the general definition on nZEB renovation can have one or more of following characteristics:
    a. *The energy performance of the building after renovation fulfils the nZEB requirements for new buildings as they are defined at level of the EU MS and regions or/and*
    b. *The primary energy consumption of the building after renovation is reduced by 75% as comparing to the pre-renovation status or/and*
    c. *Potentially an additional primary energy minimum requirement of not more than 50-60kWh/(m^2year) energy consumption (GBPN, 2013) for heating/cooling, domestic hot water, ventilation energy consumption of auxiliary building's systems and*
    d. *Potentially an additional minimum requirement for renewable energy share (proposed to be at least 50% of the remaining energy demand of the building as it is suggested in (BPIE, 2011) taking into account the nZEB definition from EPBD and*
    e. *Potentially an additional minimum CO_2 requirement of no more than 3kg CO_2/(m^2year) as it is suggested in (BPIE, 2011) based on the needs to meet the long-term decarbonisation goals for residential and services sectors as resulted from the EU 2050 Roadmap for a low-carbon economy.*

Country-specific definitions are using one or more of the listed requirements.

COUNTRY-SPECIFIC DEFINITIONS

Criteria for nZEB renovation of buildings have been identified in 13 jurisdictions, but definitions have so far only been set in 8 (Austria, Cyprus, the Czech Republic, Denmark, France, Latvia, Lithuania, Brussels Capital Region). Of these, Austria, France, and Brussels Capital Region have set primary energy use requirements for renovation at a less strict limit compared to new buildings. Germany, Ireland and Slovenia envisage doing likewise, though the renovation definitions for these countries have not yet been set. Denmark and Lithuania have the same nZEB definition for new and existing buildings, as do Bulgaria, Cyprus, Italy and Latvia, where the nZEB definition for new buildings is also applied for deep renovations. For more information on the nZEB requirements for existing buildings are described below.

Austria

The requirements for nZEB renovation were defined together with requirements for new buildings (COHERENO, 2013). They include interim targets for heating energy demand, final energy demand, total efficiency factor, primary energy demand and CO_2-emissions for the years 2014, 2016, 2018 and 2020. The most ambitious requirement for the year 2020 can be considered as nZEB renovation standard for residential buildings. It is defined with following parameters:

- Heating demand $Q_H \leq 17 \times (1+2.5 \bullet (A/V))$ kWh/(m^2 year), A/V – shape factor,
- Final energy demand defined using energy demand for technical appliances used for heating systems,
- Primary energy demand $Q_P \leq 200$ kWh/(m^2 year),
- CO_2-emissions ≤ 32 kg/(m^{2}year),

or

- Heating demand $Q_H \leq 25 \times (1+2.5 \bullet (A/V))$ kWh/(m^2 year),
- Total energy efficiency factor ≥ 0.95,
- Primary energy demand $Q_P \leq 200$ kWh/(m^2 year),
- CO_2-emissions ≤ 32 kg/(m^{2}year).

The requirements for new nZEB residential buildings are following:

- Heating demand $Q_H \leq 10 \times (1+3.0 \bullet (A/V))$ kWh/(m^2 year), A/V – shape factor,
- Final energy demand defined using energy demand for technical appliances used for heating systems,
- Primary energy demand $Q_P \leq 160$ kWh/(m^2 year),
- CO_2-emissions ≤ 24 kg/(m^{2}year),

or

- Heating demand $Q_H \leq 16 \text{ x } (1+3.0 \bullet (A/V))$ kWh/(m^2 year),
- Total energy efficiency factor ≥ 0.75,
- Primary energy demand $Q_P \leq 160$ kWh/(m^2 year),

CO_2-emissions ≤ 24 kg/(m^{2}year). The example of Austria shows that the requirements for renovated buildings are lower, what seems reasonable.

Belgium (Brussels)

nZEB renovation in the Brussels Capital Region is defined with the same parameters as for new buildings (COHERENO, 2013). They apply whenever the renovation activities cover at least 75% of the heat loss surface area and the HVAC system. In such cases all parameters for new buildings are multiplied with a factor 1.2, except for indoor temperate overrun. The requirements include energy demand for heating, primary energy demand, overheating risk and air tightness. For renovated residential buildings the requirement is following:

- Heating demand $Q_H \leq 18$ kWh/(m^2 year)
- Primary energy demand $Q_P \leq 54$ kWh/(m^2 year), the requirements apply to the total sum of the heating, hot water, auxiliary electricity,
- Summertime comfort, excessive temperature frequency (> 25 °C) ≤5%,
- Air tightness, limit value: $n_{50} \leq 0.72$ h^{-1}.

Germany

There is no official definition of nZEB renovation in Germany. However two voluntary standards can be listed - KfW Efficiency House 55 and Quality-Approved Energy Retrofit with Passive House Components. In the first case the Efficiency House 55 indicates a reduction of 45% of primary energy compared to the requirement for new construction. Houses renovated before 2009 are considered nZEB if their primary energy demand is below 40 kWh/(m^{2}year) and transmission heat loss coefficient is below 0,28 W/(m^{2}K) for external walls.

The Quality-Approved Energy Retrofit with Passive House Components (PHI, 2010) is defined with following requirements:

- Heating demand $Q_H \leq 25$ kWh/(m^2 year)
- Primary energy demand $Q_P \leq 120$ kWh/(m^2 year) + ((Q_H - 15 kWh/(m^2 year)) • 1.2), the requirements apply to the total sum of the heating, hot water, cooling, auxiliary and household electricity,
- Summertime comfort, excessive temperature frequency (> 25 °C) ≤10%,
- Moisture protection,
- Air tightness, limit value: $n_{50} \leq 1.0$ h^{-1}, target value: $n_{50} \leq 0.6$ h^{-1},
- Windows with low thermal transmittance (popular known as “U-value”)..

There is no official definition of nZEB for new buildings in Germany. For comparison requirements for new Passive Buildings were shown:

- Heating demand $Q_H \leq 15$ kWh/(m² year)
- Primary energy demand $Q_P \leq 120$ kWh/(m² year), the requirements apply to the total sum of the heating, hot water, cooling, auxiliary and household electricity,
- Summertime comfort, excessive temperature frequency (> 25 °C) ≤10%,
- Air tightness, limit value: $n_{50} \leq 0.6\ h^{-1}$,
- Buildings elements with low thermal transmittance, lack of thermal bridges.

Ireland

Irish definition of nZEB for existing dwellings can be found in the document "Towards nearly zero energy buildings in Ireland" (Ireland, 2012). By 2020 the target energy load for space heating, water heating, fixed lighting and ventilation in existing dwellings will be of the order of 125 to 150 kWh/(m² year) with a reasonable proportion of the remaining energy use of the dwelling coming from renewable energy sources onsite or nearby. According to the document "Towards nearly zero energy buildings in Ireland" (Ireland, 2012) the definition of nZEB standard for new buildings is still developed.

Part L (Conservation of Fuel and Energy) of the Irish Building Regulations also sets out the statutory minimum energy performance standards for existing dwellings undergoing extension, material alteration or which are undergoing conversion from a building previously used for non-residential purposes. U-values for key fabric elements in particular are now set at a reasonably good performance levels.

Slovenia

Slovenia definition of nZEB renovation can be found in the document "National plan for increasing the number of nearly zero-energy buildings" (Slovenia, 2013). The requirements envisage a tightening of the minimum requirements regarding the highest permitted heating, cooling or air-conditioning needs, the preparation of hot water, lighting within the building, and the growth in the headline share of renewable energy sources in total energy use for the operation of buildings. For renovated buildings they are following:

Table 1. Regulations concerning the thermal transmittance for different building elements and air tightness

Fabrics Elements	Maximum Thermal Transmittance W/(m²K)
Pitched roof • Insulation at ceiling • Insulation on slope	 0.16 0.16
Flat roof	0.20
Walls	0.21
Ground floors	0.21
Other exposed floors	0.21
External doors, windows and roof lights	1.6
Maximum permitted Air Permeability 7 m³/h/m²	

Ireland, 2012.

- Heating demand $Q_H \leq 25$ kWh/(m^2 year), where the restriction for primary energy may be met only by means of a considerably higher actual share of renewable energy sources (an envisaged increase from 25 to 50%) in the overall end-use energy for the operation of the systems in the building,
- Heating demand $Q_H \leq 15$ kWh/(m^2 year), if the technical possibilities for the use of renewable energy sources at the building's location are insufficient.
- Primary energy demand $Q_P \leq 95$ kWh/(m^2 year) - family houses, the requirements apply to the total sum of the heating, ventilation, hot water and cooling,
- Primary energy demand $Q_P \leq 90$ kWh/(m^2 year) - multi-dwelling buildings, primary energy demand $Q_P \leq 65$ kWh/(m^2 year) - non-residential buildings. For new buildings different requirements apply regarding the primary energy demand:
- Primary energy demand $Q_P \leq 75$ kWh/(m^2 year) - family houses, the requirements apply to the total sum of the heating, ventilation, hot water and cooling,
- Primary energy demand $Q_P \leq 80$ kWh/(m^2 year) - multi-dwelling buildings,
- Primary energy demand $Q_P \leq 55$ kWh/(m^2 year) - non-residential buildings.

Additional regulations are concerning the heat transfer coefficient U for different building elements.

Portugal

The requirements for nZEB renovation are in Portugal the same as for new buildings (Portugal, 2013). There are no specific regulations concerning energy demand for heating, cooling or primary energy demand. The minimum requirements are defined with regard to the minimum efficiency of the systems or building elements. Their scope includes:

Table 2. Regulations concerning the thermal transmittance for different building elements

Fabrics Elements	Maximum Thermal Transmittance W/(m^{2}K)
Pitched roof, flat roof	0.20
External walls	0.28
Walls that abut onto heated neighboring buildings	0.50
External wall of heated areas abutting onto the ground	0.35
Ground floors, floor above an unheated cellar, unheated area or a garage	0.35
Floor above ambient air	0.30
Vertical windows or balcony doors and heated winter gardens with wooden or synthetic frames	1.30
Roof windows, glazed roofs	1.40
Entrance doors	1.60
Garage doors	2.00

Slovenia, 2013.

- Surface heat transfer reference coefficients for opaque elements and glass filled spans,
- Minimum efficiency requirements on thermal production units,
- Performance classification of split, multi-split, VRF and compact units, with air to air exchangers,
- Performance classification of split, multi-split and compact units with air to water exchanger,
- Performance classification of rooftop type units,
- Performance classification of compression heat pump chillers' type units,
- Minimum energy efficiency requirements for boilers,
- Nominal efficiency for boilers and water heaters,
- Minimum energy efficiency requirements for air handling units, in accordance with standard EN 13053,
- Energy efficiency requirements for pumps and fans,
- Maximum lighting power density (LPD) values.
- Minimum energy efficiency requirements for lifts in accordance with standard VDI 4707

Slovak Republic

According to Slovak regulations (Slovak, 2013) deep renovation is renovation of a building at the ultra-low-energy construction level. An extensively renovated building must meet the nearly zero-energy requirement (the same as for new buildings) if it is technical, functionally and economically feasible. The standard is defined with following requirements:

- Heating demand $Q_H \leq 50$ kWh/(m^2 year) depending on the building shape factor - single-family house $Q_H \leq 40.7$ kWh/(m^2 year), apartment blocks $Q_H \leq 25$ kWh/(m^2 year), office buildings $Q_H \leq$ 26.8 kWh/(m^2 year),
- Primary energy demand - single-family house $Q_P \leq 54$ kWh/(m^2 year), apartment blocks $Q_P \leq 32$ kWh/(m^2 year), office buildings $Q_H \leq 60$ kWh/(m^2 year), the requirements apply to the total sum of the heating, hot water, cooling, ventilation and household lightning,

Additional regulations are concerning the heat transfer coefficient U for different building elements for ultra-low-energy buildings.

Bulgaria

In Bulgaria the requirements for renovated buildings are the same as for new nZEB (Bulgaria, 2013). They have to meet the minimum energy efficiency requirements insofar as it is technical feasible and

Table 3. Regulations concerning the thermal transmittance for different building elements for ultra-low-energy buildings

Fabrics Elements	Maximum Thermal Transmittance W/(m^{2}K)
Pitched roof, flat roof	0.10
External walls	0.22
Vent construction	0.90

Slovak, 2013.

economically viable. The legislation specifies that in the construction of new buildings or the reconstruction, major renovation, major repair or conversion of existing buildings, systems for generating energy from renewable sources shall be incorporated where technically feasible and financially viable: for public service buildings from 1 January 2012 and for other buildings from 31 December 2014. The possibility of using renewable energy forms part of the assessment of the annual energy consumption indicators for the building.

Italy

In Italy the requirements for renovated buildings are the same as for new nZEB (Italy, 2013). The Presidential Decree No 59 of 2 April 2009 set out the general criteria, calculation methods and minimum requirements for the energy performance of buildings and heating and hot water systems, as well as for air conditioning and, only for the services-commercial sector, the artificial lighting of buildings. These provisions apply to new buildings and to renovations of existing buildings, both public and private. Additionally if the building is undergoing major renovation the renewable energy sources should be integrated. The individual building elements and technical systems must comply with the minimum performance values set out by law.

France

The requirements for renovated buildings are described in Action plan for increasing the number of nearly zero-energy buildings (France, 2013). The nZEB renovation is not defined directly but the document is describing the renovation to low-consumption standard. The standard is defined with following requirements:

- **Primary Energy Demand:** Residential buildings $Q_p \leq 80$ kWh/(m^2 year), the requirements apply to the total sum of the heating, domestic hot water, cooling, ventilation, lightning, ancillary system elements,
- **Primary Energy Demand:** Office buildings $Q_p \leq 60\%$ of primary energy demand of the same building equipped with the reference insulation materials and energy systems.

The required consumption levels are subject to variations, depending at least on geographical areas and altitudes.

For new buildings different requirements apply regarding the primary energy demand:

- **Primary Energy Demand:** Residential buildings $Q_p \leq 60$ kWh/(m^2 year), the requirements apply to the total sum of the heating, domestic hot water, cooling, ventilation, lightning, ancillary system elements,
- **Primary Energy Demand:** Office buildings $Q_p \leq 110$ kWh/(m^2 year).

Luxembourg

The process of buildings renovation is supported with aid programme for energy-efficient new and existing buildings (Luxembourg, 2013). The aim of it is to provide incentives to implement more demanding energy standards and to bring forward modernization measures. According to the legislation the scale of support depends from the energy class achieved after the renovation. The most demanding is energy class A, define also as nZEB standard for new buildings. For single family houses undergoing energy class A renovation the requirement is following:

- Heating demand $Q_H \leq 22$ kWh/(m^2 year),
- Primary energy demand $Q_P \leq 45$ kWh/(m^2 year), the requirements apply to the total sum of the heating, cooling, ventilation, hot water, humidifying and dehumidifying, lighting and ventilation, and the auxiliary energy demand of the building systems,
- CO_2 emission ≤ 11 kgCO_2/(m^2 year) - environmental impact assessment.

for apartment bock:

- Heating demand $Q_H \leq 14$ kWh/(m^2 year),
- Primary energy demand $Q_P \leq 45$ kWh/(m^2 year),
- CO_2 emission ≤ 10 kgCO_2/(m^2 year) - environmental impact assessment.

In case of non-residential buildings assessment is made in terms of the percentage difference from the reference building.

Finland

The definition of nZEB renovation in Finland is not totally clear and rather general. From one site the target in buildings constructed, repaired and leased after 2015 is the "passive house" (Finland, 2012). But at the same time energy efficiency requirements for repair and renovation work, state that economically cost-effective measures should be implemented. The process of renovation should additionally increase the exploitation of renewable energy in the building stock.

Lithuania

According to Lithuania regulations (Lithuania, 2012):

- From 31 December 2014, new and existing buildings that are subject to major renovation shall comply with the requirements for the use of renewable energy resources; the compliance with such requirements may be ensured by using centrally supplied heat and cooling energy, where renewable energy resources are largely used for the production of such energy,
- From 1 January 2012 new and existing buildings belonging to state and municipal bodies and companies and requiring major renovation must comply with the requirements for the use of renewable energy resources.

Existing support programmes are requiring that after the renovation the buildings should achieve at least energy class C and the energy consumption should be reduces by at least 20%. The aim of the programmes is to modernize most residential buildings built under construction permits issued prior to

1993, to reduce heating energy costs by up to 30%, compared to the heating energy costs prior to renovation (modernization) and to reduce carbon dioxide emissions by about 400 thousand tons per year.

Sweden

In Sweden the requirements for renovated buildings are the same as for new nZEB (Sweden, 2012). For nZEB standard the requirement is following:

- Primary energy demand for residential buildings $Q_p \leq$ 30-75 kWh/(m^2 year) - depending on the reference building and location,
- Primary energy demand for non-residential buildings $Q_p \leq$ 30-105 kWh/(m^2 year) - depending on the reference building and location.

The Swedish legislation promotes the improvement of energy efficiency in existing buildings to nearly zero-energy level where reasonable. Where an improvement of energy efficiency to this level is not reasonable, the building should be brought up to the highest energy performance level that is reasonable. The legislation already provides flexibility in this regard.

POLISH DEFINITION OF NZEB RENOVATION STANDARD-

Poles live in homes that are inadequately insulated against heat loss. Heating technology is outdated and the most popular fuel is highly polluting coal, burned in old coal-fired boilers. It is estimated that more than 70% of detached single-family houses in Poland (3.6 million) have no, or inadequate, thermal insulation (IEE, 2014). Only 1% of all houses in Poland can be considered energy efficient, primarily those that have been built in the last few years (IEE, 2014; Gajewski, Węglarz, & A. Żmijewski, K., 2007). Most of the buildings without thermal insulation had been built before 1989 (Pierzchalska & Węglarz, 2016).

Data from the Central Statistical Office (Central Statistical Office, 2013, Bank Gospodarstwa Krajowego, 2016) indicates that about 50% of residential buildings in Poland have been insulated, but in the majority of cases to a sub-optimal level. Given that the economic case for improving the insulation of these partially insulated buildings is not favourable, it can be concluded that the remaining 50% of buildings should be prioritized for renovation.

Summary of the Polish Building Stock

According to data from the Central Statistical Office (Central Statistical Office, 2013), in 2011 the total number of buildings in Poland now exceeds 6 million. For comparison, in Britain there exist approximately 26 million homes that account for 27% of total UK greenhouse gas emissions (Bernier, Fenner, & Ainger, 2010).

A breakdown according to location (urban/rural) and type is provided in Table 4. Non-residential refers to property other than housing such as offices, retail, industrial, hotels, hospitals and education.

A breakdown of the Polish non-residential building stock is provided in Table 5.

Table 4. Breakdown of Polish residential building stock in 2011

<table>
<tr><th rowspan="4" colspan="2"></th><th rowspan="4">Total</th><th colspan="7">Including</th></tr>
<tr><th rowspan="3">Inhabited</th><th colspan="5">of which</th><th rowspan="3">Uninhabited</th></tr>
<tr><th rowspan="2">Residential</th><th colspan="2">of which</th><th rowspan="2">Collective accommodation</th><th rowspan="2">Non-Residential</th></tr>
<tr><th>Single-Family</th><th>Multi-Family</th></tr>
<tr><td>Total</td><td rowspan="3">Thousands</td><td>6,047.1</td><td>5,567.6</td><td>5,542.6</td><td>5,007.5</td><td>535.1</td><td>3.3</td><td>21.0</td><td>479.5</td></tr>
<tr><td>Urban areas</td><td>2,285.6</td><td>2,189.2</td><td>2,176.4</td><td>1,738.2</td><td>438.2</td><td>1.8</td><td>10.8</td><td>96.4</td></tr>
<tr><td>Rural areas</td><td>3,761.5</td><td>3,378.4</td><td>3,366.2</td><td>3,269.3</td><td>96.9</td><td>1.4</td><td>10.3</td><td>383.1</td></tr>
</table>

Central Statistical Office, 2013.

Table 5. Estimated breakdown of Polish non-residential building stock in 2010

Non-Residential Building Type	Total in Thousands
Warehouse	123.7
Hotels/Restaurants	82.5
Educational facilities	38.9
Cultural facilities	11.4
Health facilities	33.4
Office buildings	18.5
Total	308.4

Build Up Skills, 2012.

For residential buildings, the largest sectors by surface area are multi-family houses (MFH) in urban areas (37%) and detached single-family houses (SFH) in rural areas (36%). For non-residential buildings, three-quarters of the total surface area is accounted for by office buildings, educational facilities, and retail buildings, each representing around one-quarter of the total.

Building Energy Consumption

Another analysis drawn up by the Central Statistical Office (2013) shows the age structure of buildings and housing resources in Poland. The vast majority of buildings have a very high level of demand for final energy and thus are key targets for thermo-renovation. The primary energy indicator (E_P) concerns the index of primary non-renewable energy for the purpose of heating, ventilation and domestic hot water. The final energy indicator (E_F) concerns the delivered energy for heating and ventilation and domestic hot water.

Table 6 contains information about changes in the level of primary and final energy to the end of 2010. The primary energy indicator (E_P) concerns the index of primary non-renewable energy for the purpose of heating, ventilation and domestic hot water. The final energy indicator (E_F) concerns the delivered energy for heating and ventilation and domestic hot water.

Table 6. Age structure and energy demand of Polish housing stock in 2010

Year of Construction	Buildings		Dwellings		Primary Energy (E_P)	Final (Delivered) Energy (E_F)
	Thousands	%	Mln.	%	kWh/(m²a)	kWh/(m²a)
Before 1918	413.30	7.71	1.21	9.01	> 350	> 300
1918 – 1944	828.20	15.44	1.54	11.46	300 – 350	260 – 300
1945 – 1970	1367.50	25.50	3.71	27.62	250 – 300	220 – 260
1971 – 1978	676.50	12.61	2.16	16.08	210 – 250	190 – 220
1979 – 1988	763.50	14.24	2.20	16.38	160 – 210	140 – 190
1989 – 2002	698.40	13.02	1.52	11.31	140 – 180	125 – 160
2003 – 2010	616.02	11.48	1.09	8.14	100 – 150	90 – 120
All	5,363.42	100.0	13.43	100.0		

Mańkowski S. & Szczechowiak E., 2012.

For non-residential buildings, Figure 1 shows the highest delivered energy consumption (EK, represented by the red bars) is in healthcare.

In non-residential buildings, most energy is consumed for heating, ventilation and air conditioning (HVAC) (37%), followed by lighting (32%) and electrical appliances (24%). In residential buildings, space heating is the dominant energy use, accounting for over two thirds (69%) of the total, followed by DHW (15%), cooking (8%), electrical appliances (7%) and lighting (2%).

High demand for energy in buildings is due to low historic (and to a degree, existing) energy performance standards. This refers especially to single-family buildings located in the countryside. According to a report on energy efficiency in Poland (IEE, 2014), 72% of single-family buildings have a low or very

Figure 1. Energy consumption in various types of non-residential buildings
BuildDesk 2012.

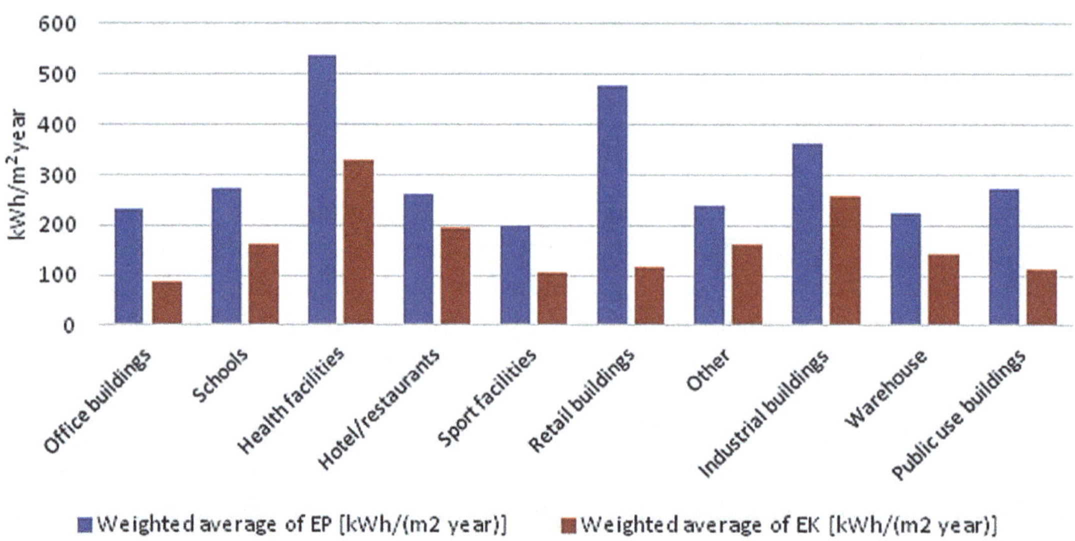

low energy standard. At the same time, 70% of single-family buildings in Poland use coal, amounting to 3.5 million coal-fired boilers (which collectively consume more than 9 million tonnes of coal per year). 28.8% of buildings have boilers that are more than 10 years old. About 3 million of these installations are based on manually fed boilers, an outdated technology which leads to significant air pollution.

Polish energy consumption in households according to study (Central Statistical Office, 2014), by energy carrier, differs significantly from other EU countries, notably with the highest per capita coal consumption. The consumption of coal per inhabitant in Poland was ten times higher than the EU-27 average. At 30%, hard coal accounts for the largest share of the 821.3 PJ of energy consumed in Polish households annually.

From an EU perspective (Central Statistical Office, 2014), the household share of total energy consumption ranged from 6% (Malta) to 31% (Latvia). The average rate for the 27 EU countries was 16%. Poland, at 19%, is slightly above the average and comparable with the proportion in Ireland, the UK, Austria, Italy and Greece. The second indicator is the energy consumption per capita in the household sector. Poland, at 21 GJ/inhabitant, is slightly below average compared with other EU countries. As noted earlier, energy in Polish residential buildings is used mainly in order to meet space heating requirements. While this has fallen slightly in recent years, total energy use has increased by 3% over the last 10 years largely due to an increase in electrical appliance energy use.

Energy Performance Standards

Amendments to the Polish regulations ought to result in the reduction of heat loss from buildings through the envelope and thus reduce energy consumption for heating. Table 7 specifies the minimum level of insulation of the building envelope thermal transmittance W/(m^2K), while Table 8 shows the maximum permissible demand for primary energy E_p kWh/(m^2year).

Table 7. Regulations concerning the thermal transmittance for different building elements over time

Period of Validity	External Wall	Roof	Ceiling Above Unheated Basement	Ceiling Under Unheated Attic	External Windows and Doors
	Thermal Transmittance (W/m²K)				
1957-1964	1.16-1.42	0.87	1.16	1.04-1.16	
1964-1974	1.16	0.87	1.16	1.04-1.16	
1974-1982	1.16	0.70	1.16	0.93	
1982-1991	0.75	0.45	1.16	0.40	2.0-2.6
1991-2002	0.55-0.70	0.30	0.60	0.30	2.0-2.6
2002-2008	0.30-0.50	0.30	0.60	0.30	2.0-2.6
2009-2013	0.30	0.25	0.45	0.25	1.7-1.8
2014-2016	0.25	0.20	0.25	0.20	1.3-1.5
2017-2020	0.23	0.18	0.25	0.18	1.1-1.3
From 2021	0.20	0.15	0.25	0.15	0.9-1.1

Firląg, S., 2015.

The present legislation, which imposes the obligation to maintain adequately high energy standards, applies only to buildings currently being designed and constructed. As for older buildings, built before 1991, when the rules regarding insulation of the building envelope were less rigorous, thermal insulation was a rarity and thus these buildings have a very high energy demand.

Possible Scope of the nZEB Renovation Definition: Results of Survey

In order to support nZEB renovation it is needed to determine what projects are cost-effective from state point of view and not cost-effective from the perspective of the investor and support them. The method and principles of support should give investors the incentive to self-incur the costs of carrying out economically viable activities. For example, there should be no support for energy efficiency projects with short payback time, they can be financed by building owners or using external capital.

It is necessary to introduce a new definition of nZEB renovation standard and define technical requirements for individual energy efficiency measures, such as the insulation of external building elements. They must be formulated in such a way that in the near future there will be no need for renovation of the currently modernized buildings.

According to the majority of experts (on basis of survey results carried out by BPIE among experts at the meeting of Efficient Poland), the definition of nZEB renovation should include only two parameters:

- Heating demand index Q_H expressed in kWh/(m^2 year),
- Percentage reduction of the primary, non-renewable energy Q_P demand (including in case of residential buildings heating, ventilation, domestic hot water, cooling and auxiliary systems) determined in relation to the building before renovation.

These two parameters would make the definition very flexible but at the same time very demanding. The value of heating demand index Q_H depends among other on thermal transmittance of building elements, thermal bridges, air tightness and solutions in regard to ventilation system. Achieving the low value of the index will be possible on different ways depending on the buildings situation. For example in case when insulation of grand floor is not possible other elements of the building can have better ther-

Table 8. Permissible level of primary energy indicator (E_P) for newly designed buildings

Type of Building	Indicator E_P for Heating, Ventilation and DHW Heating [kWh/(m^{2}year)]		
	From 1st Jan 2014	**From 1st Jan 2017**	**From 1st Jan 2021***
Residential building: Detached Multifamily	 120 105	 95 85	 70 65
Collective residential building	95	85	75
Public utility building Health care Others	 390 65	 290 60	 190 45
Farm building, warehouse, production	110	90	70

Building Code 2013.

* From 1st Jan 2019 in the case of buildings occupied and owned by public authorities

mal transmittance or ventilation with heat recovery can be implemented. Such solution gives flexibility and opportunity to choose the best renovation measures. The requirement referring to the reduction of primary energy includes other aspects like efficiencies of the heating and DHW system, energy source type and use of RES. Depending on the situation the different solutions can be chosen. Use of requirement referring to the heating demand prevent against situation in which only RES will be implemented in order to reduce the primary, non-renewable energy demand.

The results of the survey were the basis for starting work on the definition of nZEB renovation for single and multi-family buildings.

NZEB Definition Based on Cost Optimum Calculations

Process of defining the Polish nZEB renovation standard was divided into two stages. First of all the optimum U-value for building envelope were calculated and then on this basis the energy demand for heating and reduction of non-renewable primary energy demand.

Stage I: Cost-optimal heat transfer coefficients for renovated elements of building envelope. The aim of the calculation was to determine the optimal U-values for the fabrics elements of renovated single-family house. The results are depending on the type of the element, its initial insulation and the cost of energy. As an optimizing criterion the minimum cumulative cost (investment + cost of energy losses) was used, calculated for the 30-year time-scale.

Stage II: Cost-optimal renovation standard of the single-family building. The second part of the analysis consisted finding the optimal variant of the renovation of a reference building. As an optimizing criterion the minimum cumulative cost (investment + energy cost of heating, DHW and auxiliary

Table 9. Results of survey among polish experts on the possible scope of nZEB definition for renovated residential buildings

Should the Definition of Renovation to nZEB Standard Include Requirements for?	Yes	No	Importance (Yes-No)
Index of non-renewable, primary energy demand Q_P	12	5	7
Index of final (delivered) energy demand Q_F	10	6	4
Heating demand index Q_H	15	2	**13**
CO_2 emission index	10	5	5
Share of renewable energy sources	10	7	3
Thermal transmittance (U-value) of different building elements	12	4	8
Air tightness	7	7	0
Ventilation systems including the efficiencies of heat recovery	9	6	3
Efficiencies of heating and domestic hot water systems	9	6	3
Summertime comfort – risk of overheating	5	9	-4
Energy demand of auxiliary systems	6	8	-2
Energy efficiency of improvement - percentage reduction of the primary, non-renewable energy Q_P demand	13	3	**10**
Index of final (delivered) energy demand Q_F for cooling	1	0	1

electricity) was used, calculated for the 30-year time-scale.12 variants of renovation were defined, based on the heat transfer coefficient of the building elements and ventilation, central heating and domestic hot water solutions. For each variant energy demand for heating and ventilation was calculated with computer program. The reduction of primary non-renewable, energy demand for heating, ventilation, domestic hot water and auxiliary systems was also determined.

Cost-Optimal U-Values of Renovated Building Envelope

The Table 10 and Figure 2 show the example calculation results for renovated external wall. Depending on the heating energy cost per 1 GJ the cost-optimum thermal transmittance in for 0.20 to 0.12 W/(m²K).

Cost-optimal thermal transmittance for renovated building elements is presented in Table below. The optimal thermal transmittance depends strongly on energy price. That is way two values were presented for lowest and the highest energy price per GJ. The limit values were the basis for defining the renovation variants used in second part of the analysis:

- First variant (W1) – the building envelope was renovated in accordance with cost-optimum thermal transmittance specified for energy price of 20 PLN per GJ,
- First variant (W1) – the building envelope was renovated in accordance with cost-optimum thermal transmittance specified for energy price of 60 PLN per GJ.

Table 10. The discounted, cumulative cost (k_{Rd}) per m² including the cost of energy loss and the unit cost of renovation with tax (k_{VAT}), depending on the renovation variant and the cost of 1 GJ of energy (1 EUR=4,2PLN). Insulation of external walls with ETICS using EPS with λ = 0.033 W/(mK).

U-Value, Insulation Cost and Energy Loss per m² of External Wall Depending on Insulation Thickness					Heating Energy Cost in PLN per 1 GJ				
					20	**30**	**40**	**50**	**60**
d	**Thermal Transmittance**	**K**	**K_{VAT}**	**Energy Loss**	**Discounted, Cumulative Cost in PLN per m²**				
cm	**W/(m²K)**	**PLN/m²**	**PLN/m²**	**GJ/(m²year)**	**K_{Rd_20}**	**K_{Rd_30}**	**K_{Rd_40}**	**K_{Rd_50}**	**K_{Rd_60}**
0	0.533	0.00	0.00	0.170	86.1	129.2	172.2	215.3	258.4
10	0.204	99.93	107.92	0.065	**141**	157	174	190	207
12	0.181	103.56	111.84	0.058	141	156	170	185	200
14	0.163	107.19	115.76	0.052	142	**155**	169	182	195
16	0.149	110.81	119.68	0.048	144	156	168	180	192
18	0.136	114.44	123.60	0.044	146	157	**168**	179	190
20	0.126	118.07	127.52	0.040	148	158	168	**178**	189
22	0.117	122.04	131.80	0.037	151	160	170	179	**189**
24	0.109	125.58	135.63	0.035	153	162	171	180	189
26	0.103	129.99	140.38	0.033	157	165	173	182	190
28	0.097	133.53	144.21	0.031	160	168	175	183	191
30	0.091	137.07	148.03	0.029	163	170	177	185	192

Figure 2. The discounted, cumulative cost (k_{Rd}) per m² for renovated external wall depending on thermal transmittance and energy cost

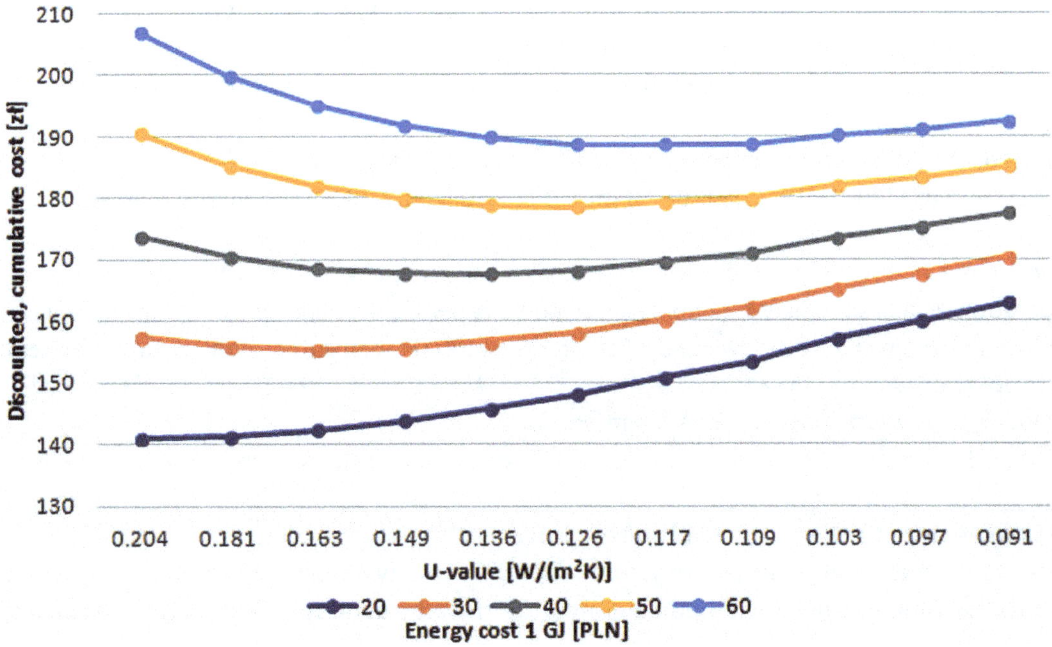

Table 11. The cost-optimum renovation variants (depending on the energy price) of buildings envelope used in the II stage on the analysis

Building Element	W1 - 20 PLN per 1 GJ				W2 - 60 PLN per 1 GJ			
	d	Thermal Transmittance	Renovation Cost	Unit	d	Thermal Transmittance	Renovation Cost	Unit
	cm	W/(m²K)			cm	W/(m²K)		
External wall	10	0.204	107.92	PLN/m²	22	0.117	131.80	PLN/m²
Floor above an unheated cellar	7	0.247	52.92	PLN/m²	10	0.187	75.60	PLN/m²
Flat roof	12	0.206	92.62	PLN/m²	26	0.117	120.20	PLN/m²
Windows	-	0.9	562.36	PLN/m²	-	0.9	562.36	PLN/m²
External doors	-	1.3	4087.73	PLN/pcs.	-	0.9	4347.35	PLN/pcs.

Cost-Optimal Building Renovation Standard

On basis of defined U-values of buildings envelope and solutions referring to modernization of ventilation system and use of RES 12 different renovation variants were identified. All of them include modernization of existing heating and domestic hot water system with exchange of heat source. The symbols used to distinguish the variants are explained below:

W0: Base variant of reference building – before renovation,
W1: I variant of buildings envelope renovation, energy price 20 PLN per 1 GJ,
W2: II variant of buildings envelope renovation, energy price 60 PLN per 1 GJ,
G: Natural ventilation – base case before renovation,
H: Hybrid ventilation – it was assumed that the ventilation energy loss will be reduced by 20% (Firlag, S. & Rucińska, J., 2007),
R: Balanced ventilation with heat recovery – efficiency of heat recovery 90%,
S: Solar system used for DHW heating – assumed coverage 60% of DHW energy demand.

Table 12 and Figure 3 show the calculation results for energy cost of 40 PLN per GJ.

Calculation results show that for energy price 40 PLN per GJ the cost-optimum variant of renovation is W1/G for Szczecin and Warszawa and W2/G for Suwałki. It seems that modernization of ventilation system is not justified from the economic point of view. Energy demand for heating of renovated single-family house is between 46,8 – 51,2 kWh/(m^{2}year).

Based on the results the Polish definition of nZEB renovation of single-family house can be following:

- Heating demand index $Q_H \leq 50$ kWh/(m^2 year),
- Percentage reduction of the primary, non-renewable energy Q_P demand (including in case of residential buildings heating, ventilation, domestic hot water, cooling and auxiliary systems) $\leq$70%.

With regard to existing multifamily residential buildings, the optimal level of demand for non-renewable primary energy and final energy for heating, ventilation and DHW heating (Mańkowski S. & Szczechowiak E., 2012) is the following:

- Existing buildings, built after 1970:
  - E_P = 50-75 kWh/(m^{2}year); E_F = 45-68 kWh/(m^{2}year);
- Older buildings, built before 1970:
  - E_P = 75-100 kWh/(m^{2}year); E_F = 65-90 kWh/(m^{2}year);
  - Taking into account the share of RES, EP could be reduced to 50-75 kWh/(m^{2}year).

Table 12. Summary of energy demand for heating (Q_H) and discounted cumulated cost (K_{CZ}) - energy cost 40 PLN per GJ, for different variants and Polish climates

Variants	Warszawa		Szczecin		Suwałki	
	Q_H	K_{CZ}	Q_H	K_{CZ}	Q_H	K_{CZ}
	kWh/(m^{2}year)	PLN	kWh/(m^{2}year)	PLN	kWh/(m^{2}year)	PLN
W0	169.9	149 125	159.5	141 737	205.3	174 338
W1/G	51.2	**117 399**	46.8	**115 139**	67.6	125 811
W2/G	35.8	117 785	32.3	115 969	48.9	**124 506**
W1/H	47.3	120 632	43.1	118 475	62.8	128 556
W2/H	32.1	121 075	28.7	119 338	44.5	127 444
W1/R	29.4	136 750	26.0	134 967	42.5	143 436
W2/R	16.4	138 294	13.4	136 784	25.6	143 017

Warszawa – the average number of heating degree days, Szczecin – the smallest, Suwałki – the highest.

Figure 3. Energy demand for heating (Q_H) in relation to discounted cumulated cost (K_{CZ}) - energy cost 40 PLN per GJ, for different variants and Polish climates
Warszawa – the average number of heating degree days, Szczecin – the smallest, Suwałki – the highest.

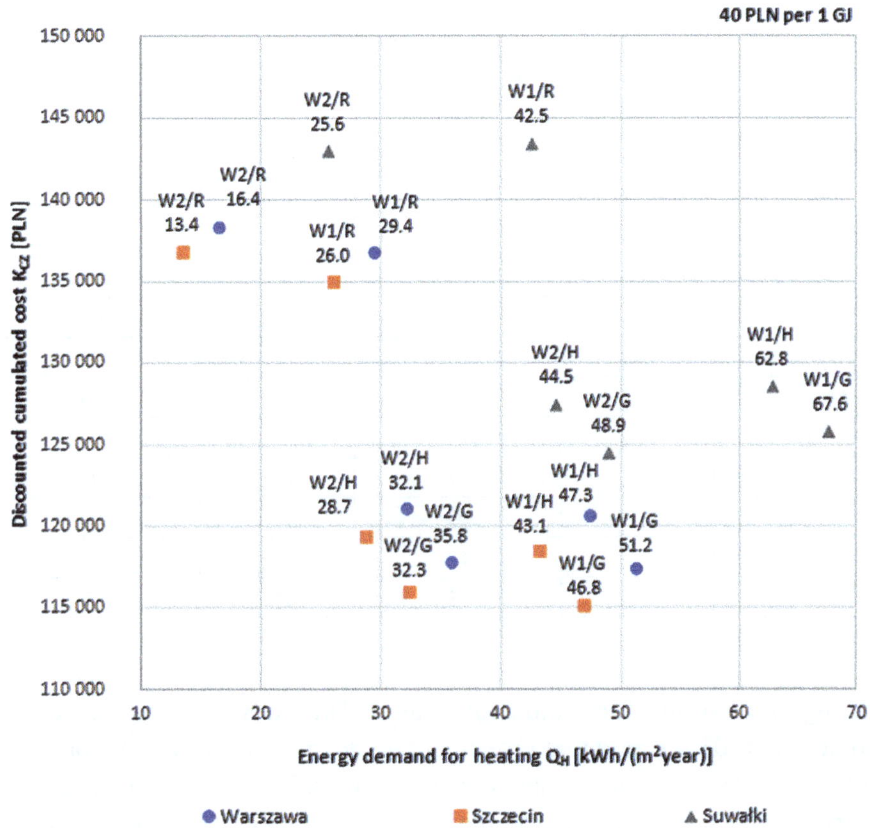

Achieving the indicated energy performance parameters requires deep renovation of buildings.

CONCLUSION

The research shows that not all countries in EU have definitions of nZEB renovation standard. Also an approach to defining it is different. Some countries have the same requirements for new and existing buildings, others not. Calculations made for polish conditions show that the second solution is probably better. In not all cases the implementation of the same requirements as for new buildings is technical, functionally and economically feasible. For example maximum thickness of additional insulation layer is can be limited by the height of the rooms or law. Also the only economically cost-effective measures should be implemented. The nZEB renovation definition itself should be simple, clear and demanding but at the same time flexible. Existing buildings have different standards, construction, function and location. It is not possible to proposed detailed list of renovation measures that can be implemented in every case.

In order increase the rate of nZEB renovation in EU, the following actions should be undertaken:

Table 13. Reduction of primary energy demand Q_P in regard to base variant (W0) depending on the location of the building and renovation variant

Variant	Primary Energy Demand Q_p (Heating, Ventilation, DHW and Auxiliary Systems)					
	Warszawa		Szczecin		Suwałki	
	kWh/(m²year)	Reduction	kWh/(m²year)	Reduction	kWh/(m²year)	Reduction
W0	360	0%	342	0%	420	0%
W1/G	110	69%	105	69%	130	69%
W2/G	91	75%	87	75%	107	74%
W1/H	106	70%	101	70%	125	70%
W2/H	88	76%	83	76%	103	76%
W1/R	93	74%	88	74%	109	74%
W2/R	77	79%	73	79%	88	79%
W1/G/S	90	75%	85	75%	110	74%
W2/G/S	71	80%	67	80%	87	79%
W1/H/S	86	76%	81	76%	105	75%
W2/H/S	67	81%	63	81%	83	80%
W1/R/S	73	80%	68	80%	89	79%
W2/R/S	57	84%	53	85%	68	84%

- Public funding for renovation needs to be increased. The study of Pikas, E., Kurnitski, J., Lilias, R. & Thalfeldt, M. (2015) shows that investment in energy efficiency is not only environmentally important but provides economic benefits on an individual and government budget level, e.g. tax return is between 32–33% depending on the renovation project. Promotion of economic benefits can help in convincing the key stakeholders.
- Levels of gearing, i.e. proportion of funding from third parties such as building owners and other investors, need to be improved. One of the solutions can be ESCo working according to an integral renovation approach like it is in Italy (Bonacina, Masera, & Pavan, 2015) or Russia (Paiho, Abdurafikov, Hoang, & Kuusisto, 2015). Investing in renovation can be a way of earning money because of relatively high rate of return. The analyzed scenarios show that greatest return on investment is achieved under the Ambitious scenario. According to study of Morelli, Harrestrup, & Svendsen, (2014) it is also better to invest in renovation than in new buildings.

REFERENCES

Bank Gospodarstwa Krajowego. (2016). *Figures of the Thermomodernisation and Renovation Fund [Dane liczbowe Funduszu Termomodernizacji i Remontów]*. Warsaw: National Economy Bank. (in Polish)

Bernier, P., Fenner, R. A., & Ainger, C. (2010) Assessing the sustainability merits of retrofitting existing homes. *Proceedings of the Institution of Civil Engineers - Engineering Sustainability, 163*, 197-207. doi:10.1680/ensu.2010.163.4.197

Bonacina, C. F., Masera, G., & Pavan, A. (2015). Investment Grade Energy Audit: A Financial Tool for the Cost-effective Renovation of Residential Buildings. *Energy Procedia, 70*, 709–718. doi:10.1016/j.egypro.2015.02.180

BPIE (2011). *Principles for nearly zero-energy buildings. Paving the way for effective implementation of policy requirements.* Brussels: Buildings Performance Institute Europe.

BPIE. (2014). *Renovation strategies of selected eu countries. A status report on compliance with article 4 of the energy efficiency directive.* Brussels: Buildings Performance Institute Europe.

Build Up Skills. (2012). *The professional qualifications and staff for the green building.* Warszawa: NAPE.

BuildDesk. (2012). *REPORT: Energy performance of buildings in Poland [RAPORT: Stan energetyczny budynków w Polsce].* BuildDesk Polska. (in Polish)

Building Code. (2013). *Regulation of the Minister of Transport, Construction and Maritime Economy dated 5 July 2013r. amending the regulation on technical conditions to be met by buildings and their location* [Rozporządzenie Ministra Transportu, Budownictwa i Gospodarki Morskiej z dnia 5 lipca 2013r. zmieniające rozporządzenie w sprawie warunków technicznych, jakim powinny odpowiadać budynki i ich usytuowanie]. Academic Press.

Bulgaria. (2013). *Outline national plan to increase the number of nearly zeroenergy buildings in accordance with article 9 of directive 2010/31/eu on the energy performance of buildings.* Ministry of Regional Development and Public Works.

Central Statistical Office. (2013). *Inhabited buildings, the National Census of Population and Housing 2011.* Warsaw: Central Statistical Office.

Central Statistical Office. (2014). *Energy consumption in households in 2012.* Warsaw: Central Statistical Office.

COHERENO. (2013). *NZEB criteria for typical single-family home renovations in various countries.* Retrieved from http://www.cohereno.eu/fileadmin/media/Dateien/D2_1_BPIE_WP2_12092013_3_5_-final.pdf

EED. (2012). *Directive 2012/27/EU of the European Parliament and of the Council of 25 October 2012 on energy efficiency, amending Directives 2009/125/EC and 2010/30/EU and repealing Directives 2004/8/EC and 2006/32/EC Text with EEA relevance.* Brussels: European Parliament, Council of the European Union. EEA relevance, COD 2011/0172.

EPBD. (2010). *Directive 2010/31/EU of the European Parliament and of the Council of 19 May 2010 on the energy performance of buildings (recast).* Brussels: European Parliament, Council of the European Union. COD 2008/0223.

EPBD. (2016). *Proposal for a Directive of the European Parliament and of the Council amending Directive 2010/31/EU on the energy performance of buildings.* Brussels: European Commission. Com(2016) 765 Final, 2016/0381 (Cod).

Finland. (2012). *Finland's national plan to increase the number of nearly zero-energy buildings. Report to the European Commission pursuant to Article 9 of Directive 2010/31/EU on the energy performance of buildings.*

Firląg, S. (2015). How to meet the minimum energy performance requirements of Technical Conditions in year 2021? *Procedia Engineering, 111*, 202–208. doi:10.1016/j.proeng.2015.07.077

Firlag, S., & Rucińska, J. (2007). *Simplified method of designing an air-ground heat exchange.* CESB 2007 PRAGUE International Conference - Central Europe Towards Sustainable Building 1.

France: ш Ref. Ares (2013) 54186 - 16/01/2013.

France. (2013). *Action plan for increasing the number of nearly zero-energy buildings.*

Gajewski R. R., Węglarz, A., & Żmijewski, K. (2007). *Poly-Optimal Method of Designing Environmentally Friendly Buildings.* IOS Press BV.

GBPN. (2013). *What is a deep renovation definition?* Technical report. GBPN.

IEE. (2014). *Energy Efficiency in Poland 2013 Review, Single family houses energy efficiency and air quality*. Kraków, Poland: Institute of Environmental Economics.

Ireland. (2012). *Towards nearly zero energy buildings in Ireland, planning for 2020 and beyond.* Department of the Environment, Community and Local Government.

Italy. (2013). *National plan for increasing the number of "nearly-zero energy buildings" Guidelines and development framework.*

Lithuania. (2012). *National plan for increasing the number of nearly-zero energy buildings.* Ministry of the Environment of the Republic of Lithuania.

Luxembourg. (2013). *National Plan for Luxembourg to increase the number of nearly zero-energy buildings.* Luxembourg: Le Gouvernement du Grand-Duche de Luxembourg, Ministere de l'Economie et du Commerce exterieur.

Mańkowski, S., & Szczechowiak, E. (2012). *Strategic research project entitled "Integrated system for reducing operating energy consumption in buildings" Research Task No. 2 Volume I, Part A: Conditions of transformations in construction.* Academic Press.

Morelli, M., Harrestrup, M., & Svendsen, S. (2014). Method for a component-based economic optimisation in design of whole building renovation versus demolishing and rebuilding. *Energy Policy, 65*, 305–314. doi:10.1016/j.enpol.2013.09.068

Paiho, S., Abdurafikov, R., Hoang, H., & Kuusisto, J. (2015). An analysis of different business models for energy efficient renovation of residential districts in Russian cold regions. *Sustainable Cities and Society, 14*, 31–42. doi:10.1016/j.scs.2014.07.008

PHI. (2010). *Quality-Approved Energy Retrofit with PH Components - Criteria for residential-use refurbished buildings.* Darmstadt: Passive House Institute.

Pierzchalska, D. & Węglarz, A. (2016). *Praktyczna realizacja idei głębokiej termomodernizacji.* Materiały Budowlane 1/2016 str. 23-25.

Pikas, E., Kurnitski, J., Liias, R., & Thalfeldt, M. (2015). Quantification of economic benefits of renovation of apartment buildings as a basis for cost optimal 2030 energy efficiency strategies. *Energy and Building*, *86*, 151–160. doi:10.1016/j.enbuild.2014.10.004

Portugal. (2013). *National plan for increasing the number of nearly zero-energy buildings in Portugal.* Direcao Geral de Energia e Geologia.

Slovak. (2013). *National plan for increasing the number of nearly zero-energy buildings.* Academic Press.

Slovenia. (2013). *National plan for increasing the number of nearly zero-energy buildings in accordance with Article 9 of the Recast EPBD.* Republika Slovenija, Ministrstvo za Infrastrukturo in Prostor.

Sweden. (2012). *Towards nearly zero-energy buildings.* Communication 2011/12:131. Ministry of Industry, Employment and Communications.

ZEBRA 2020. (2015). *Nearly Zero-Energy Building Strategy 2020.* Author.

Chapter 2
Selection of Design Methods in the Modernization Process of Buildings to the nZEB Standard

Hanna Irena Jędrzejuk
Warsaw University of Technology, Poland

ABSTRACT

Typical methods of improving energy efficiency of buildings to the nZEB standard, such as better external envelope insulation, using better windows, or introducing a proper ventilation system, may prove inadequate. Often, it is necessary to apply unconventional systems and renewable energy sources. Each case has to be analyzed considering the location, building technology, usage of the building, and even conservation officer's requirements. To verify the effectiveness of standard activities in various climates, the following locations in Europe were chosen: Athens, Stockholm, and Warsaw. For evaluation, two methods were applied: the monthly balance method and the simplified hourly method 5R1C.

INTRODUCTION

The modernization of buildings to the 'nearly zero energy' standard requires identification of all conditions that affect energy consumption, such as local climate, environmental impact, building technology and type, technical equipment, purpose, and operating conditions.

The choice of technology and then specific technical solutions to achieve the nZEB standard must precede any modernization initiative designed to rationally reduce energy demand. For this purpose, an energy assessment of an existing building and an analysis of the energy demand structure are necessary.

A preliminary energy analysis of a building can be carried out using the monthly balance method and the simple hourly method (ISO, 2008). An advantage of this approach is the ability to perform relatively simple calculations in any spreadsheet, but, on the other hand, it provides only an approximate representation of dynamic processes. Simulation tools for building performance analyses, such as an open-source project ESP-r, or EnergyPlus, offer a complex analysis, but using them requires expert knowledge of these programs and is time-consuming.

DOI: 10.4018/978-1-5225-4105-9.ch002

The final evaluation should take into account not only energy-related and economic aspects but also the internal comfort (thermal, acoustic and lighting conditions).

This chapter focuses on energetic effects which can be achieved by improving thermal insulation of building's external partitions. The analysis was conducted for various climatic conditions for selected European capitals.

BACKGROUND

Directive 2010/31/EU defines a 'building' as a roofed construction having walls, for which energy is used to condition the indoor climate. That definition can be applied to isolate the analysed system from the surrounding environment and identify its components to be analyzed by defining system boundaries (Figure 1).

Energy in buildings is essential to provide:

Figure 1. Energy flow in a building system
Kurnitski, et al. 2013.

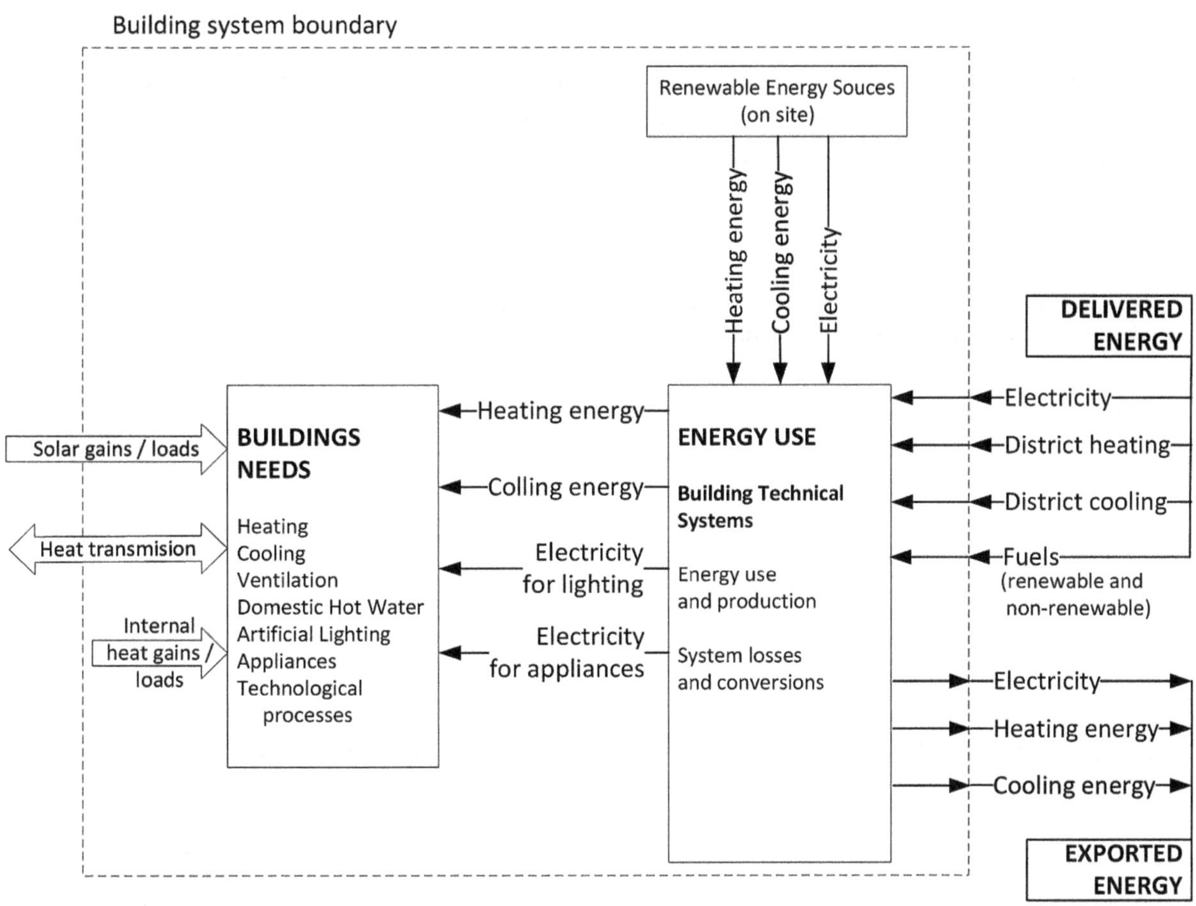

- Proper internal air temperature (heating, cooling - as needed),
- Air humidification or dehumidification (if required),
- Indoor air exchange (providing fresh air, removing contaminants, air treatment processes: humidification, dehumidification, etc.),
- Artificial lighting,
- Hot water preparation,
- Power supply of various types of electrical appliances (wide range: electric kettle, washing machine, dishwasher, television set, elevators, fans, pumps, control systems, etc.), and
- Technological processes or preparing meals.

The required energy supply depends also on external determinants, such as climate, landform features, interaction with other buildings, and local vegetation.

The energy required to meet human or technological needs must be supplied, but renewable sources can be applied instead of conventional sources. Balancing energy needs and energy on-site production has led to the concept of zero-energy buildings.

A nearly zero-energy building is defined as a building that has a very high energy performance (Directive 2010/31/EU). The supply of nearly zero or a very low amount of energy required to match particular building requirements can be provided by the use of energy from renewable sources. Producing the energy from renewable sources on-site or nearby makes it possible to decrease distribution losses and contributes to the creation of a decentralized energy system.

A modernization process of a building towards the nZEB standard should include a detailed identification of its actual state in terms of construction, internal installations, control systems, and finally the actual mode of operation, with emphasis on the behaviour of inhabitants.

Then, the next step is possible, including:

- Identification of the heat flow and energy processes,
- Finding weak points in terms of the heat and energy flow,
- Evaluating the possibility of heat or energy recovery within the system, and
- Analysing profits resulting from the use of renewable energy sources.

Identifying the state of an existing building must consider:

- Thermal insulation of all external partitions of the building (walls, windows, roof, doors, ground floor slab, etc.),
- The ventilation system,
- The lighting system,
- The electric power system, and
- The hot water preparation system.

Directive 2010/31/EU has triggered the development of the nZEB standard. The technology has advanced since then. Apart from new buildings, there are also examples of modernised buildings. A report prepared by Erhorn et al. (2014) presents 32 pilot projects in European countries, of which seven were renovation projects. The modernisation activities included improving thermal insulation quality of the building's envelope, applying mechanical ventilation with heat recovery, and using photovoltaic modules for local electricity production. A district heating system, a heat pump (variable refrigerant flow - VRF) with the heating and cooling mode, or even a modulating condensing, weather-compensated gas boiler were chosen as heat sources. Thermal comfort was estimated with the aid of thermodynamic and CFD analyses.

Barbolini et al. (2017) presented a design strategy that can be introduced to meet the nZEB requirements in Europe's temperate climates. A social housing complex in Modena (Italy) was substantially renovated. The shape of the building was optimised. Green surfaces for climate mitigation were increased. Thermal insulation of the envelope was improved. Passive cooling was applied. Solar gains through greenhouses were optimised. Air-tight windows with additional trickle ventilators were used. A completely new HVAC system was introduced, consisting of a mechanical ventilation system equipped with earth tubes, heat recovery and a heating unit, PV and solar thermal collectors, and a combined heating power system (CHP).

Becchio et al. (2015) provided a cost optimality assessment of a single family house (Northern Italy). The global cost of the nearly zero energy building was established to be higher by 212–313 EUR/m^2 than in the case of a typical house.

The energy analysis was performed in the EnergyPlus program. It was shown that in Italian conditions the nZEB standard can be reached: either with a conventional technical system (a condensing gas boiler with water terminals) and an average level of insulation but with a large number of PV panels, or with advanced technical systems (mechanical ventilation with heat pumps), high insulation level, and a large number of PV panels.

METHODOLOGY

The first step in modernizing an existing building to meet the nZEB standard is an analysis of the actual state of energy use for heating, cooling, hot water preparation, lighting, etc. Then, the impact of the following basic thermomodernization activities on energy efficiency can be assessed:

- Improving thermal quality of external partitions of the envelope,
- Application of external blinds to decrease solar gains during days in the summer,
- Application of external blinds to decrease heat transfer loss during nights in the winter, and
- Application of mechanical ventilation to control the airflow rate.

The monthly balance method was chosen as the simplest tool for building energy analysis that can be implemented in any spreadsheet. This method is based on the energy balance, i.e. the comparison of heat losses and heat gains. In the heating mode the heat losses include heat transfer by transmission and heat transfer by ventilation, while heat gains include solar heat gains and internal heat gains. The heat gain component is taken into account with the utilization factor. In the cooling mode the heat balance is formulated similarly. To calculate the heat balance, some data should be provided. A bottom-up algorithm of data processing is shown in Figure 2.

To express heat transfer characteristics of the envelope, it is necessary to analyse thermal insulation of the building, its partitions and thermal bridges. The thermal characteristic of ventilation can be determined as a function of the air tightness of the building and depends on the air flow through the conditioned area.

Figure 2. General algorithm of the monthly heat balance method

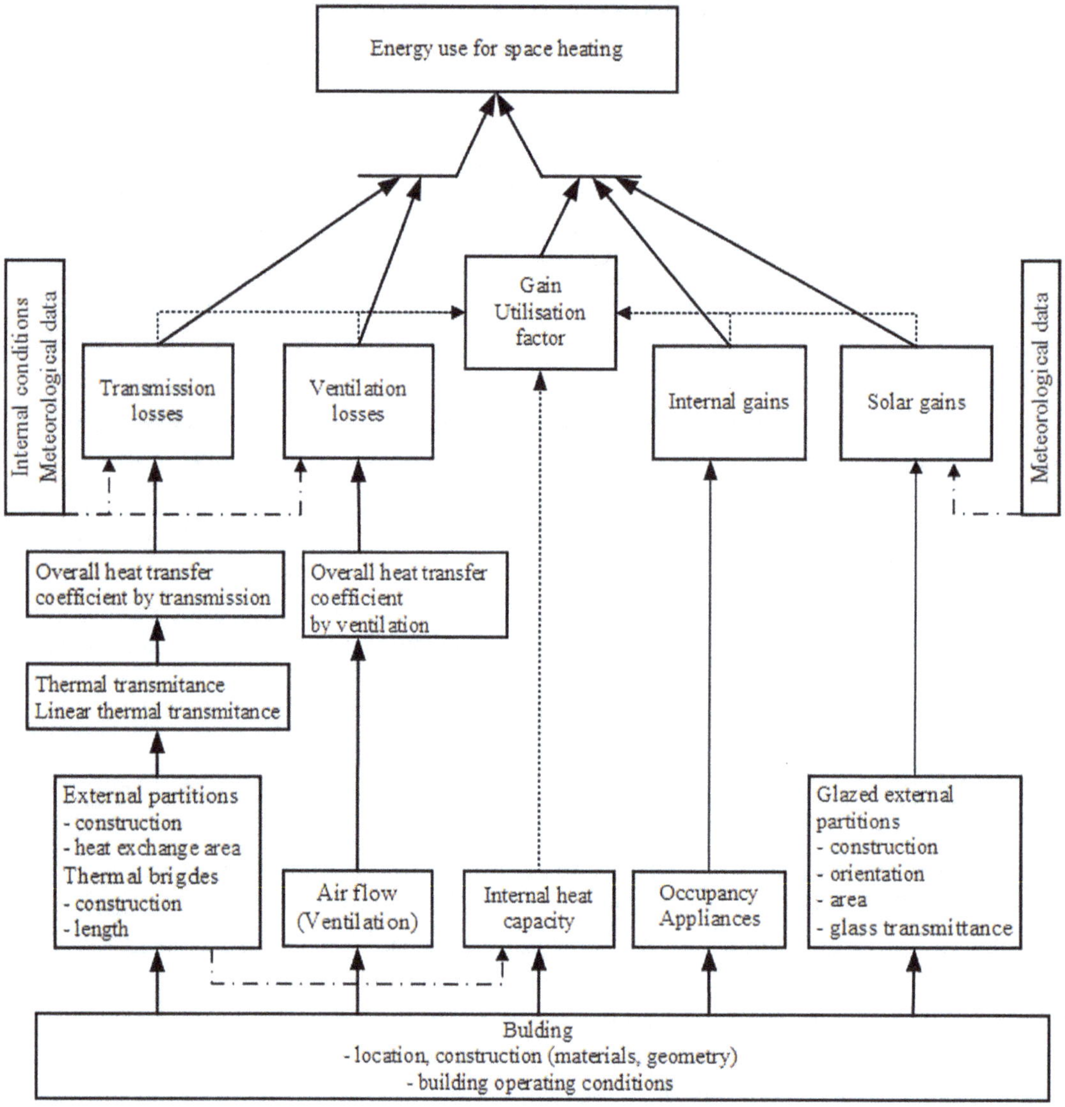

The method applied is described in detail further in the chapter.

Thermal Insulation of the Building Envelope and Its Partitions

A building envelope is the physical separator between the conditioned and unconditioned environment of a building and it has various functions. It protects the internal space from heat, noise, water, and air transfer and it must feature adequate mechanical strength and fire resistance.

To meet these requirements, a typical building partition consists of several layers. Thermal insulation serves to decrease the heat flow through a specific building partition.

The most important thermal insulation feature is the coefficient of thermal conductance.

The coefficient of thermal conductance (λ, W m^{-1} K^{-1}) is defined as the amount of heat (J) conducted in one second through the area of 1 m^2 of material with a thickness of 1 m, when a temperature change through the material in steady heat flow conditions is 1 K.

In addition to the coefficient of thermal conductance, the following properties are important:

- Density,
- Specific heat capacity,
- Surface emissivity (if radiation heat exchange is present),
- Water vapour diffusion resistance coefficient,
- Mechanical resistance,
- Dimensional stability,
- Acoustic performance,
- Non-flammable properties, and
- Economics.

The hydrothermal properties of building materials can be found in the ISO (2007c) standard. Most common thermal insulation materials for buildings include: wood fibre, cellulose, glass wool, stone wool, expanded polystyrene (EPS), extruded polyethylene foam (XPE), polyurethane foam (PUF), phenolic foam (PF), vacuum insulation panels (VIP), and aerogel. The preferred technology, its limits and actual requirements are provided by manufacturers. Some thermal insulation materials are characterized in Table 1.

The simplified method of calculating thermal transmittance of thermally homogeneous elements of the envelope is described in detail in the ISO (2007a) standard.

This method is used to determine the total thermal resistance of a partition as the sum of thermal resistances of the component processes:

Table 1. Basic properties of thermal insulation materials (Pando A.R. et al., 2014)

Material	Density, [kg m^{-3}]	Thermal Conductivity, [W m^{-1} K^{-1}]	Specific Heat, [J kg^{-1} K^{-1}]	Vapour Resistance Factor [-]
Cellulose	27 to 65	0.035 in lofts; 0.038-0.040 in walls	2020	No data
Expanded polystyrene (EPS)	15 to 30	0.034 to 0.038	1300	No data
Extruded polystyrene (XPS)[)]	20 to 40	0.029 to 0.035	1300	100
Glass mineral wool	20	0.035	1030	No data
Rock mineral wool	40	0.031 – 0.050	No data	No data
Wood fibre rigid [1)]	240	0.046	2100	1 to 2
Wood fibre flexible [1)]	50	0.038	2100	5
Cork	100 - 200	0.045 – 0.060	No data	No data
Aerogel		0.012 – 0.013	No data	100
VIP		0.005 to 0.007	No data	No data

[1)]STEICO flex

- Convection and radiation on the internal side,
- Conduction of heat through the material layers, and
- Convection and radiation on the external side.

The total thermal resistance (R_{tot}) of a building component consisting of homogeneous layers can be expressed as:

$$R_{tot} = R_{si} + \sum_{i=1}^{n} R_i + R_{se}, \qquad m^2KW^{-1} \tag{1}$$

where:

R_{si} = Internal surface resistance, $m^2\ K\ W^{-1}$,
R_i = Design thermal resistances of layer *i*:

$$R_i = \frac{d_i}{\lambda_i}, \qquad m^2KW^{-1} \tag{2}$$

d_i = Thickness of layer *i*, m,
λ_i = Design thermal conductivity of the material in layer *i*, $W\ m^{-1}\ K^{-1}$,
R_{se} = External surface resistance, $W\ m^{-1}\ K^{-1}$.

Thermal transmittance (known as a '*U*-value') characterizes the thermal insulation of a partition and is expressed as:

$$U = \frac{1}{R_{tot}}, \qquad Wm^{-2}K^{-1} \tag{3}$$

Surface resistances (Table 2) depend on the direction and surroundings conditions.

The surface resistances only apply to surfaces in contact with air. The internal surface resistance applies to all surfaces on the internal site of the building partition. The point is that the air flow on the internal site is a natural convection. The low speed of air flow in contact with the partition makes heat

Table 2. Conventional surface resistances (ISO 2007a.)

Surface Resistance	Direction of Heat Flow		
	Upwards (↑)	Horizontal (↔)	Downwards (↓)
	Conventional Surface Resistances, [$m^2 \cdot K\ W^{-1}$]		
R_{si}	0.10	0.13	0.17
R_{se}	0.04	0.04	0.04

transfer conditions worse, which is why the surface resistance on the internal site is higher than on the external site. It is worth to add that the internal surface resistance is applied to unheated spaces as well.

If unventilated air layers with high emissivity surfaces are present, their thermal resistance is assumed to be constant (Table 3).

Two simple examples of calculations are given in Table 4.

This type of external wall construction (Table 4) was commonly used in Poland in the first half of the 20th century. Its thermal insulation quality ($U = 1.23$ W m^{-2} K^{-1}) is nowadays unacceptable.

An external wall (Table 5) was widely used in 1960s in single family houses in Poland. Its thermal insulation quality ($U = 1.17$ W m^{-2} K^{-1}) is nowadays unacceptable. The unventilated air layer served as the thermal insulation.

The methodology in more complicated cases, for example in the case of a building component consisting of inhomogeneous layers or having a slightly or well ventilated air layer, are described in the ISO (2007a) standard.

Table 3. Thermal resistance of unventilated air layers with high emissivity surfaces (ISO 2007a)

Thickness of Air Layer	Thermal Resistance, [m²·K W⁻¹]		
	Direction of Heat Flow		
	Upwards (↑)	Horizontal (↔)	Downwards (↓)
0	0.00	0.00	0.00
5	0.11	0.11	0.11
7	0.13	0.13	0.13
10	0.15	0.15	0.15
15	0.16	0.17	0.17
25	0.16	0.18	0.19
50	0.16	0.18	0.21
100	0.16	0.18	0.22
300	0.16	0.18	0.23

Table 4. A wall of ceramic bricks: Calculation of thermal resistance and thermal transmittance

No.	Layer Description	Thickness, [m]	Thermal Conductivity, [W m⁻¹ K⁻¹]	Thermal Resistance, [m² K W⁻¹]	
i	internal surface resistance	-	-	0.130	
1	lime plaster	0.015	0.700	0.021	
2	wall of ceramic bricks	0.48	0.770	0.623	
e	external surface resistance	-	-	0.040	
			R_{tot} =	0.814	m² K W⁻¹
			U =	1.229	W m⁻² K⁻¹

Table 5. A wall of cellular concrete blocks on cement-lime mortar with an unventilated air layer: Calculation of thermal resistance and thermal transmittance

No.	Layer Description	Thickness [m]	Thermal Conductivity, [W m^{-1} K^{-1}]	Thermal Resistance, [m^{2}K W^{-1}]	
i	internal surface resistance	-	-	0.130	
1	cement-lime plaster	0.015	0.820	0.018	
2	cellular concrete blocks on cement-lime mortar	0.12	0.770	0.156	
3	unventilated air layer *)	0.03	-	0.180	
4	cellular concrete blocks on cement-lime mortar	0.24	0.770	0.312	
5	cement-lime plaster	0.015	0.820	0.018	
e	external surface resistance	-	-	0.040	
			R_{tot} =	0.854	m^2 K W^{-1}
			U =	1.171	W m^{-2} K^{-1}

*)When the thickness of the unventilated air layer is not included in the table, intermediate values should be obtained by linear interpolation.

Thermal Bridges

Thermal bridges are identified as areas of the envelope in which heat exchange is intensified. The reasons could be: a change in the construction (e.g. a connection between the wall and the frame of a window, a connection of a window sill with the wall, additionally with the influence of the window frame), a change of dimensions (e.g. a change in wall thickness or in a corner of the wall), or the use of different materials (e.g. a concrete pillar inside the wall). These examples can be classified as linear thermal bridges and their length can be specified. The heat exchange through a thermal bridge is characterized by linear thermal transmittance. In some cases it is possible to use a simplified method and apply values given in the ISO (2007e) standard, but the values provided in the standard are limited to the cases considered, e.g. for external wall thickness 300 mm, internal wall thickness 200 mm, thermal transmittance of insulated walls 0.343 W m^{-2} K^{-1}, and thermal resistance of the insulation layer 2.5 m^2 K W^{-1}. A consequence of the above mentioned assumptions is that the applicability of this method is very limited.

In the modernization process of a building to the nZEB standard the value of the linear thermal transmittance should be determined by numerical methods verified according to the ISO (2007b) standard. Some examples of numerical programs for calculating heat exchange through thermal bridges are listed in Table 6.

An example of temperature distribution in the corner of a ceramic brick wall (a connection of two walls, Table 4) is shown in Figure 3 and Figure 4. In this case the linear thermal transmittance is negative and is equal to -0.47 W m^{-1} K^{-1}. Such case is not included in the ISO (2007e) standard. The lowest temperature on the internal surface is 8.0 °C. That value can be applied to avoid the mould growth (ISO 13788:2012).

Some constructional solutions cause the creation of point thermal bridges (e.g. a connection point of a non-insulated column with the ceiling). Their impact on heat exchange is characterized by a point thermal transmittance (χ, W K^{-1}), which can be determined using 3D numerical programs.

Table 6. Numerical programs for calculating linear thermal transmittance of thermal bridges

Program	Web Page	Pricing
AnTherm	www.antherm.at	commercial
BISCO, TRISCO	http://www.physibel.be	commercial
CYPE	http://www.cype.com/en/cype/	commercial
Heat 2D, Heat 3D	www.buildingphysics.com	commercial
HTflux	www.htflux.com	commercial
Psi Therm	www.psitherm.uk	commercial
Therm	windows.lbl.gov/software/therm/therm.html	freeware

Figure 3. Temperature field in the corner of a ceramic brick wall: Isotherms
LBNL. Therm 7.5.

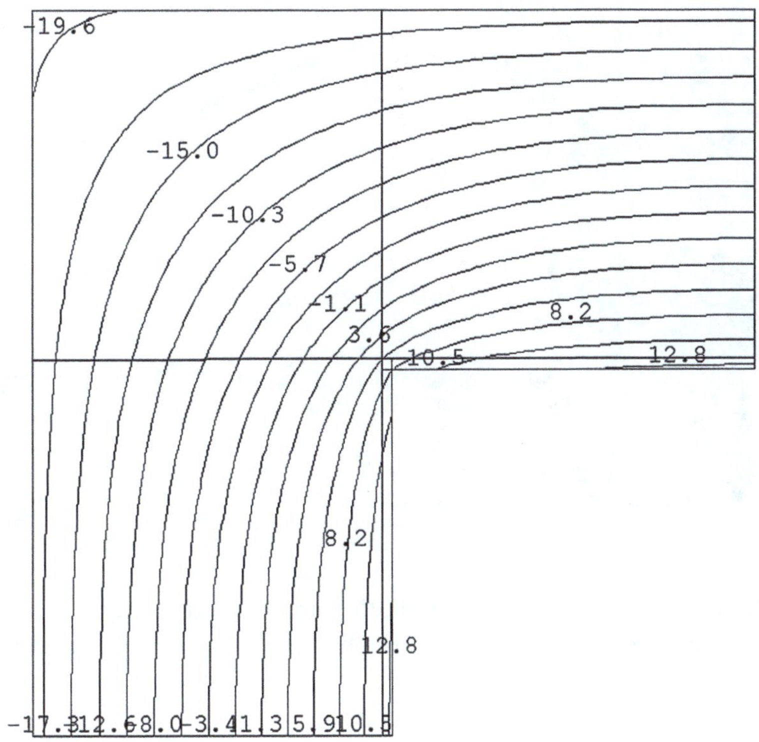

Heat Transfer Characteristics of the Envelope

The building envelope separates a space from the environment. Its parts, such as walls, roof, floor, windows, and doors, must fulfil construction, durability, thermal, and acoustic requirements and must also offer adequate fire resistance for safety reasons.

Technologies needed to improve thermal insulation quality depend on the type of the wall.

The description of heat flow through the envelope is simplified. It is assumed that the total heat transfer is a sum of heat transfers through each element under steady-state conditions. Then the heat exchange

Figure 4. Temperature field in the corner of a ceramic brick wall: Colour infrared
LBNL. Therm 7.5.

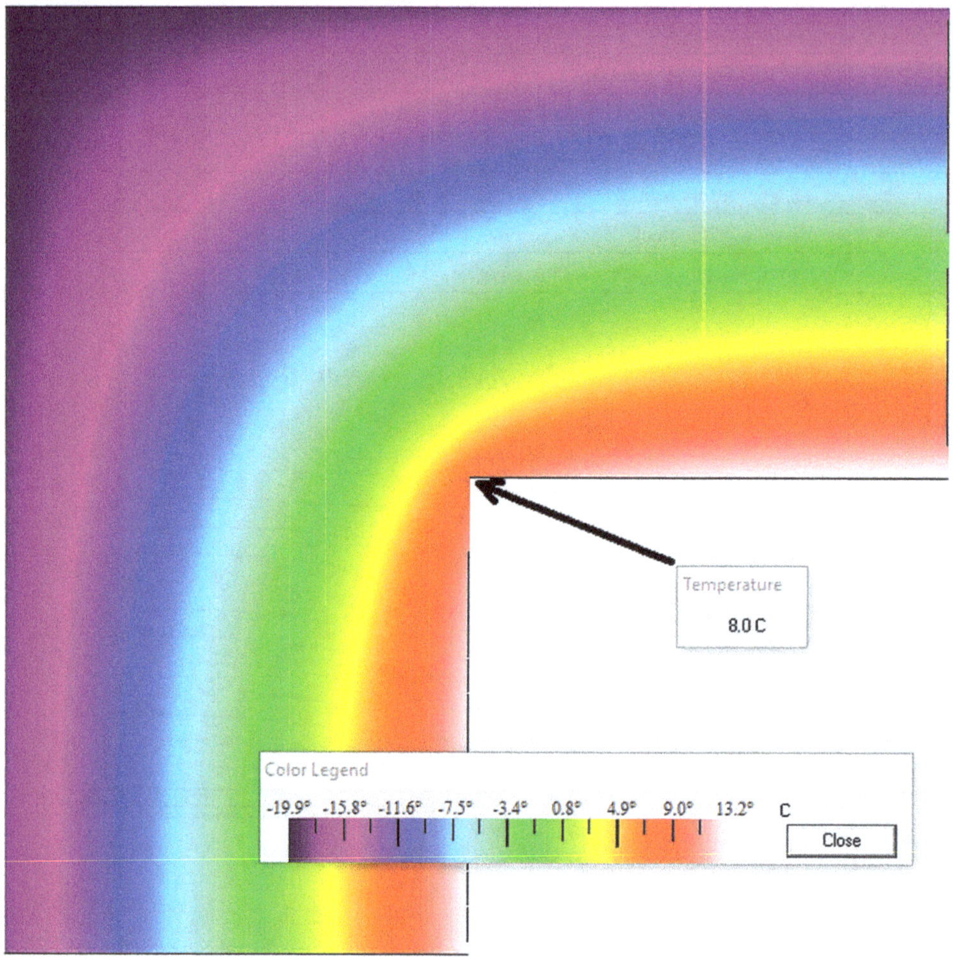

is proportional to thermal transmittance (*U*-value), the area through which the heat exchange occurs, the temperature difference between both sides of the building component, and the period considered, for example a month.

The following elements of the envelope are considered:

- Wall (or walls),
- Windows,
- Door (doors),
- Roof (or roofs),
- Ceiling (ceilings) under unheated room,
- Floor on the ground (or floors), and
- Ceiling (or ceiling above the basement).

The heat flow is forced by the temperature difference (Figure 5):

Figure 5. A diagram of the temperature-forced heat flow between the building and the environment

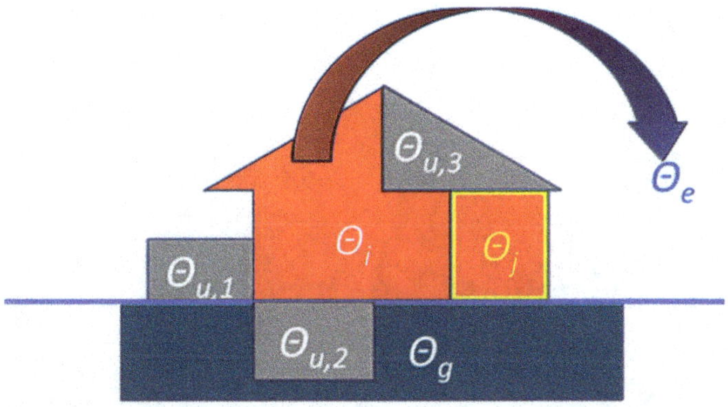

- Between internal (θ_i) and external temperature (θ_e): $(\theta_i - \theta_e)$, K,
- Between internal (θ_i) and another internal temperature of the conditioned space (θ_j): $(\theta_i - \theta_j)$, K,
- Between internal (θ_i) and unheated spaces temperature (θ_u): $(\theta_i - \theta_u)$, K,
- Between internal (θ_i) and ground temperature (θ_g): $(\theta_i - \theta_u)$, K.

Let us assume that all temperatures are monthly averages $\left(\bar{\theta}_i, \bar{\theta}_e, \bar{\theta}_j, \bar{\theta}_u, \bar{\theta}_g\right)$ and consider all partitions of the external envelope (walls, roofs, windows, and thermal bridges). Then, the heat exchange between the internal conditioned space and the environment over a month (the number of hours: t_M, h) can be described as:

$$
\begin{aligned}
Q_{tr,M} = & \sum_{k}\left(U_{ie,k}\cdot A_{ie,k} + \sum_{k1}\Psi_{ie,k1}\cdot L_{ie,k1} + \sum_{k2}\chi_{ie,k2}\cdot n_{ie,k2}\right)\cdot\left(\bar{\theta}_i - \bar{\theta}_e\right)\cdot t_M \\
& + \sum_{l}\left(U_{ije,l}\cdot A_{ije,l} + \sum_{l1}\Psi_{ije,l1}\cdot L_{ije,l1} + \sum_{l2}\chi_{ije,l2}\cdot n_{ije,l2}\right)\cdot\left(\bar{\theta}_i - \bar{\theta}_{j,l}\right)\cdot t_M \\
& + \sum_{m}\left(U_{iue,m}\cdot A_{iue,m} + \sum_{m1}\Psi_{iue,m1}\cdot L_{iue,m1} + \sum_{m2}\chi_{iue,m2}\cdot n_{iue,m2}\right)\cdot\left(\bar{\theta}_i - \bar{\theta}_{u,m}\right)\cdot t_M \\
& + \sum_{n}\left(U_{ige,n}\cdot A_{ige,n} + \sum_{n1}\Psi_{ige,n1}\cdot L_{ige,n1} + \sum_{n2}\chi_{ige,n2}\cdot n_{ige,n2}\right)\cdot\left(\bar{\theta}_i - \bar{\theta}_{g,n}\right)\cdot t_M, \quad \text{Wh,}
\end{aligned}
\tag{4}
$$

where:

k, l, m, n = Designation of partitions of the envelope,
$k1, l1, m1, n1, k2, l2, m2, n2$ = Designation of linear and point thermal bridges.

The equation above can be converted into the form:

$$Q_{tr,M} = \Bigg[\sum_k \left(U_{ie,k} \cdot A_{ie,k} + \sum_{k1} \Psi_{ie,k1} \cdot L_{ie,k1} + \sum_{k2} \chi_{ie,k2} \cdot n_{ie,k2}\right) \cdot \frac{\left(\bar{\theta}_i - \bar{\theta}_e\right)}{\left(\bar{\theta}_i - \bar{\theta}_e\right)}$$
$$+\sum_l \left(U_{ije,l} \cdot A_{ije,l} + \sum_{l1} \Psi_{ije,l1} \cdot L_{ije,l1} + \sum_{l2} \chi_{ije,l2} \cdot n_{ije,l2}\right) \cdot \frac{\left(\bar{\theta}_i - \bar{\theta}_{j,l}\right)}{\left(\bar{\theta}_i - \bar{\theta}_e\right)}$$
$$+\sum_m \left(U_{iue,m} \cdot A_{iue,m} + \sum_{m1} \Psi_{iue,m1} \cdot L_{iue,m1} + \sum_{m2} \chi_{iue,m2} \cdot n_{iue,m2}\right) \cdot \frac{\left(\bar{\theta}_i - \bar{\theta}_{u,m}\right)}{\left(\bar{\theta}_i - \bar{\theta}_e\right)} \tag{5}$$
$$+\sum_n \left(U_{ige,n} \cdot A_{ige,n} + \sum_{n1} \Psi_{ige,n1} \cdot L_{ige,n1} + \sum_{n2} \chi_{ige,n2} \cdot n_{ige,n2}\right) \frac{\left(\bar{\theta}_i - \bar{\theta}_{g,n}\right)}{\left(\bar{\theta}_i - \bar{\theta}_e\right)}\Bigg]$$
$$\cdot\left(\bar{\theta}_i - \bar{\theta}_e\right) \cdot t_M, \quad \text{Wh}$$

The expression in square brackets is the thermal characteristic of the building's envelope, i.e. the heat transfer coefficient:

$$H_{tr} = \sum_k \left(U_{ie,k} \cdot A_{ie,k} + \sum_{k1} \Psi_{ie,k1} \cdot L_{ie,k1} + \sum_{k2} \chi_{ie,k2} \cdot n_{ie,k2}\right)$$
$$+\sum_l \left(U_{ije,l} \cdot A_{ije,l} + \sum_{l1} \Psi_{ije,l1} \cdot L_{ije,l1} + \sum_{l2} \chi_{ije,l2} \cdot n_{ije,l2}\right) \cdot b_{r,j}$$
$$+\sum_m \left(U_{iue,m} \cdot A_{iue,m} + \sum_{m1} \Psi_{iue,m1} \cdot L_{iue,m1} + \sum_{m2} \chi_{iue,m2} \cdot n_{iue,m2}\right) \cdot b_{r,u} \tag{6}$$
$$+\sum_n \left(U_{ige,n} \cdot A_{ige,n} + \sum_{n1} \Psi_{ige,n1} \cdot L_{ige,n1} + \sum_{n2} \chi_{ige,n2} \cdot n_{ige,n2}\right) \cdot b_{r,g}, \quad \text{WK}^{-1}$$

where

$$b_{r,x} \frac{\left(\bar{\theta}_i - \bar{\theta}_x\right)}{\left(\bar{\theta}_i - \bar{\theta}_e\right)} \tag{7}$$

denotes the temperature adjustment factor which takes into account the change in temperature difference that forces the heat flow. In fact, the heat flow is caused by the actual temperature but it is referred to the external air temperature.

Such a mathematical notation makes it possible to formulate a very simple relationship describing the heat transfer between the building and its surrounding in the period of one month (compare: Equations 18 and 19):

$$Q_{tr,M} = H_{tr} \cdot \left(\bar{\theta}_i - \bar{\theta}_e\right) \cdot t_M, \qquad \text{Wh} \tag{8}$$

Thermal Characteristics of Ventilation

Ventilation is needed to supply fresh air into the conditioned space and to remove moisture, contaminants and odours produced in that space. There are three main types of ventilation systems: natural, mechanical, and hybrid (being a combination of the natural and mechanical ones).

In the case of natural ventilation the air flows through gaps or openings in the external envelope and is removed through ventilation ducts (a chimney). The main advantage of this kind of ventilation is that it requires no energy to run, but its reliability and quality can be unacceptable, and it is difficult to control the airflow.

In mechanical systems the airflow is forced by a fan and can be controlled by changing the rotational speed of the fan.

Hybrid ventilation has advantages of both systems; the fan is only used when the effectiveness of the natural ventilation is insufficient.

In all types of the ventilation systems the total amount of airflow is treated as a sum of the required airflow and the infiltrated airflow. The amount of the infiltrated airflow depends on air tightness and height of the building, temperature difference and wind conditions. In an existing building the air infiltration rate can be estimated by the fan pressurization method (ISO 9972:2015).

A simplified method of calculating the airflow in a building uses an air change rate (air change per hour, ACH), being a measure of the air volume added to or removed from a space (q_{ve}, $m^3\ h^{-1}$) referred to the volume of the space (V_B, m^3):

$$n = \frac{q_{ve}}{V_B}, \qquad h^{-1} \tag{9}$$

Note: The air exchange rate cannot be used to design HVAC systems.

The air change rate should be determined based on building air tightness tests (so-called n_{50}), and under this condition it is justified to assume specific values. For most buildings, however, the value of n_{50} is not known. Real air change rates in residential buildings in Poland are in the range of 0.1 h^{-1} to 40.0 h^{-1} (Laskowski. 2008). Suggested air change rates in residential buildings in Polish conditions are shown in Table 7. It should be noted that mechanical ventilation is required in airtight buildings.

Based on the concept of the air exchange rate, the thermal characteristic of the ventilation system can be determined by the overall ventilation heat transfer coefficient:

Table 7. Air change rates in residential buildings in Polish conditions (RMI. 2008)

Type of Building	Air Change Rate, [h^{-1}]
Passive residential buildings	0.6
Buildings with low energy demand for heating	1.5 (1.0)
Buildings with a gravitational system (new)	4.0
Buildings with a gravitational system (old)	from 7.0

$$H_{ve,adj} = \rho_a \cdot c_a \cdot n \cdot V_B, \quad WK^{-1} \tag{10}$$

In a more advanced model used for energy analysis, different kinds of airflows are distinguished; hence the overall ventilation heat transfer coefficient can be expressed as:

$$H_{ve,adj} = \rho_a \cdot c_a \cdot \left(\sum_k b_{ve,k} \cdot q_{ve,k,mn} \right), \qquad WK^{-1} \tag{11}$$

where:

$\rho_a \cdot c_a$ = Heat capacity of air per volume, $\rho_a \cdot c_a = 1200 \quad Jm^{-3}K^{-1}$,
$q_{ve,k,mn}$ = Time-average airflow rate of air flow element k, m^3 s^{-1},
$b_{ve,k}$ = Temperature adjustment factor for air flow element k, with value $b_{ve,k} \neq 1$ if the supply temperature, $\theta_{sup,k}$, is not equal to the temperature of the external environment, such as in the case of pre-heating, pre-cooling or heat recovery:

$$b_{ve,k} = \frac{\theta_{int,set} - \theta_{sup,k}}{\theta_{int,set} - \theta_e} \tag{12}$$

$\theta_{int,set}$ = Set-point temperature of the building zone for heating or cooling, °C,
$\theta_{sup,k}$ = Supply temperature, equal to the internal temperature of the adjacent building, °C,
θ_e = Temperature of the external environment, °C.

In the case of a heat recovery unit, the temperature adjustment factor is calculated as

$$b_{ve,k} = \left(1 - f_{ve,frac,k} \cdot \eta_{hru}\right) \tag{13}$$

where:

η_{hru} = Thermal efficiency of the heat recovery unit:

$$\eta_{hru} = \frac{\theta_{sup,k} - \theta_e}{\theta_{int,set} - \theta_e} \tag{14}$$

$\theta_{sup,k}$ = Supply temperature from the heat recovery unit, °C,
$f_{ve,frac,k}$ = Fraction of the considered air flow element k that goes through the heat recovery unit.

In the case of a series of two heat recovery provisions, the resultant efficiency of heat recovery is:

$$\eta_{ve,k} = \eta_{ve,k1} + \eta_{ve,k2} - \left(\eta_{ve,k1} \cdot \eta_{ve,k2}\right) \tag{15}$$

$\eta_{ve,k1}$ = Average efficiency of heat recovery by provision 1, -,

$\eta_{ve,k2}$ = Average efficiency of heat recovery by provision 2, - .

Energy Analysis

Internal Conditions

The intended use of the building determines the required internal conditions and makes it easier to define occupancy conditions. Thus, the actual behaviour of inhabitants or employees and actual internal heat gains can be reflected more adequately.

National regulations may impose requirements on indoor climate parameters and specify, for example, internal heat gains per unit area. In addition, standard scheduling of building occupancy, operating modes of appliances or load profiles can be set. In the energy analysis the internal temperature (θ_{int}) is assumed.

Meteorological Data

For energy analysis, representative meteorological data are needed.

Depending on the applied model of the energy and/or heat flow, different sets of meteorological data should be known, for example:

- The monthly balance method requires:
  - Dry bulb temperature, °C,
  - Solar radiation (Direct normal, Diffuse, Global Horizontal), Wh m^{-2},
- The simplified model 5R1C requires hourly data mentioned above and:
  - sky temperature, °C;
- A detailed simulation (depending on the model) requires additionally:
  - Wet bulb temperature, °C,
  - Wind speed, m s^{-1},
  - Wind direction, -.

An important problem concerning the comparison of the proposed solutions for different countries is to apply a consistent set of meteorological data appropriate for the building energy performance analysis. For example, EUROSTAT publishes only Heating Degree Days for EU-28 countries, and Norway is not included in the list (EUROSTAT).

Monthly Balance Method

Effective Energy Demand for Heating and Cooling

To determine the effective energy demand for heating and cooling, the simplified monthly method can be used (ISO, 2008). The main advantage of this method is its simplicity and that it can be implemented in a spreadsheet, but this method can be applied only at the first step when evaluating energy quality of a particular building.

The monthly balance method (ISO, 2008) is one of the simplest methods known that can be used to estimate energy use for space heating and cooling and that can be applied to non-dynamic cases. In this method the energy demand is described as a sum of energy for heating and/or cooling (if used) during the chosen period of the year (the heating or cooling season).

In each month considered:

- The energy needed for space heating is expressed as a function of heat losses by transmission (Q_{tr}, Equation 18), heat losses by ventilation (Q_{ve}, Equation 22), internal heat gains (Q_{int}, Equation 25), solar heat gains (Q_{sol}, Equation 24), and the gain utilization factor ($\cdot_{H,gn}$, Equation 30):

$$Q_{H,nd} = \left(Q_{tr} + Q_{ve}\right) - \eta_{H,gn} \cdot \left(Q_{\text{int}} + Q_{sol}\right), \qquad \text{kWh / m-th} \tag{16}$$

- The energy needed for space cooling is expressed in a similar form:

$$Q_{C,nd} = \left(Q_{sol} + Q_{\text{int}}\right) - \eta_{C,ls} \cdot \left(Q_{tr} + Q_{ve}\right), \qquad \text{kWh / m-th} \tag{17}$$

where:

$\eta_{C,ls}$ = Utilization factor for heat losses.

Total Heat Transfer by Transmission

The total heat transfer by transmission is calculated for each month:

- For heating:

$$Q_{tr} = H_{tr,adj} \cdot \left(\theta_{init,H,set} - \theta_{e}\right) \cdot t_{M} \cdot 10^{-3}, \quad \text{kWh / m-th} \tag{18}$$

- For cooling:

$$Q_{tr} = H_{tr,adj} \cdot \left(\theta_{init,C,set} - \theta_{e}\right) \cdot t_{M} \cdot 10^{-3}, \quad \text{kWh / m-th} \tag{19}$$

where:

$H_{tr,adj}$ = Overall heat transfer coefficient by transmission of the building (zone), adjusted for the indoor-outdoor temperature difference (Equation 6), W K^{-1},

$\theta_{int,H,set}$ = Set-point temperature of the building (zone) for heating, calculated as a weighted average value for the heated space:

$$\theta_{init,H,set} = \frac{\sum_s A_{f,s} \cdot \theta_{init,s,H,set}}{\sum_s A_{f,s}}, \quad °C \tag{20}$$

$\theta_{int,s,H,set}$ = Set-point temperature for heating of space *s*, °C,

$\theta_{int,C,set}$ = Set-point temperature of the building zone for cooling:

$$\theta_{init,C,set} = \frac{\sum_s A_{f,s} \cdot \theta_{init,s,C,set}}{\sum_s A_{f,s}}, \quad °C \tag{21}$$

$\theta_{int,s,C,set}$ = Set-point temperature for cooling of space *s*, °C,

$A_{f,s}$ = Conditioned floor area of space *s*, m^2,

θ_e = Temperature of the external environment, °C,

t_M = Duration of the month, h.

Total Heat Transfer by Ventilation

The total heat transfer by ventilation is calculated for each month:

- For heating:

$$Q_{ve} = H_{ve,adj} \cdot \left(\theta_{init,H,set} - \theta_e\right) \cdot t_M \cdot 10^{-3}, \quad \text{kWh / m-th} \tag{22}$$

- For cooling:

$$Q_{ve} = H_{ve,adj} \cdot \left(\theta_{init,C,set} - \theta_e\right) \cdot t_M \cdot 10^{-3}, \quad \text{kWh / m-th} \tag{23}$$

where:

$H_{ve,adj}$ = Overall heat transfer coefficient by ventilation, adjusted for the indoor-outdoor temperature difference (compare: Equation 11), W K^{-1}.

Solar Heat Gains

Depending on the location, solar gains can significantly contribute to the increase in energy demand for cooling (e.g. in the south of Europe). On the other hand, they can also reduce the energy demand for heating (e.g. in central and northern Europe in transitional months).

$$Q_{sol} = \sum_{k} A_{G,k} \cdot I_{sol,k} \cdot g_{g,k} \cdot F_{sh,k} \cdot 10^{-3}, \quad \text{kWh / m-th} \tag{24}$$

where:

$A_{G,k}$ = The effective collecting area (glass) of surface *k* with a given orientation and tilt angle, m^2,
$I_{sol,k}$ = Energy of the solar irradiation over a month, per square metre of collecting area of surface *k* with a given orientation and tilt angle, Wh m^{-2},
$g_{g,k}$ = Total solar energy transmittance of a building element, -,
F_{sh} = Shading reduction factor for external obstacles for the solar effective collecting area of surface *k*, -.

Internal Heat Gains

Heat gains from internal heat sources, including negative heat gains, comprise any heat generated in the conditioned space by internal sources other than the energy intentionally utilized for space heating, space cooling or hot water preparation.

The internal heat gains include:

- Metabolic heat from occupants and dissipated heat from appliances,
- Heat dissipated from lighting devices,
- Heat dissipated from, or absorbed by, hot and mains water and sewage systems,
- Heat dissipated from, or absorbed by, heating, cooling and ventilation systems, and
- Heat from processes.

In the simplified method the monthly value of internal heat gains can be calculated as:

$$Q_{int} = q_{int} \cdot A_f \cdot t_M \cdot 10^{-3}, \qquad \text{kWh / m-th} \tag{25}$$

where:

q_{int} = Unit heat gains, W m^{-2},
A_f = Area of the conditioned space, m^2,
t_M = Number of hours in a particular month, h.

In the case of low energy losses from a building, not all heat gains can reduce the energy demand for heating, simply because some heat is removed through open windows from the space to avoid high internal temperature. This is reflected in the mathematical model by the gain utilization factor for heating ($\eta_{H,gn}$), which reduces total monthly heat gains.

In some cases this factor reflects dynamic properties of a building and is expressed as a relationship between heat losses, heat gains and the internal heat capacity.

Internal Heat Capacity

The internal capacity is calculated by summing heat capacities of all the building elements in direct thermal contact with internal air:

$$C_m = \sum_j \kappa_j \cdot A_j, \qquad \mathrm{JK^{-1}} \tag{26}$$

where:

κ_j = Internal heat capacity per area of building element j, J m^{-2} K^{-1},
A_j = Area of element j, m^2.

The internal heat capacity per area of building element j can be in a simplified way calculated as:

$$\kappa_j = \sum_j \sum_i \left(c_{ij} \cdot \rho_{ij} \cdot d_{ij}\right), \qquad \mathrm{J\,m^2K^{-1}} \tag{27}$$

and for given partition j:

$$\sum_i d_{ij} \leq 0.10, \quad \mathrm{m} \tag{28}$$

where:

c_{ij} = Specific heat of the material in layer i of partition j, J kg^{-1} K^{-1},
ρ_{ij} = Density of the material in layer i of partition j, kg m^{-3},
d_{ij} = Thickness of the material in layer i of partition j, m.

More detailed description of calculation methods concerning the dynamic thermal characteristics of a building component is given in the ISO (2017) standard.

In some instances the internal heat capacity can be calculated as a function of the conditioned area (A_f), in which case Table 8 is useful (ISO, 2008), since a traditional construction of external walls, i.e. a brickwork wall or a hollow clay block wall, characterized by a very high internal heat capacity, is the most common one in Poland. Therefore, the 'F' class, comprising buildings of an 'ultra-heavy' construction, was added (Table 8).

Table 8. Internal heat capacity (ISO 2008)

Type	Building Class	Internal Heat Capacity C_m, [J K^{-1}]
A	Very light	80,000 A_f
B	Light	110,000 A_f
C	Medium	165,000 A_f
D	Heavy	260,000 A_f
E	Very heavy	370,000 A_f
F	Ultra-heavy (PL)	600,000 A_f

Time Constant of the Building

Dynamic properties of a building are described by the time constant of the building (τ), which is a function of the internal heat capacity (C_m), the transmission heat transfer coefficient (H_{tr}), and the ventilation heat transfer coefficient (H_{ve}):

$$\tau = \frac{\frac{C_m}{3600}}{H_{tr} + H_{ve}}, \qquad \text{h} \tag{29}$$

The time constant is proportional to the internal heat capacity and inversely proportional to the thermal characteristic of the external envelope and ventilation. A higher time constant indicates slower changes in internal temperature.

Gain Utilization Factor for Heating

The gain utilization factor ($\eta_{H,gn}$) for heating is calculated in each month in the following way:

$$if\ \left(\gamma_H > 0\ and\ \gamma_H \neq 1\right),\ then\ \eta_{H,gn} = \frac{1-\gamma_H^{a_H}}{1-\gamma_H^{a_H+1}},$$

$$if\ \left(\gamma_H = 1\right),\ then\ \eta_{H,gn} = \frac{a_H}{a_H+1}, \tag{30}$$

$$if\ \left(\gamma_H < 1\right),\ then\ \eta_{H,gn} = \frac{1}{\gamma_H},$$

- Dimensionless heat gains to heat losses ratio:

$$\gamma_H = \frac{Q_{H,gn}}{Q_{H,ht}} = \frac{Q_{H,sol} + Q_{H,int}}{Q_{H,tr} + Q_{H,ve}} \quad (31)$$

- Dimensionless numerical parameter a_H depends on the time constant (τ):

$$a_H = a_{H,0} + \frac{\tau}{\tau_{H,0}} \quad (32)$$

where:

$a_{H,0}$ = Dimensionless reference numerical parameter (Table 9),
$\tau_{H,0}$ = Reference time constant, h.

Loss Utilization Factor for Cooling

The loss utilization factor for cooling ($\eta_{C,ls}$) is calculated in each month in the following way:

$$if\left(\gamma_C > 0\ and\ \gamma_C \neq 1\right),\ then\ \eta_{C,ls} = \frac{1-\gamma_C^{-a_C}}{1-\gamma_C^{-(a_C+1)}},$$

$$if\left(\gamma_C = 1\right),\ then\ \eta_{C,ls} = \frac{a_C}{a_C+1}, \quad (33)$$

$$if\left(\gamma_C < 1\right),\ then\ \eta_{C,ls} = 1,$$

- Dimensionless heat gains to heat losses ratio:

$$\gamma_C = \frac{Q_{C,gn}}{Q_{C,ht}} = \frac{Q_{C,sol} + Q_{C,int}}{Q_{C,tr} + Q_{C,ve}}, \quad (34)$$

Table 9. Reference parameters: $a_{H,0}$, $\tau_{H,0}$ (ISO 2008)

Type of Method	$a_{H,0}$	$\tau_{H,0}$
Monthly calculation method	1.0	15

- The dimensionless numerical parameter depending on the time constant:

$$a_C = a_{C,0} + \frac{\tau}{\tau_{C,0}}, \tag{35}$$

where:

$a_{C,0}$ = Dimensionless reference numerical parameter (Table 10),
$\tau_{C,0}$ = Reference time constant, h.

Heating Season

The heating season can be defined in various ways according to national regulations:

- As a given and established period,
- According to dynamic properties of the building and operational conditions,
- Including all months with positive heating (cooling) demand.

The length of heating season L_H to determine the number of hours of operation of certain season-length-dependent provisions is calculated as:

$$L_H = \sum_{M=1}^{12} f_{H,M} \cdot t_M, h \tag{36}$$

where:

$f_{H,M}$ = Fraction of the month that is part of the heating season.

Two methods are provided in the ISO (2008) standard, and method ‘b’ is explained below:

- Determine the limit value of the dimensionless heat-balance ratio for the heating mode:

$$\gamma_{H,\lim} = \frac{a_H + 1}{a_H} \tag{37}$$

- Determine for each month:
    - The value of $\gamma_{H,B}$ at the beginning of the month as the mean value of γ_H for the month considered and the previous month (for January, the previous month is December),
    - The value of $\gamma_{H,E}$ at the end of the month as the mean value of γ_H for the month considered and the next month (for December, the next month is January),

Table 10. Reference parameters: $a_{C,0}$, $\tau_{C,0}$ (ISO, 2008)

Type of Method	$a_{C,0}$	$\tau_{C,0}$
Monthly calculation method	1.0	15

- When applicable: negative values of γ_H shall be replaced by the value from the nearest month with a positive value of γ_H:
- Find $\gamma_{H,1} = \min\left(\gamma_{H,B}; \gamma_{H,E}\right)$,
- Find $\gamma_{H,2} = \max\left(\gamma_{H,B}; \gamma_{H,E}\right)$.

- Consider cases:
  - If $\gamma_{H,2} < \gamma_{H,\lim}$, then $f_H = 1$ (the whole month is part of the heating period),
  - If $\gamma_{H,1} > \gamma_{H,\lim}$, then $f_H = 0$ (the whole month is outside the heating period),
  - Otherwise:

$$if\ \gamma_H > \gamma_{H,\lim}, then\ f_H = 0.5\frac{\gamma_{H,\lim} - \gamma_{H,1}}{\gamma_H - \gamma_{H,1}},$$

$$if\ \gamma_H \leq \gamma_{H,\lim}, then\ f_H = 0.5 + 0.5\frac{\gamma_{H,\lim} - \gamma_H}{\gamma_{H,2} - \gamma_H}.$$

Cooling Season

The length of cooling season L_C to determine the number of hours of operation of certain season-length-dependent provisions is calculated as:

$$L_C = \sum_{M=1}^{12} f_{C,M} \cdot t_M, \qquad h, \tag{38}$$

where:

$f_{C,M}$ = Fraction of the month that is part of the cooling season.

Two methods are provided in the ISO (2008) standard, and method 'b' is explained below:

- Determine the limit value of the dimensionless heat-balance ratio for the cooling mode:

$$\left(1/\gamma_C\right)_{\lim} = \frac{a_C + 1}{a_C} \tag{39}$$

- Determine for each month:
  - The value of $\left(1/\gamma_C\right)_B$ at the beginning of the month as the mean value of $1/\gamma_C$ for the month considered and the previous month (for January, the previous month is December),
  - The value of $\left(1/\gamma_C\right)_E$ at the end of the month as the mean value of $1/\gamma_C$ for the month considered and the next month (for December, the next month is January),
  - When applicable: negative values of $1/\gamma_C$ shall be replaced by the value from the nearest month with a positive value of $1/\gamma_C$,
  - Find $\left(1/\gamma_C\right)_1 = \min\left[\left(1/\gamma_C\right)_B;\left(1/\gamma_C\right)_E\right]$,
  - Find $\left(1/\gamma_C\right)_2 = \max\left[\left(1/\gamma_C\right)_B;\left(1/\gamma_C\right)_E\right]$.
- Consider cases:
  - $if\ \left(1/\gamma_C\right)_2 < \left(1/\gamma_C\right)_{\lim},\ then\ f_C = 1$, (the whole month is part of the cooling period),
  - $if\ \left(1/\gamma_C\right)_1 > \left(1/\gamma_C\right)_{\lim},\ then\ f_C = 0$, (the whole month is outside the cooling period),
  - Otherwise:

$$if\ \left(1/\gamma_C\right) > \left(1/\gamma_C\right)_{\lim},\ then\ f_C = 0.5\frac{\left(1/\gamma_C\right)_{\lim} - \left(1/\gamma_C\right)_1}{1/\gamma_C - \left(1/\gamma_C\right)_1},$$

$$if\ \left(1/\gamma_C\right) \le \left(1/\gamma_C\right)_{\lim},\ then\ f_C = 0.5 + 0.5\frac{\left(1/\gamma_C\right)_{\lim} - \left(1/\gamma_C\right)}{\left(1/\gamma_C\right)_2 - \left(1/\gamma_C\right)}.$$

Simple Hourly Method 5R1C

In the case of dynamic solutions, such as:

- A construction comprising additional unheated buffer spaces with glazed walls to gain solar energy (solar passive solution),
- Using exterior blinds during certain periods,
- Using variable-volume forced ventilation with heat recovery, and
- Decreasing the ventilation air flow rate during the absence of occupants,

the simple hourly method 5R1C (ISO, 2008) or 6R1C (Narowski et al., 2010) is more appropriate.

The schematic diagram (Figure 6) shows the main quantities needed when applying the 5R1C method (a lumped-parameter method):

- Temperatures:
  - The actual, resulting room air temperature $\left(\theta_{air}\right)$,
  - Temperature on the internal surface of the envelope $\left(\theta_{S}\right)$,
  - External temperature $\left(\theta_{e}\right)$,
  - Temperature of the air supplied by the ventilation system $\left(\theta_{\sup}\right)$,
  - Temperature of heavy elements $\left(\theta_{m}\right)$,
- Thermal characteristics of the building envelope:
  - Heavy Partitions: Walls, roof, floors, ceilings (their heat capacity C_m is important in the dynamic process of changing internal temperature),
  - Light Partitions (their heat capacity is not important in the dynamic process of changing internal temperature),
- Heat fluxes:
  - Solar gains flow rate $\left(\phi_{sol}\right)$,
  - Internal gains flow rate $\left(\phi_{int}\right)$,
  - Internal heating or cooling power $\left(\phi_{HC,nd}\right)$.

Thermal characteristics of the envelope comprise four parts ($H_{tr,w}$, $H_{Tr,em}$, $H_{Tr,ms}$, $H_{Tr,is}$) and a characteristic of ventilation (H_{ve}). These five quantities represent 'resistances' (hence 5R).

As the next step, the sum of streams of internal heat gains and the heat gain from solar radiation is divided into three auxiliary heat fluxes: the first one $\left(\phi_{m}\right)$ affects the temperature of heavy partitions $\left(\theta_{m}\right)$, the second one $\left(\phi_{st}\right)$ affects the temperature $\left(\theta_{S}\right)$, and the third one $\left(\phi_{ia}\right)$ reflects air temperature $\left(\theta_{air}\right)$.

In comparison to the monthly balance method, the 5R1C method requires, in addition to the internal heat capacity, the internal heat transfer surface area (A_m) (Table 11).

The 5R1C method was applied to determine the annual effective energy demand for heating ($Q_{H,nd}$, kWh) and the annual effective energy demand for cooling ($Q_{C,nd}$, kWh).

Modernization of Existing Buildings

To evaluate the possibility of modernization, the actual energy performance of an existing building should be estimated. The evaluation can be executed based on:

- A simplified analysis of the numerical model,
- A detailed simulation of the building.

The final energy consumption should be verified on the basis of information taken from energy bills.

Possible changes which lead to improving the energy performance of an existing residential building include:

Figure 6. Schematic diagram of the 5R1C method
ISO, 2008.

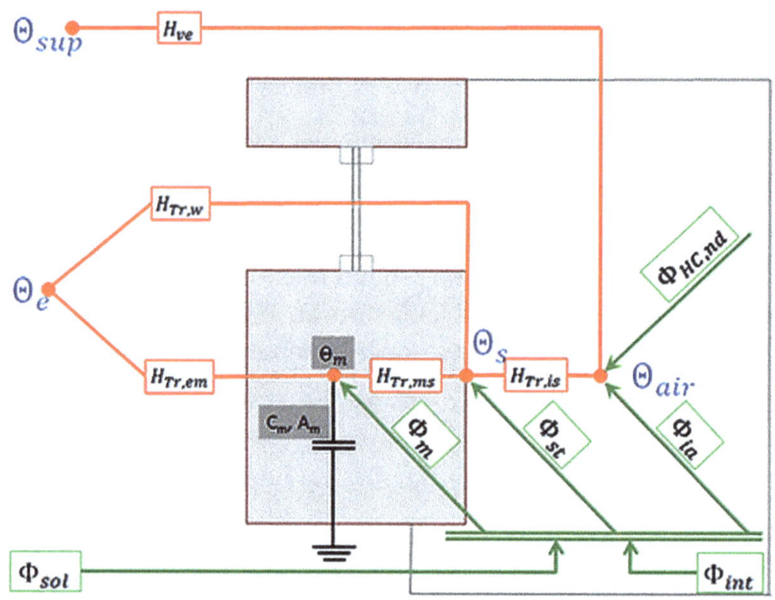

Table 11. Dynamic parameters for the 5R1C method (ISO, 2008)

Variant	Type	Building Class	Internal Heat Transfer Surface Area, A_m, [m^2]	Internal Heat Capacity, C_m, [J/K]
B1	A	Very light	2.5 A_f	80,000 A_f
B2	B	Light	2.5 A_f	110,000 A_f
B3	C	Medium	2.5 A_f	165,000 A_f
B4	D	Heavy	3.0 A_f	260,000 A_f
B5	E	Very heavy	3.5 A_f	370,000 A_f
B6	F	Ultra-heavy (PL)	3.5 A_f	600,000 A_f

- Permanent improvement of thermal performance of the envelope, including walls, a roof under a heated area, and ceilings over unheated spaces (e.g. the basement),
- Replacing windows and external doors,
- Increasing the internal heat capacity of the building,
- Construction of additional unheated buffer spaces, treated as a thermal insulation,
- Construction of additional unheated buffer spaces, treated as passive solar direct or indirect gain components of the envelope.

Thermal Insulation and Energy Indicators

To estimate the energy performance of a building and to compare the influence of modernization activities on energy efficiency, the following indicators are defined:

- Average thermal transmittance, as a simplified assessment of thermal insulation quality:

$$U_{AV,E} = \frac{H_{Tr}}{A_E}, \quad \mathrm{Wm^{-2}K^{-1}}, \tag{40}$$

- Useful energy indicator for heating:

$$EU_H = \frac{Q_{H,nd}}{A_f}, kWhm^{-2}yr^{-1}, \tag{41}$$

- Useful energy indicator for cooling:

$$EU_C = \frac{Q_{C,ls}}{A_f}, \quad \mathrm{kWhm^{-2}yr^{-1}}, \tag{42}$$

where:

H_{tr} = Thermal characteristic of the building's envelope (Equation 6), W K^{-1},
A_E = External area of the building's envelope where heat exchange between the building and the environment occurs, m^{-2},
$Q_{H,nd}$ = Energy needed for space heating (Equation 16), kWh yr^{-1},
$Q_{C,ls}$ = Energy needed for space cooling (Equation 17), kWh yr^{-1},
A_f= Area of the conditioned space, m^{-2}.

When using the average thermal transmittance of the building's envelope, it is assumed that heat losses by transmission are the same as in the case of a building with various external partitions. This very simple indicator shows the quality of thermal insulation, and its lower values characterize better solutions.

ENERGY ANALYSIS: CASE STUDY

Locations and Meteorological Data

For a simplified analysis, three different European locations were chosen:

Athens (Greece) – South Europe, Stockholm (Sweden) – North Europe, and Warsaw (Poland) – Central Europe.

The meteorological data were taken according to (METEO ASHRAE).

Description of the Building

A two-storey single-family building was analysed with a conditioned space area of 200 m^2 (100 m^2 per each storey), with a height of the storey equal to 2.80 m.

The area of transparent partitions was taken according to the minimum technical requirements as one eighth of the useful floor area. It was assumed that the areas of the south and north elevations each make up 30%, and the areas of the east and west ones each make up 20% of the area of external walls. The area of windows was assumed to make up 40% of the area of the south elevation, 10% of the north elevation, and 25% of the area of each of the east and west elevations. The building is occupied by four inhabitants.

The required indoor temperature during the heating season was assumed to be 20 °C; during summer it cannot exceed 26 °C in the cities in the south of Europe (Athens) or 24 °C in the cities in the northern and central Europe (Stockholm and Warsaw). According to the Köppen climate classification, the climate is classified for Athens as Csa, for Stockholm as Cfb, and for Warsaw as Dfb (WIKIPEDIA. Climate).

Description of Variants

For standard quality of thermal insulation, current Polish legal requirements were assumed (Table 12).

The characteristics of ventilation were described using the simplified method (Equation 10), which uses the air change rate. The values of the air flow rate considered are collected in Table 13.

The assumed values of heat gains are 3.5 W m^{-2} for variant D1 and 5.0 W m^{-2} for variant D2.

Tables 14 to 18 summarize the selected data variants taken for calculations: Table 14 - location, Table 15 - internal heat capacity, Table 16 - thermal insulation quality, Table 17 - internal heat gains, Table 18. - exchange rate.

Variants of thermal insulation quality (C) were evaluated according to the assumed building data to obtain thermal characteristics of the external envelope (Table 19).

Table 12. Assumed thermal performance of selected partitions (RMI. 2002)

Type of Partition	Thermal Performance of Selected Partitions, [W m^{-2} K^{-1}]					
	Very Low	Low	Standard (PL)	High	Very High	Future
Variant	**C1**	**C2**	**C3**	**C4**	**C5**	**C6**
External walls $\left(\theta_{int} \geq 16°C\right)$	0.55	0.30	0.23	0.20	0.10	0.05
Roofs, deck roofs and floors under unheated attic spaces and over passages $\left(\theta_{int} \geq 16°C\right)$	0.30	0.25	0.18	0.15	0.10	0.05
Floors over unheated rooms and closed crawl spaces $\left(\theta_{int} \geq 16°C\right)$	0.60	0.45	0.25	0.25	0.10	0.05
Windows $\left(\theta_{int} \geq 16°C\right)$	2.60	1.80	1.10	0.90	0.60	0.30
External doors	3.00	2.60	1.50	1.30	0.60	0.30

Table 13. Thermal characteristics of ventilation

Variant	E1	E2	E3
Air flow rate, h^{-1}	0.5	1.5	4.0
Ventilation heat transfer coefficient, $W\ K^{-1}$	93.3	280.0	746.7

Table 14. Variants of location (A)

Variant	A1	A2	A3
Location	Athens	Stockholm	Warsaw

Table 15. Variants of internal heat capacity (B)

Variant	B1	B2	B3	B4	B5	B6
Internal heat capacity	Very light	Light	Medium	Heavy	Very heavy	Ultra-heavy (PL)

Table 16. Variants of thermal insulation quality (C)

Variant	C1	C2	C3	C4	C5	C6
Thermal insulation quality	Very low	Low	Standard	High	Very high	Future

Table 17. Variants of internal heat gains (D)

Variant	D1	D2
Internal heat gains, $W\ m^{-2}$	3.5	5.0

Table 18. Air exchange rate variants (E)

Variant	E1	E2	E3
Air exchange rate, h^{-1}	0.5	1.5	4.0

Table 19. Thermal characteristics of the external envelope

Variant	C1	C2	C3	C4	C5	C6
Thermal quality	Very low	Low	Standard (PL)	High	Very high	Future
Transmission heat transfer coefficient, $W\ K^{-1}$	455.1	304.3	196.68	168.4	97.7	48.9
Average thermal transmittance, $W\ m^{-2}\ K^{-1}$	0.86	0.58	0.37	0.32	0.19	0.09

The transmission heat transfer coefficient (Equation 6) is a function of the external area of a building and heat transmittances of external partitions. However, when this coefficient is used, it is difficult to assess whether or not thermal insulation quality is good. It is much easier to evaluate the quality based on the average thermal transmittance (Equation 40); e.g. at present, in Polish climatic conditions only variant C5 is reasonable from a technical and economic point of view.

Based on thermal characteristics of the external envelope (Equation 6), the structure of transmission heat losses from a building can be checked. Depending on the share of external walls, the roof, windows, linear bridges, and other components considered in thermal characteristics of the external envelope (Equation 6), the structure of heat losses can vary.

In the variant characterized by the worst thermal insulation quality (C1) and the highest airflow rate (E3), ventilation generates more than 60% of the losses. Lower thermal insulation of windows accounts for almost 18% of the losses (Figure 7).

Decreasing the airflow (E2) changes the structure, but still ventilation is responsible for 38% of heat losses (Figure 8).

The last example refers to the best thermal insulation quality (C6) and the lowest airflow rate (E1). In that case (Figure 9), the share of ventilation in heat losses grows to 65% and to decrease the energy demand a mechanical ventilation system with heat recovery can be applied. Another way to decrease the energy demand is to preheat the ventilation air, for instance in a ground heat exchanger.

The value of the time constant (Equation 29) was calculated as a combination of the following variants: internal heat capacity (B), thermal insulation quality (C), and the air flow rate (E). A higher airflow rate decreases the time constant of the building and reduces the heat accumulation in the building.

The smallest value of the time constant for the building is only 3.7 h. It is observed when the internal heat capacity is the lowest (B1), thermal insulation quality of the building is the worst (C1), and the air flow rate is the highest (E3) (Figure 10).

Figure 7. Structure of transmission heat losses: Variant (C1, E3), %

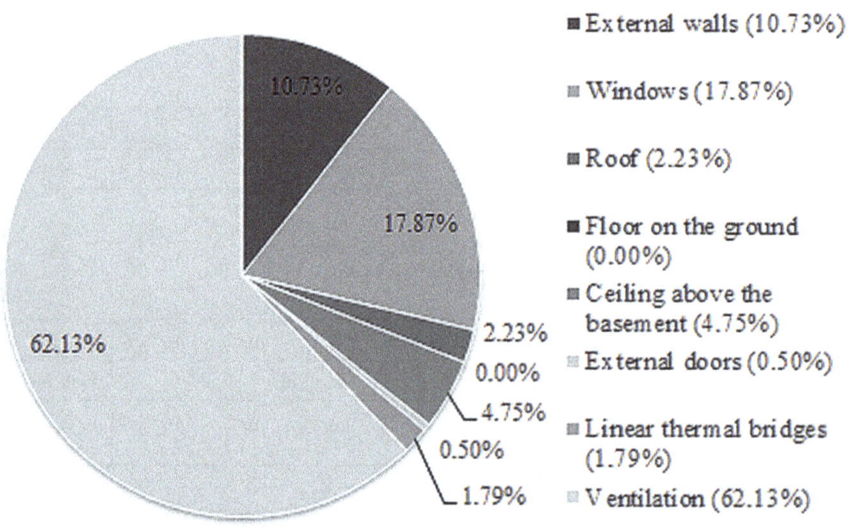

Figure 8. Structure of transmission heat losses: Variant (C1, E2), %

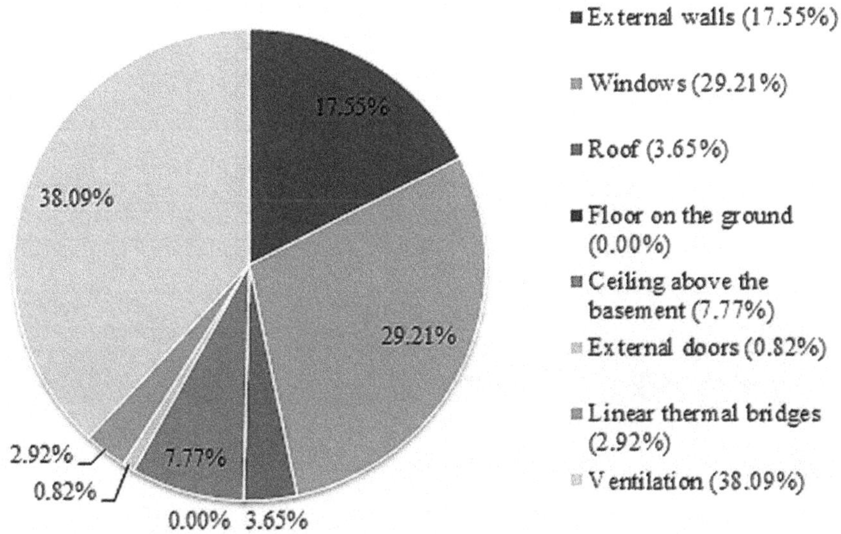

Figure 9. Structure of transmission heat losses: Variant (C6, E1), %

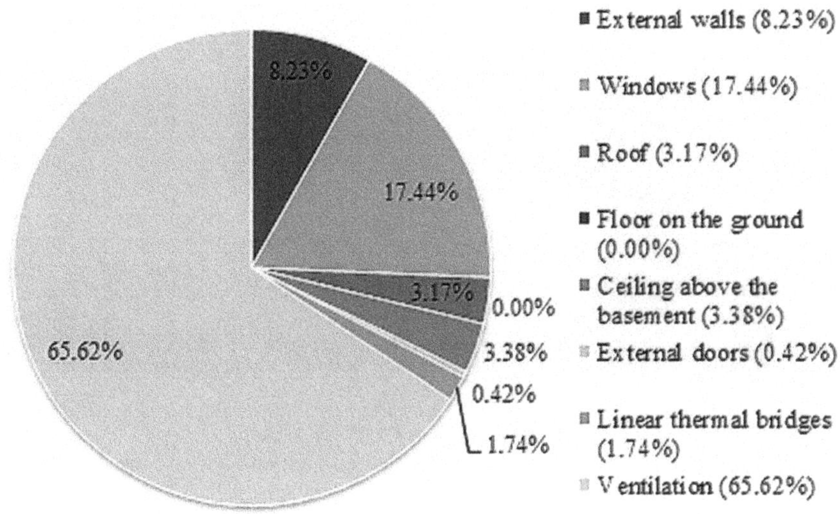

The highest value of the time constant of the building is equal to 234.4 h and it occurs for the highest internal heat capacity (B2), the highest thermal insulation quality (C6), and the lowest air flow rate (E1) (Figure 11).

Results

Calculations for all (162) combinations of variants were performed in a spreadsheet, and selected results are presented in Figures 12 to 23.

Figure 10. Time constant variation for buildings with the lowest heat capacity (B1)

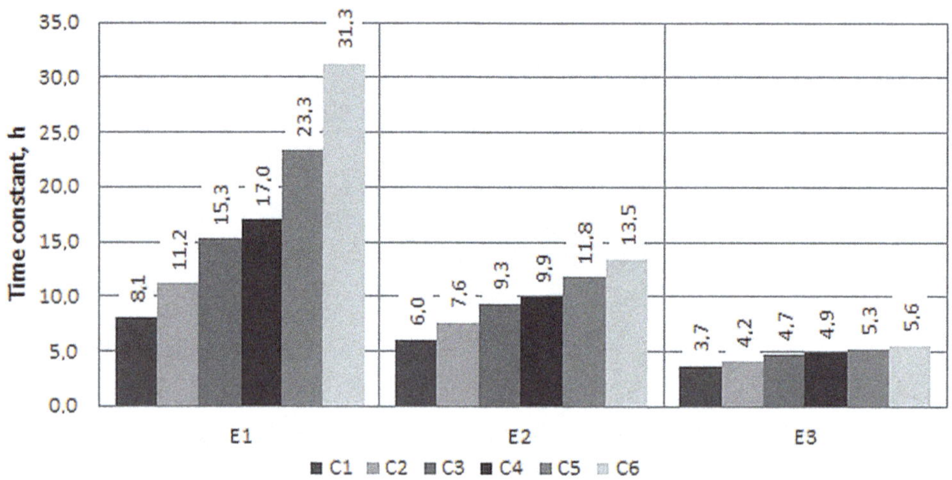

Figure 11. Time constant variation for buildings with the highest heat capacity (B6)

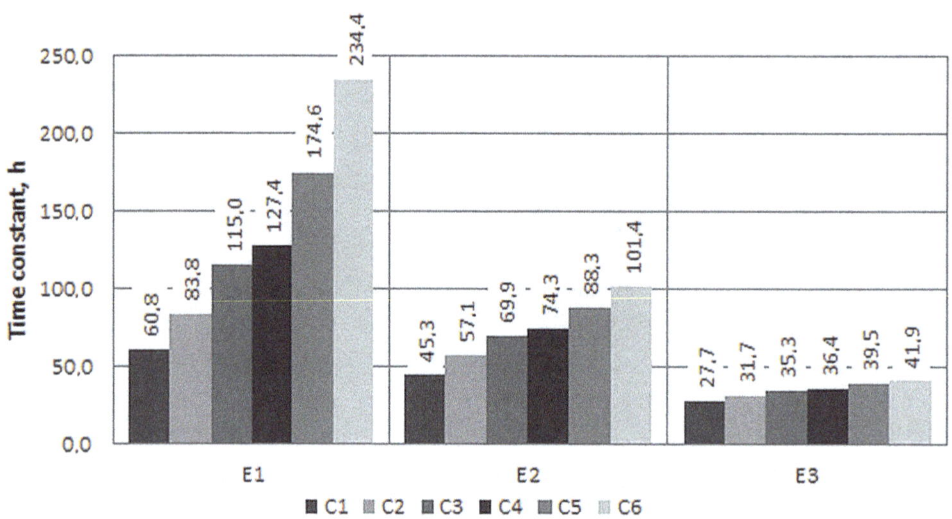

Athens

The longest heating season, 3796 h, occurs when the airflow is the highest (E3) and internal heat gains are the smallest (D1) (Figure 12). In such cases the influence of the internal heat capacity on the duration of the heating season is negligible (less than 2%).

In the case of Athens it is possible to avoid energy demand for heating with at least a 'Heavy' construction (e.g. B4, B5, and B6), at least a 'Standard' thermal insulation (e.g. C3, C4, C5, and C6), and the lowest air exchange rate (E1).

The biggest energy demand for heating, 99.5 kWh m^{-2} yr^{-1}, occurs with the lowest internal capacity (B1), the lowest thermal insulation quality (C1), and the highest air exchange rate (E3) (Figure 13).

Figure 12. Athens: Heating season

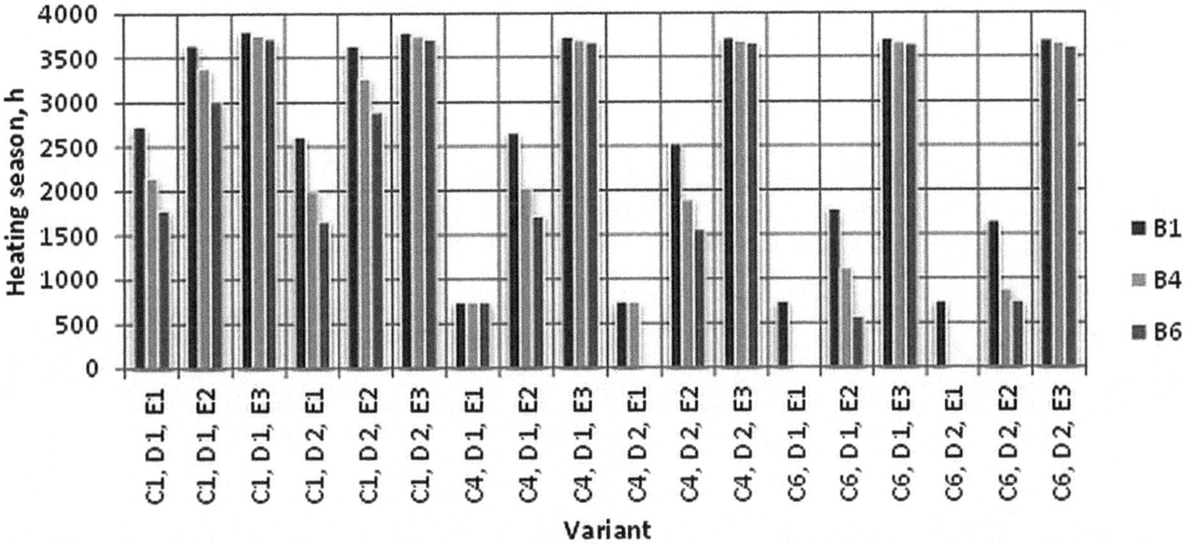

Figure 13. Athens: Energy demand for heating

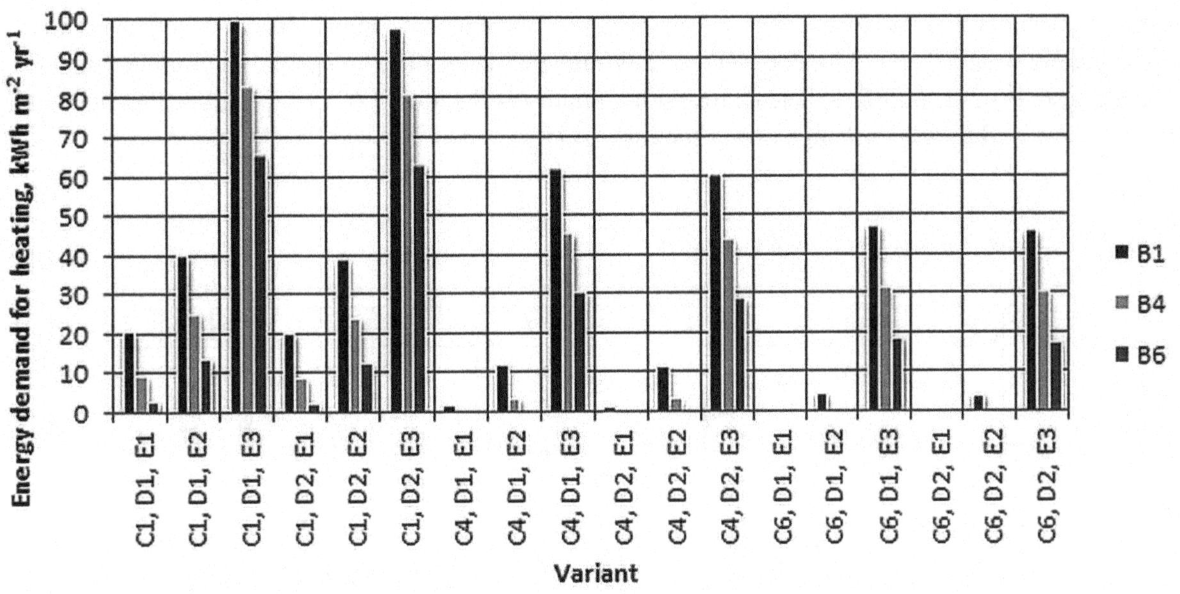

The longest cooling season lasts the whole year, i.e. 8760 h, and occurs when the airflow is the lowest (E1) and thermal insulation is at least 'High' (e.g. C4, C5, and C6) (Figure 14). In such cases there is no influence of the internal heat capacity on the duration of the cooling season. In all other cases, the greater the internal heat capacity, the shorter the cooling season is.

The shortest cooling season lasts 3223 h when the internal heat capacity is 'Heavy' (B4), thermal insulation is very low (C1), and the airflow rate is the highest (E3).

Figure 14. Athens: Cooling season

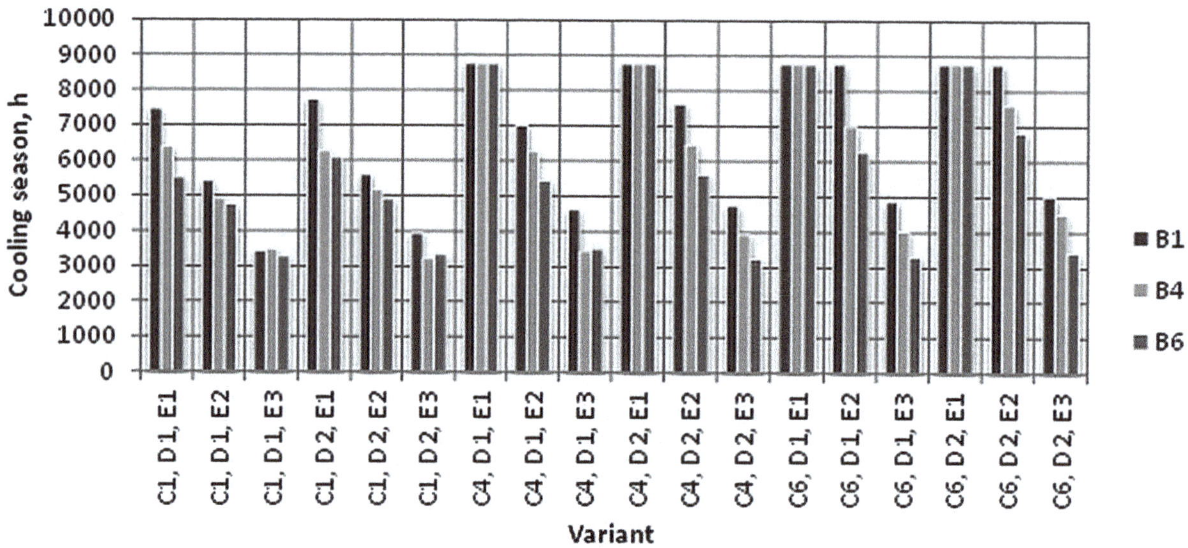

The energy demand for cooling is the highest (459 kWh m^{-2} yr^{-1}) when thermal insulation is the highest (C6) and the airflow rate is the lowest (E1) (Figure 15). Certainly, the biggest internal heat gains (D2) increase the energy demand for cooling. In this case, the internal heat capacity has no significant impact.

The lowest energy demand for cooling, 264 kWh m^{-2} yr^{-1}, is observed when the internal heat capacity is the highest (B6), the building thermal insulation quality is the lowest (C1), internal heat gains are the smallest (D1), and the air exchange rate is the highest (E3).

Figure 15. Athens: Energy demand for cooling

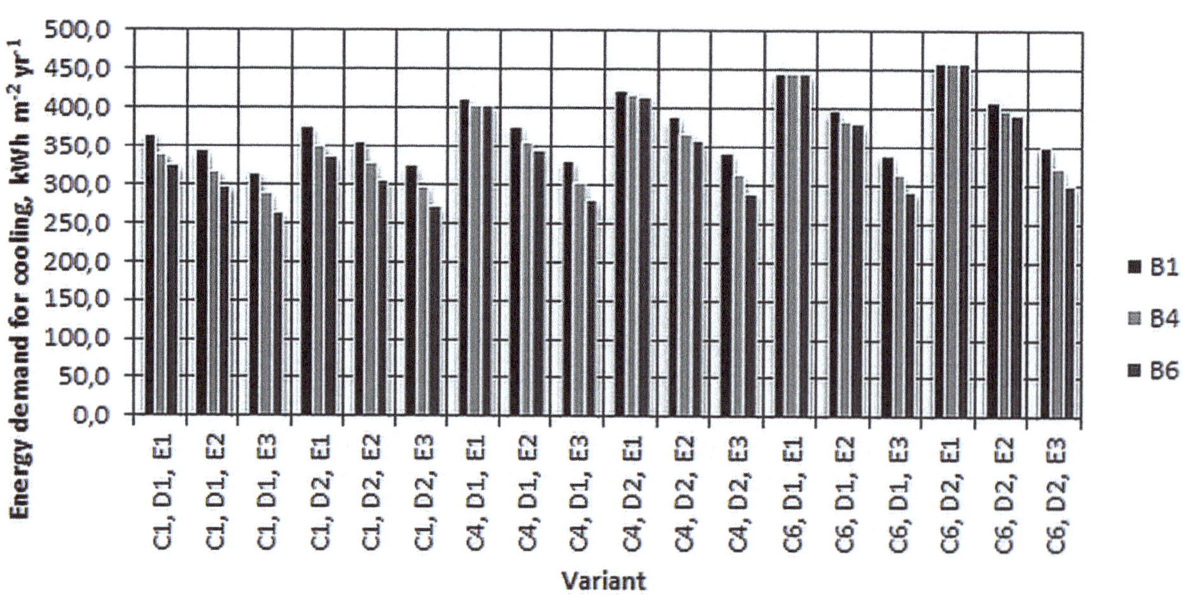

Stockholm

The longest heating season lasts the whole year, i.e. 8760 h, and occurs when thermal insulation is the worst (C1) and airflow is the highest (E3) (Figure 16). In this case, in a building with the highest internal heat capacity (B6), the heating season is 11% shorter compared to the building with the lowest heat capacity (B1).

The shortest heating season lasts 2362 h and occurs when the internal heat capacity is the smallest (B1), thermal insulation is the best (C6), and internal gains are the highest (D2).

The greater the internal heat capacity, the lower the energy required for heating is.

The lowest energy demand for heating, 7.1 kWh m^{-2} yr^{-1}, occurs with the highest internal capacity (B6), the highest thermal insulation quality (C6), and the lowest air exchange rate (E1).

The highest energy demand for heating, 519 kWh m^{-2} yr^{-1}, occurs with the lowest internal capacity (B1), the lowest thermal insulation quality (C1), and the highest air exchange rate (E3) (Figure 17).

The longest cooling season lasts 6004 h and occurs when the internal heat capacity is the lowest (B1), thermal insulation is the best (C6), airflow is the lowest (E1), and internal heat gains are the highest (Figure 18). In these cases, the greater the internal heat capacity, the shorter the cooling season is.

The shortest cooling season lasts 7 h when the internal heat capacity is the smallest (B1), thermal insulation is very low (C1), and the airflow rate is the highest (E3).

Energy demand for cooling is the highest (268 kWh m^{-2} yr^{-1}) when the internal heat capacity is the lowest (B1), thermal insulation is the highest (C6), and the airflow rate is the smallest (E1) (Figure 19). Certainly, the highest internal heat gains (D2) increase the energy demand for cooling.

The lowest energy demand for cooling, 52 kWh m^{-2} yr^{-1}, is observed when the internal heat capacity is the highest (B6), the quality of the building thermal insulation is the lowest (C1), internal heat gains are the lowest (D1), and the air exchange rate is the highest (E3).

Figure 16. Stockholm: Heating season

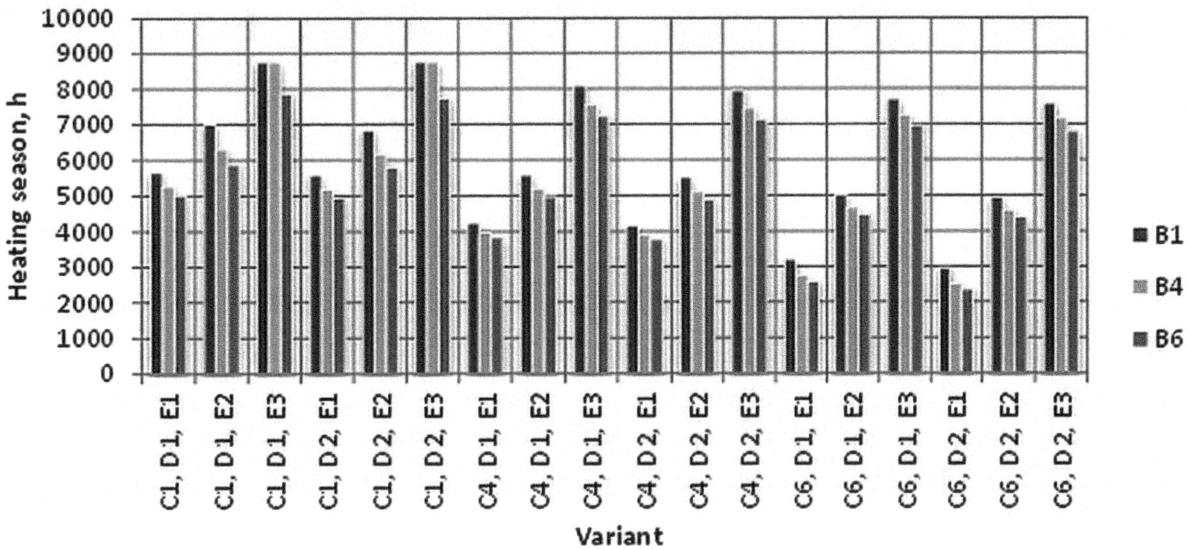

Figure 17. Stockholm: Energy demand for heating

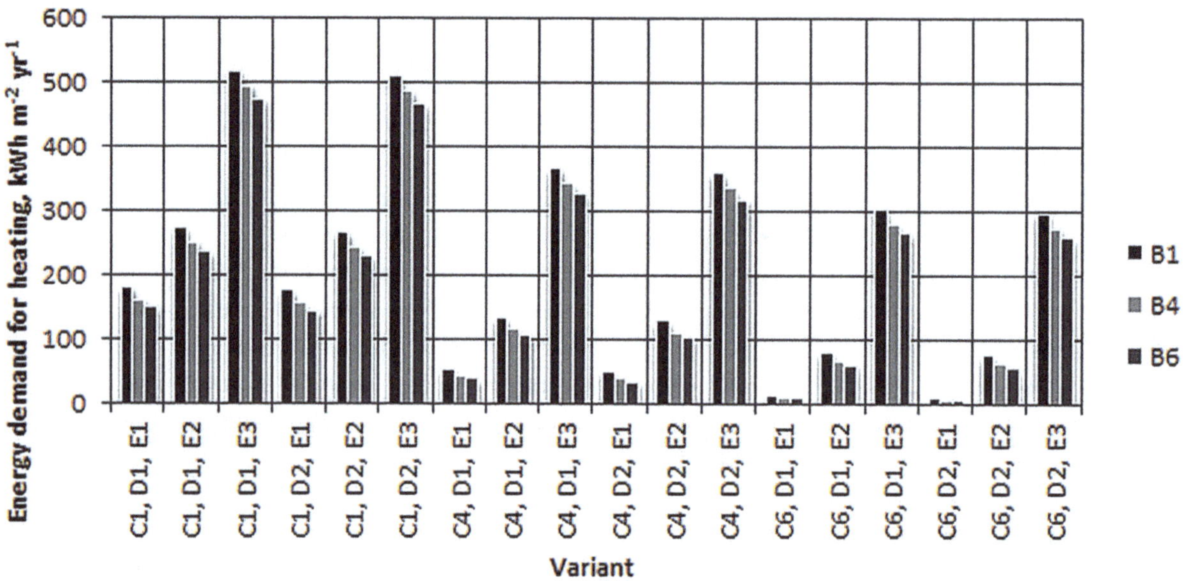

Figure 18. Stockholm: Cooling season

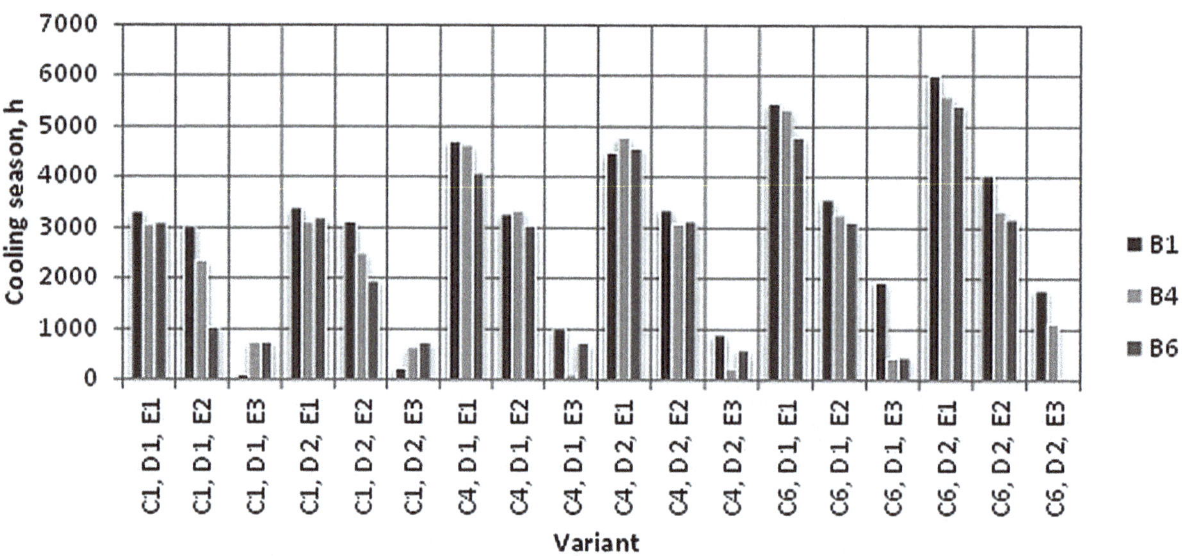

Warsaw

The longest heating season lasts 7553 h and occurs when the internal heat capacity is the smallest (B1), thermal insulation is the worst (C1), internal heat gains are the lowest (D1), and airflow is the highest (E3) (Figure 20).

The shortest heating season lasts 389 h and occurs with the internal heat capacity B5, the best thermal insulation (C6), the highest internal gains (D2), and the lowest airflow rate (E3).

Figure 19. Stockholm: Energy demand for cooling

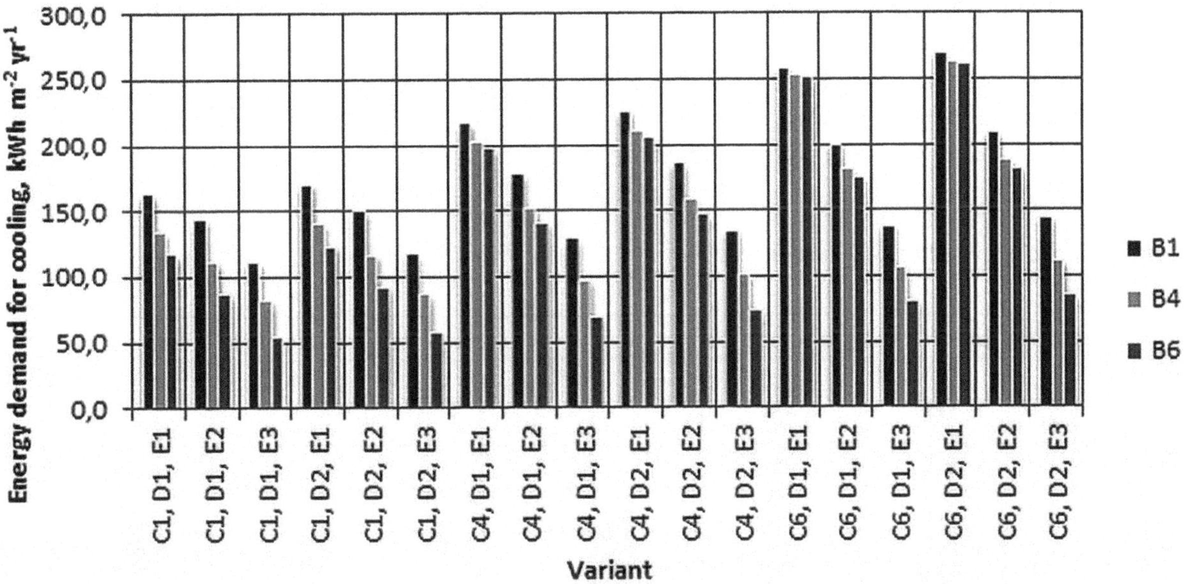

Figure 20. Warsaw: Heating season

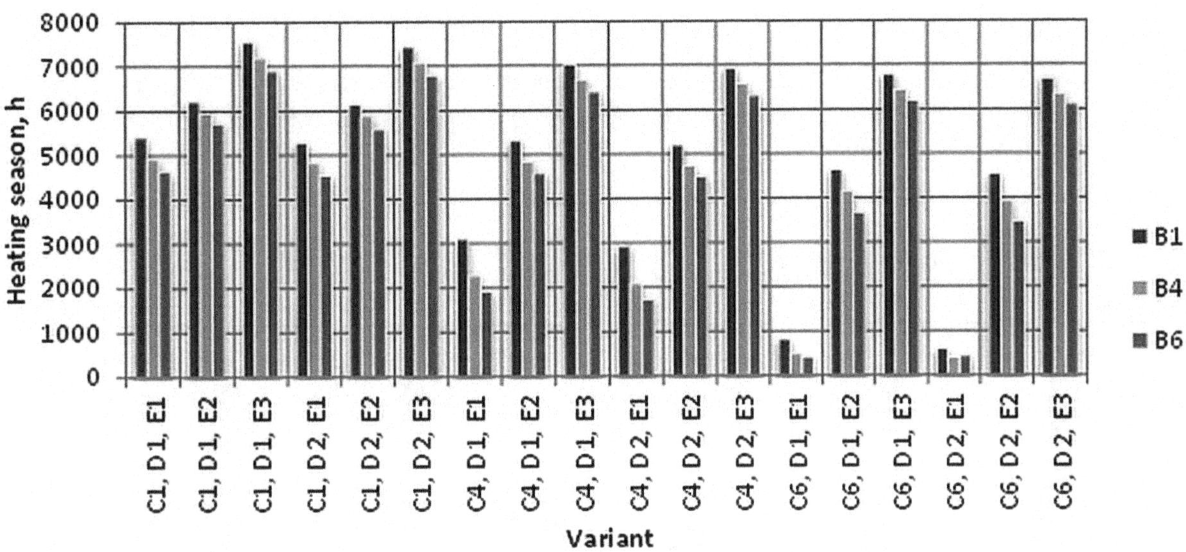

The greater the internal heat capacity, the smaller amount of energy is required for heating.

The smallest energy demand for heating, 0.1 kWh m^{-2} yr^{-1}, occurs with the internal capacity B6, the highest thermal insulation quality (C6), the highest internal gains (D2), and the lowest air exchange rate (E1).

The highest energy demand for heating, 401.9 kWh m^{-2} yr^{-1}, occurs with the lowest internal capacity (B1), the lowest thermal insulation quality (C1), the lowest internal gains (D1), and the highest air exchange rate (E3) (Figure 21).

The longest cooling season lasts 8760 h and occurs when the internal heat capacity is the lowest (B1), thermal insulation is the best (C6), airflow is the lowest (E1), and internal heat gains are the highest (D2) (Figure 22). In these cases, the greater the internal heat capacity, the shorter the cooling season is.

Figure 21. Warsaw: Energy demand for heating

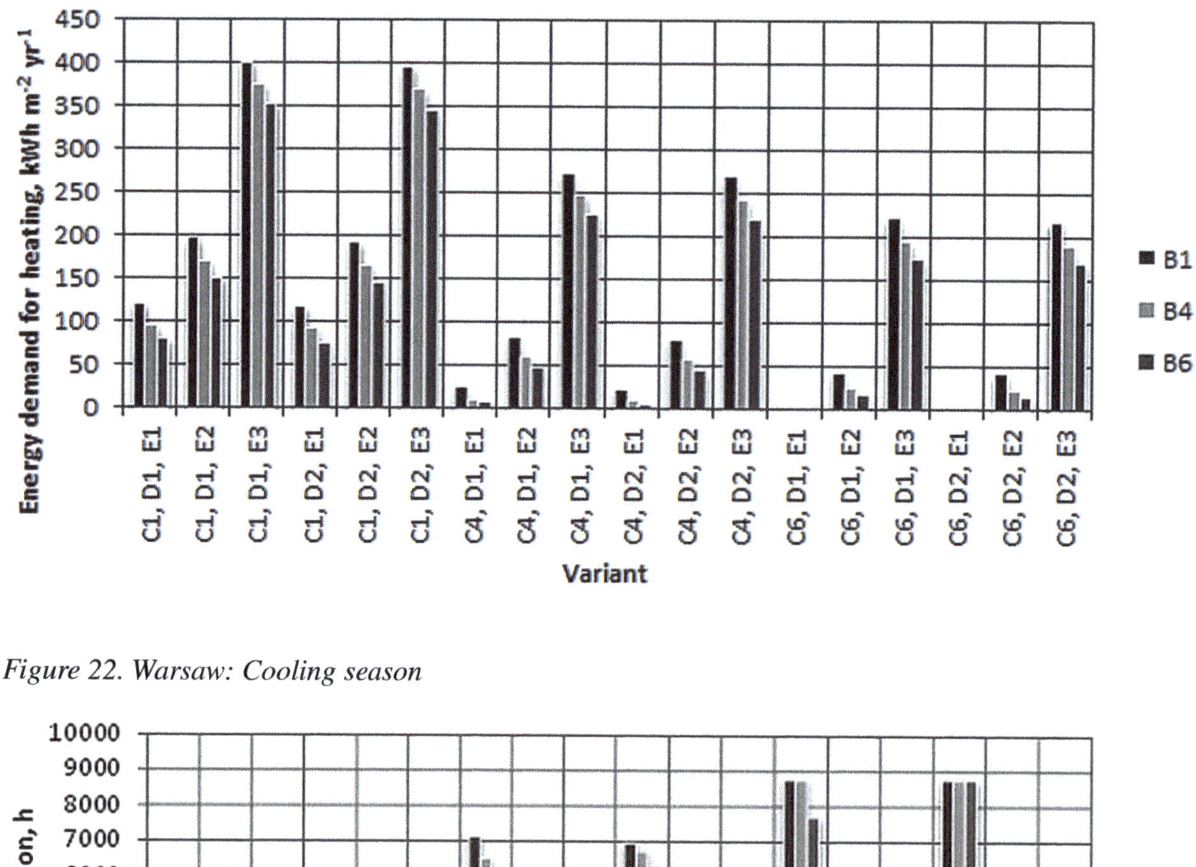

Figure 22. Warsaw: Cooling season

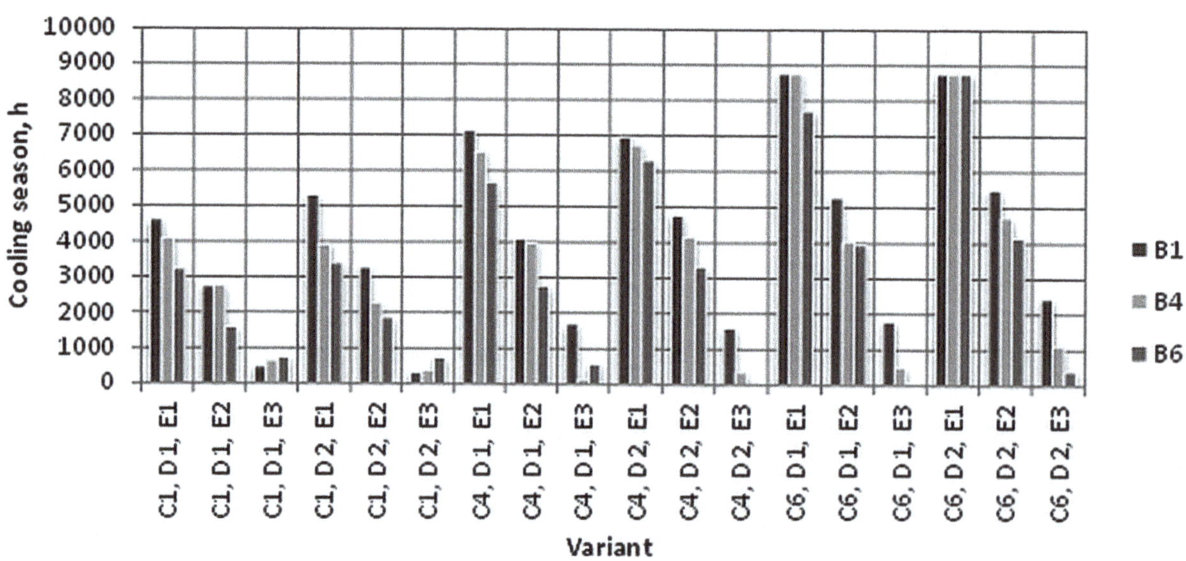

The shortest cooling season lasts 14 h when the internal heat capacity is the greatest (B6), thermal insulation is C4 and the airflow rate is the highest (E3).

Energy demand for cooling is the highest, i.e. 302 kWh m^{-2} yr^{-1}, when the internal heat capacity is the lowest (B1), thermal insulation is the highest (C6), internal gains are the biggest (D2), and the airflow rate is the smallest (E1) (Figure 23).

The lowest energy demand for cooling, 55 kWh m^{-2} yr^{-1}, is observed when the internal heat capacity is the highest (B6), the quality of the building thermal insulation is the lowest (C1), internal heat gains are the smallest (D1), and the air exchange rate is the highest (E3).

Effectiveness of External Blinds

It is known that especially in the south of Europe external blinds play an important role in protecting against overheating of internal spaces in summer. In the north of Europe during nights in winter, when the lowest external air temperatures occur, external blinds can reduce transmission heat losses. To verify energy effectiveness of external blinds, the Warsaw climate was chosen and the 5R1C method (ISO 2008) was applied. In comparison to the monthly heat balance method, additional assumptions necessary to use the hourly simplified method were made. Artificial lighting is turned on when solar irradiance is less than 70 W m^{-2} and when there are occupants in the building according to the occupancy pattern used. When solar irradiance exceeds 300 Wm^{-2}, exterior blinds, with the efficiency of 40%, can be used to prevent rooms from overheating. In the heating season, with temperatures below 0 °C, between 10 p.m. and 6 a.m. the next day, exterior foldable plastic blinds can be used (they are filled with foam characterized by medium air permeability, which improves thermal performance of windows). The results are presented in Figures 24 for the heating mode and 25 for the cooling mode. The effectiveness is taken as a percentage of the reduction in the demand for heating or cooling.

Figure 23. Warsaw: Energy demand for cooling

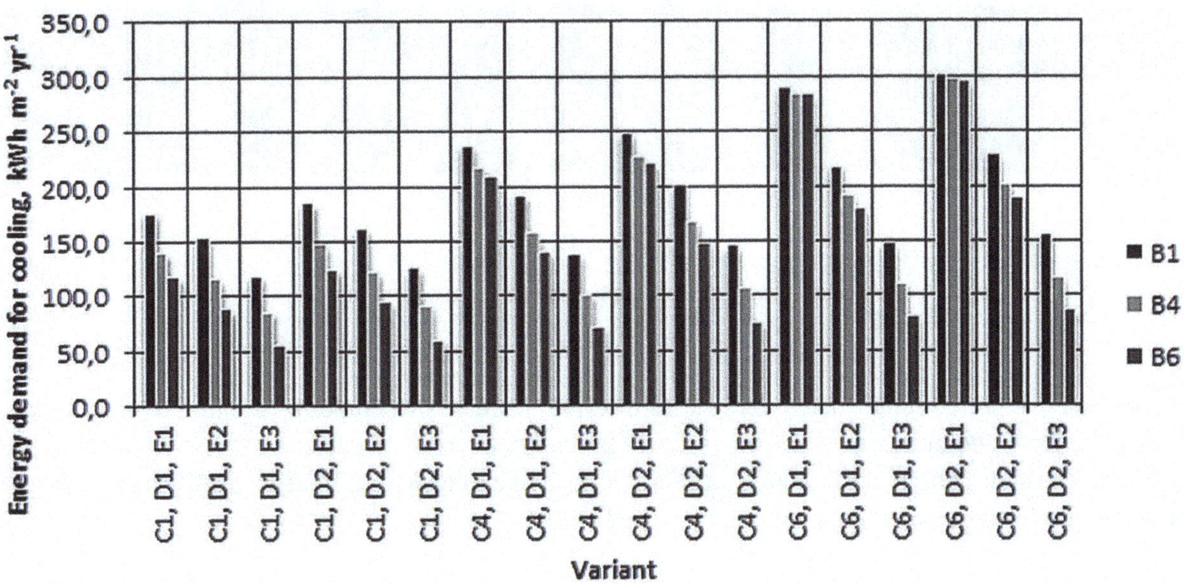

Figure 24. Reduction in demand for heating due to external blinds, %

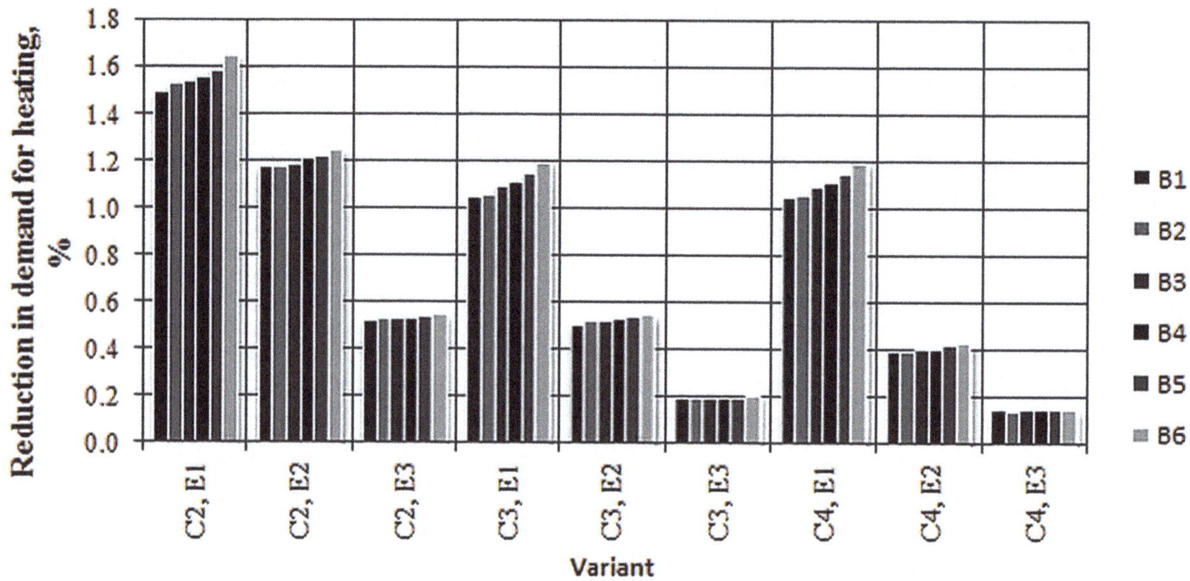

Figure 25. Reduction in demand for cooling due to external blinds, %

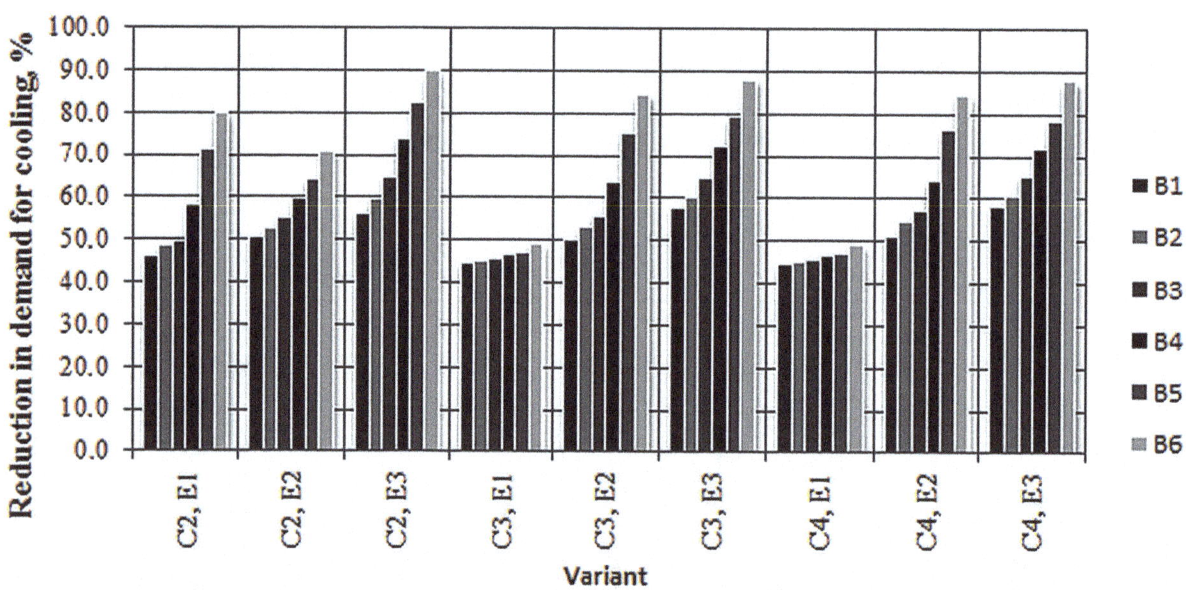

Positive influence on the energy demand for heating decreases when both the internal heat capacity and the air flow rate increase.

Given the assumptions listed above, in Warsaw in the heating season relative profits due to the use of external blinds are negligible and do not exceed 1.70% (Figure 24).

As far as cooling is concerned, the situation is different (Figure 25). A significant impact of this solution on reducing the demand for energy in the cooling mode is noticeable, and bigger with a better insulation (up to almost 90% in the case 'B6, C4, E3').

For a higher air flow rate, the efficiency of shading blinds increases (up to 44%).

The impact of the internal heat capacity is bigger when cooling is applied (up to 40%).

Better thermal insulation of building partitions only slightly improves the efficiency of shading devices (up to 10%).

FUTURE RESEARCH DIRECTIONS

As a general strategy, the future research should include the following main issues:

- Optimisation of the whole building construction (thermal insulation, dynamic properties),
- Optimisation of the HVAC, lighting, and control systems,
- Optimisation of renewable energy sources to obtain heat and electricity, and
- Application of a building management system (BMS).

Additionally, the proposed technical solutions should be verified in terms of internal comfort, economy, energy efficiency, and pollutants emission. A life cycle analysis (LCA) is also worth considering. Finally, a wider range of technical solutions should be taken into account (PCM materials, transparent partitions with selective layers, double skin facades, building-integrated photovoltaics, etc.). However, in the case of dynamic systems, the monthly balance method is inappropriate.

CONCLUSION

The chapter presents the methodology useful for analyzing the influence of typical modernisation activities on the energy demand for heating and cooling. Firstly, the mathematical model used in the monthly heat balance method (MHB) was described and applied to estimate possible energy savings in three locations in Europe. The mathematical model is very simplified, but the effects can be determined relatively quickly in a spreadsheet. However, in the case of dynamic processes its reliability is insufficient to carry out detailed analysis. This is why the methodology presented is useful only at the first stage of the analysis. The dynamic, and time-dependent behaviour of buildings requires the use of at least the simplified hourly method (5R1C or 6R1C).

A large or very large internal heat capacity of buildings, extremely high standard of thermal insulation, and a minimum air exchange rate significantly reduce energy demand for heating.

A higher internal heat capacity decreases the energy demand for both heating (Stockholm, up to 35%) and cooling (Athens, up to 16%). However, in an existing building it is difficult to change it significantly without a complex reconstruction, but to some extent it is possible by introducing PCM materials.

When a renovation of an existing building towards the nZEB standard is considered, in addition to the comfort of inhabitants or employees, technical possibilities, economical aspects, possible legal restrictions, and architectural recommendations, the following should be taken into account:

- Improving quality of thermal insulation of the envelope, and
- Rearranging ventilation to provide a mechanical system with heat recovery.

REFERENCES

ASHRAE. (2013). *ASHRAE Handbook - Fundamentals (SI Edition)*. Atlanta, GA: ASHRAE.

Barbolini, F., Cappellacci, P., & Guardigli, L. (2017). A design strategy to reach nZEB standards integrating energy efficiency measures and passive energy use. *Energy Procedia*, *111*, 205–214. doi:10.1016/j.egypro.2017.03.022

Becchio, C., Dabbene, P., Fabrizio, E., Monetti, V., & Filippi, M. (2015). Cost optimality assessment of a single family house: Building and technical systems solutions for the nZEB target. *Energy and Building*, *90*, 173–187. doi:10.1016/j.enbuild.2014.12.050

DI. (2017). *dateandtime.info*. Retrieved September 2, 2017, from http://dateandtime.info/

Directive 2002/91/EC of the European Parliament and of the Council of 16 December 2002 on the energy performance of buildings. (2002). *Official Journal of the European Communities, 4*(1), 65-71.

Directive 2010/31/EU of the European Parliament and of the Council of 19 May 2010 on the energy performance of buildings (recast). (2010). *Official Journal of the European Union, 3,* 124-146.

EnergyPlus. (n.d.). Retrieved September 2, 2017, from https://energyplus.net/

EPBD. (2015). *Implementing the Energy Performance of Buildings Directive (EPBD). Featuring Country Reports*. Lisboa: ADENE.

Erhorn, H., & Erhorn-Kluttig, H. (2014). *Selected examples of Nearly Zero-Energy Buildings. Detailed Report*. Retrieved September 2, 2017, from http://www.epbd-ca.eu/wp-content/uploads/2011/05/CT5_Report_Selected_examples_of_NZEBs-final.pdf)

ESP-r. (n.d.). Retrieved September 2, 2017, from http://www.esru.strath.ac.uk/Programs/ESP-r.htm

EUROSTAT. (n.d.). *Heating Degree Days (HDD)*. Retrieved September 2, 2017, from http://ec.europa.eu/eurostat/web/energy/data

ISO. (2006). *ISO 10077-1:2006 Thermal performance of windows, doors and shutters — Calculation of thermal transmittance — Part 1: General*. Retrieved from https://www.iso.org/standard/40360.html

ISO. (2007a). *ISO 6946:2007 Building components and building elements -- Thermal resistance and thermal transmittance -- Calculation method*. Retrieved from https://www.iso.org/standard/40968.html

ISO. (2007b). *ISO 10211:2007 Thermal bridges in building construction -- Heat flows and surface temperatures -- Detailed calculations*. Retrieved from https://www.iso.org/standard/40967.html

ISO. (2007c). *ISO 12524:2000 Building materials and products -- Hygrothermal properties -- Tabulated design values*. Retrieved from https://www.iso.org/obp/ui/#iso:std:iso:10456:ed-3:v1:en

ISO. (2007d). *ISO 13789:2007 Transmission and ventilation heat transfer coefficients—Calculation method.* Retrieved from https://www.iso.org/standard/40894.html

ISO. (2007e). *ISO 14683:2007 Thermal bridges in building construction -- Linear thermal transmittance -- Simplified methods and default values*. Retrieved from https://www.iso.org/standard/40964.html

ISO. (2008). *ISO 13790:2008 Energy performance of buildings -- Calculation of energy use for space heating and cooling*. Retrieved from https://www.iso.org/standard/41974.html

ISO. (2012). *ISO 13788:2012 Hygrothermal performance of building components and building elements -- Internal surface temperature to avoid critical surface humidity and interstitial condensation -- Calculation methods*. Retrieved from https://www.iso.org/standard/51615.html

ISO. (2015). *ISO 9972:2015 Thermal performance of buildings -- Determination of air permeability of buildings -- Fan pressurization method*. Retrieved from https://www.iso.org/standard/55718.html

ISO. (2017). *ISO 13786:2017 Thermal performance of building components -- Dynamic thermal characteristics -- Calculation methods*. Retrieved from https://www.iso.org/standard/65711.html

Kurnitski, J. (2013). *REHVA nZEB technical definition and system boundaries for nearly zero energy buildings*. Federation of European Heating, Ventilation and Air-Conditioning Associations, Report No 4. Brussels: REHVA.

Laskowski, L. (2008). *Thermal protection and energy performance of the building*. Warszawa: OWPF. (in Polish)

LBNL. (N.D.) Therm. Lawrence Berkeley National Laboratory. Therm 7.5. Retrieved September 2, 2017, from windows.lbl.gov/software/therm/therm.html

Meteo AHRAE. (2001). *Meteorological data for EnergyPlus, Copywrite: American Society of Heating, Refrigerating and Air-Conditioning Engineers*. Atlanta, GA: ASHRAE.

Narowski, P., Mijakowski, M., Panek, A. D., Rucińska, J., & Sowa, J. (2010). Integrated 6R1C Energy Simulation Method – Principles, Verification and Application. *Proceedings CLIMA 2010, 11th REHVA World Congress "Sustainable Energy Use in Buildings"*.

Pando, A.R., Capestro, I., Colarusso, A., Solari, M., Saba, A., Morando, M., Reccardo, D., Bogacz, A., Raco, F., Uriarte, A., Tellado, B. (2014). *Technologies and strategies for Public Building and District Retrofitting*. Project Funded by the European Community in the 7th Framework Programme, A2PBEER GA n°: 609060 D2.2 Technologies and strategies for Public Building and District Retrofitting.

RMI. (2008). Regulation of the Minister of Infrastructure of 6 November 2008 On the methodology for calculating the energy performance of the building and dwelling or part of the building and the technical-utilitarian independent parts and the preparation and design of energy performance certificates. *Journal of Laws 2014*, Item 888 (in Polish).

RMI. (2002). Regulation of the Minister of Infrastructure of 12 April 2002 on technical specifications for buildings and their location. *Journal of Laws 2015*, Item 1422, consolidated text (in Polish).

Wikipedia. (n.d.). *Köppen climate classification*. Retrieved September 2, 2017, from https://en.wikipedia.org/wiki/K%C3%B6ppen_climate_classification

KEY TERMS AND DEFINITIONS

Heat Transfer Characteristics of the Envelope: A physical quantity that describes thermal insulating properties of the external envelope of a building, expressed in W K^{-1}.

Internal Heat Capacity: Heat capacity of internal components and the building structure that has an influence on internal temperature changes, expressed in J K^{-1}.

Monthly Balance Method: A simplified mathematical method of calculating energy demand for heating and cooling in which a month is assumed as the balance period.

nZEB: A building characterised by high energy performance, in which energy consumption and production are almost balanced during a year (a nearly zero-energy building).

Simplified Hourly Method 5R1C: A simplified simulation method of calculating energy demand for heating and cooling in which the building behaviour is described by the first order differential equation.

Thermal Characteristic of Ventilation: A physical quantity that describes thermal properties of the ventilation system, expressed in J K^{-1}.

Thermal Upgrading of an Existing Building: All technical actions taken to improve energy performance of a building.

Chapter 3
Selection of Renewable Energy Sources for Buildings

Hanna Irena Jędrzejuk
Warsaw University of Technology, Poland

ABSTRACT

This chapter describes a general issue of selecting renewable energy sources (RES) and technical systems. To achieve the nearly zero-energy building (nZEB) standard, application of an RES (e.g., solar, wind, geothermal, hydropower, and biomass energy) is necessary. Each type of RES has specific characteristics and can be used to produce electricity and/or heat in certain systems. A short review of various systems using renewable energy sources is presented. To find the required and satisfactory solution that guaranties meeting the nZEB standard, an analysis must be carried out considering a number of aspects: local availability, structure and time-dependence of energy demand, building construction, economic conditions, legal regulations, and specific requirements. Finally, two examples of modernisation towards the nZEB standard are included.

INTRODUCTION

A rational decrease in energy demand for HVAC (heating, ventilation, and air-conditioning), hot water preparing, lighting, and other electrical appliances does not guarantee achieving the nearly zero-energy building (nZEB) standard. Any actions must be supported by the use of renewable energy sources: solar, geothermal, aerothermal, wind, hydropower, and biomass energy.

At present, the issue of greenhouse gases (GHG) emissions is mentioned mainly in the context of the global warming effect (Directive 2001/77/EC, Directive 2009/28/EC), but the problem is more complex (e.g. Tiba et al. 2017; Kim et al. 2017). Application of RES-based technical systems has a wide impact on local development and employment opportunities as well as energy diversification and independence.

DOI: 10.4018/978-1-5225-4105-9.ch003

BACKGROUND

Energy is essential for meeting various human needs. In Europe, transport, industry, households, services, fishing, agriculture, and forestry are the main sectors where energy is consumed (Figure 1). In 2010-2014, a drop in final energy consumption was observed (a total decrease of 8.8%). A similar trend can be seen in all the sectors. The smallest decrease is found in transport (3.2%), while the biggest one in fishing, agriculture, and forestry (19.6%). As far as households are concerned, a significant decrease was noticed (17.0%). In 2014, the share of households in final energy consumption was 24.8% (11.0 EJ).

Energy can be obtained from primary energy sources, such as nuclear and fossil fuels, bioenergy and geothermal, direct solar, wind, hydropower, and ocean energy.

These energy sources are classified in terms of their recovery on a human time scale and divided into renewable (RES: bioenergy, geothermal, direct solar, wind, hydropower, and ocean energy) and non-renewable sources (nuclear and fossil fuels).

Redirection of energy production from conventional to renewable energy sources results in:

- Independence from the supply of fossil fuels from countries outside the European Union,
- Decentralization of energy sources, and
- Expanding the labour market linked to renewable energy sources (production, installation, and maintenance).

Although renewable energy sources can generate different forms of energy (heat, mechanical, and electric), the aspect of generating electricity is most emphasized today (Directive 2003/54/EC).

In the period 2005–2015 (Figure 2), hydropower has the main share in the production of electricity. Geothermal power accounted for only a small share. During the period concerned, the growth in the use of wind energy is going to be significant (from 70.5 TWh in 2005 to 301.9 TWh in 2015). A

Figure 1. Final energy consumption in Europe (EU-28) by sector, Mtoe [*)]*. Period 2010–2014*
IEA. Final energy.
[*)] Mtoe - Million Tonnes of Oil Equivalent. Tonne of oil equivalent - amount of energy released by burning one tone of crude oil and it is 41.868 GJ or 11.63 MWh (IEA. UC).

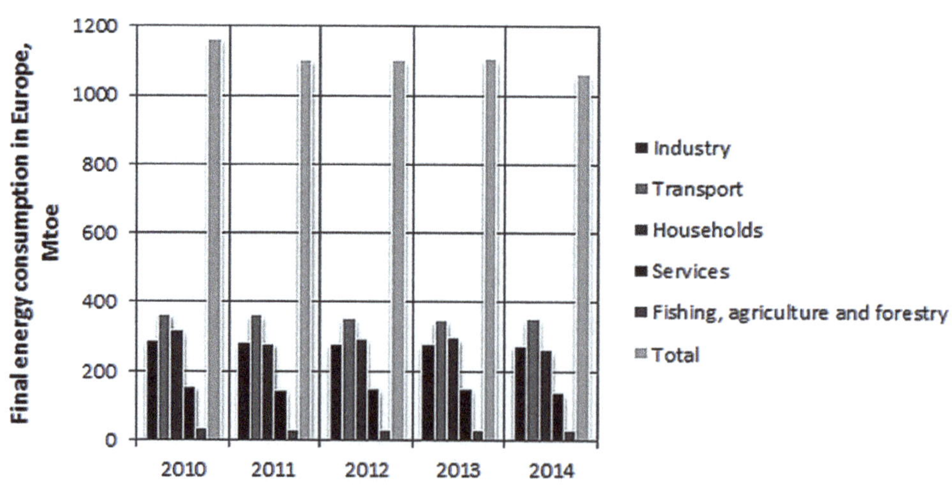

Figure 2. Electricity generated from renewable energy sources, EU-28, 2005–2015
EUROSTAT. RES.

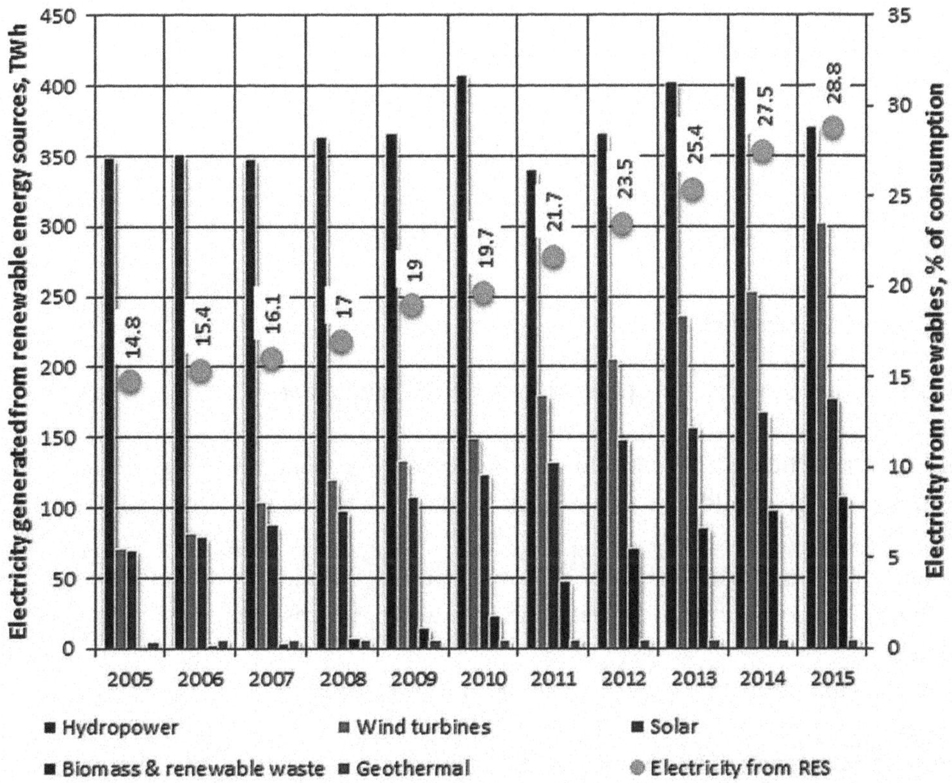

rapid growth of solar energy (from 69.9 TWh in 2005 to 177.9 TWh in 2015) as well as biomass and renewable waste (from 1.5 TWh in 2005 to 107.9 TWh in 2015) is observed. Finally, the total amount of electricity production from renewables is going to be significant (28.8%), while in 2005 it was almost two times smaller.

RENEWABLE ENERGY SOURCES

Modernisation of an existing building towards the nZEB standard requires a detailed analysis of the actual energy consumption and structure and improving the efficiency of all processes involved.

On this basis, a set of actions may be proposed to rationalise energy usage.

Further reduction in non-renewable energy consumption, needed to ensure required operating conditions in the building, is possible through the use of renewable energy sources.

In Directive 2003/54/EC, renewable energy sources are mentioned as renewable non-fossil energy sources: wind, solar, geothermal, wave, tidal, hydropower, biomass, landfill gas, sewage treatment plant gas, and biogases.

Ideal renewable energy sources should be characterised by the following features:

- Unlimited resources,
- Easy accessibility,
- Reliability of supply,
- Availability when specific needs arise (heat, electricity), and
- No harmful impact on the environment.

An RES-based system should be characterised by:

- Low investment and operating costs (including maintenance),
- High efficiency, and
- Long and trouble-free operation.

In all modernisation projects, the application of an RES must consider:

- Reliability of a particular source and its ability to cover reasonable energy needs,
- Local constraints of a building (construction, roof area, etc.) and its surroundings (e.g. shadowing effects),
- Economics of the proposed solution (investment costs, running costs before and after modernisation, bank rates, possibility of preferential financing terms, i.e. local programs, deployment subsidies),
- Environment (pollution, change of local conditions),
- Legal regulations,
- Requirements of a spatial development plan, and
- Recommendations for conservation supervision (if applicable).

The last three aspects strongly differ in EU countries, but discussing them is beyond the scope of this study.

The following main renewable non-fossil energy sources were taken into account: solar, wind, geothermal, and biomass energy. These renewable energy sources and systems are described below in a preferable order.

SOLAR ENERGY

Solar radiation is one of the most important, relatively easily accessible and environmentally friendly renewable energy sources.

To estimate the potential of solar energy, the following should be known:

- **Sunshine Duration (or Sunshine Hours):** Duration of sunshine in a given period (hour, month, year),
- **Irradiance:** The rate at which radiant energy is incident on a surface per unit area (W m^{-2}), and
- **Irradiation (Insolation - Solar Energy Irradiation):** Incident energy per unit area on a surface, found by integration of irradiance over a specified time (J m^{-2}, kWh m^{-2}).

All above mentioned quantities should be established by taking into account: orientation of the plane (azimuth - the deviation of the projection on a horizontal plane of the normal to the surface from the local meridian), its slope (horizontal tilt - the angle between the considered plane and the horizontal), and surrounding objects causing shadowing effects (buildings, mountains, etc.).

Availability of solar radiation in Europe differs in terms of sunshine hours (Table 1) and solar irradiation (Figure 3). Solar irradiation in Figure 3 is given on the assumption that the plane is inclined at an optimal angle to the horizontal surface and has a southern orientation (Huld et al. 2012). The greatest solar potential is in Southern European countries and reaches almost 2200 kWh m^{-2} yr^{-1} (Athens, Madrid). In the North, it drops to a value not greater than 600 kWh m^{-2} yr^{-1}. The longest sunshine duration (Table 1) is noticed in Athens (2,848 h), while the shortest one in Warsaw (1,571 h).

Estimation of solar energy application opportunities in buildings, especially with buildings that will be modernised into the nZEB standard, requires adequate data. If they are unavailable, mathematical models can be applied. Some models are theoretical, while others are based on measurements. Measured data must be statistically designed to be representative of a given terrain.

Generally, only global solar radiation on a horizontal surface is measured at weather stations, but for evaluating technical systems this quantity must be recalculated for sloped surfaces.

A total solar radiation on a horizontal surface is calculated as a sum of the beam (direct), diffuse, and reflected components (Duffie et al. 2013). Beam solar radiation is received from the Sun without having been scattered by the atmosphere. Diffuse solar radiation is received from the Sun after its direction has been changed by scattering by the atmosphere. The last component is the reflected solar radiation of surrounding surfaces (ground, buildings, trees, etc.).

Two main types of solar radiation sky models differ in how they describe the diffuse solar radiation. In the isotropic model, the diffuse solar radiation is received uniformly from the entire sky dome. In the anisotropic model, two other subcomponents are included: circumsolar diffuse and horizon brightening solar radiation. The most commonly applied mathematical models for calculating incident solar radiation are listed in Table 2. Reviews of mathematical models of the global solar radiation on inclined surfaces can be found in various publications (e.g. Maleki et al. 2017).

Table 1. Sunshine hours for selected cities in Europe, h (Wikipedia, 2017d.)

Location	Month												
	Jan	Feb	Mar	Apr	May	Jun	Jul	Aug	Sep	Oct	Nov	Dec	Year
Athens, Greece	158	168	189	225	304	360	384	360	252	198	144	105	2,848
Berlin, Germany	47	74	121	159	220	222	217	211	156	112	51	37	1,626
Madrid, Spain	148	157	214	231	272	310	359	335	261	198	157	124	2,769
Oslo, Norway	40	76	126	178	220	250	246	216	144	86	51	35	1,668
Paris, France	63	79	129	166	194	202	212	212	168	118	68	51	1,662
Stockholm, Sweden	40	72	135	185	276	292	260	221	154	99	54	33	1,821
Vienna, Austria	66	106	128	183	239	228	260	251	168	139	66	51	1,884
Warsaw, Poland	43	59	115	150	211	237	226	214	153	99	39	25	1,571

Figure 3. Solar global irradiation map for Europe
Huld et al., 2012.

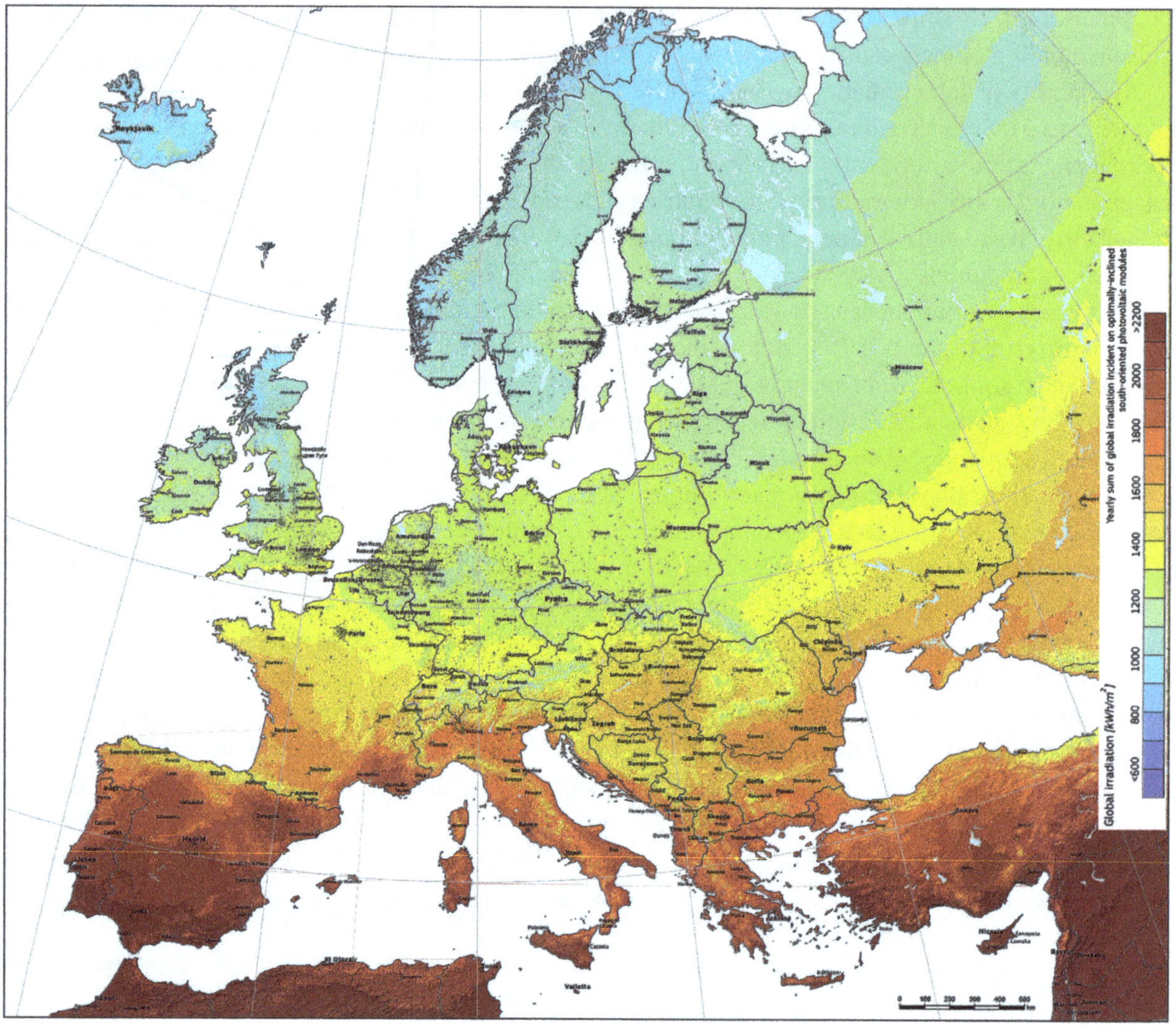

Table 2. List of mathematical models for calculating incident solar radiation

Sky Model	Model Name	Authors	Source
Isotropic	Isotropic Sky Model	Liu B.Y.H., Jordan R.C.	Liu et al. (1963)
Anisotropic	Klucher	Klucher T.M.	Klucher (1979)
	Reindl et al.	Reindl D.T., Beckman W.A., Duffie J.A.	Reindl et al. (1990a) Reindl et al. (1990b)
	Perez et al.	Perez R., Ineichen P., Seals R., Michalsky J., Stewart R.	Perez et al. (1990)
	HDKR model	Proposed Duffie and Beckman combination of models: Hay J.E., Davies J.A., Klucher T.M., Riedl D.T.	Duffie et al. (2013)

The reference Duffie et al. (2013) is recommended as a main source of the solar engineering knowledge.

Solar energy can be used for generating:

- Electricity (directly, in the process of photovoltaic conversion in photovoltaic panels systems),
- Electricity (indirectly, using mirrors or lenses to concentrate a large area of sunlight, or solar thermal energy, onto a small area; then the concentrated light is converted to heat which drives a heat engine (usually a steam turbine) connected to an electrical power generator in concentrated solar power systems, or CSPs (Wikipedia, CSP),
- Heat (directly, in the process of thermal conversion in solar collector systems, then used for preparing hot water or heating spaces, or to run absorption heat pumps or coolers), and
- Heat (directly, used to heat internal spaces as a direct or indirect passive solar heating).

Photovoltaics Systems

The process of converting solar radiation into electricity occurs in cells. Today, two types of cells are sold commercially: crystalline silicon cells and thin layer cells. The first type is represented by monocrystalline cells (efficiency of 15–18%) and polycrystalline cells (efficiency of 13–16%), while the second one by CdTe cells (cadmium telluride, efficiency of 6–9%), amorphous silicon cells (efficiency of 5–8%), and CIS cells (copper, indium, gallium, selenide, efficiency of 7.5–9.5%).

Additionally, two other categories of PV cells are under development: multi-junction (concentrator or non-concentrator) solar cells and emerging photovoltaics (WIKIPEDIA, 2017b).

As reported by the National Renewable Energy Laboratory, the efficiency of multi-junction concentrator solar cells reaches 46% and is significantly higher than that of cells currently available on the market (NREL 2017).

For estimating electrical characteristics of photovoltaic (PV) modules or arrays, several models were established (e. g. Akhsassi et al.; Chang 2015; Cuce et al. 2017; Duffie et al. 2013; Fara et al. 2017). Generally, the characteristics of solar cells are the function of solar irradiation (Figure 4) and temperature (Figure 5). The drop in efficiency with an increase in temperature results from increased resistance.

Single solar cells are connected electrically in series and/or parallel circuits into modules (panels) to produce higher voltages, currents, and power levels. To achieve the desired voltage and current, modules are connected in series and parallel into a PV array.

Usually, PV arrays are mounted on roofs, but they can also be mounted on walls, the ground, and polesor installed as part of a shading structure. In all cases, the problem of decreasing efficiency due to the shading effect should be analysed, since solar modules are groups of cells wired in series with each other, so shading one cell will essentially turn off all the cells in its group (WSUEEP. 2009). To reduce shading losses, stringing arrangements, bypass diodes, and module-level power electronics can be applied (Brown. 2016).

Three basic types of photovoltaic systems can be mentioned:

- Direct – PV modules are directly connected to the load,
- Off-grid (‘stand-alone’), and
- On-grid.

An off-grid PV system is designed for independent power generation, which is why it is ideal for remote rural areas. The system consists of:

Figure 4. Theoretical solar cell I-V (current – voltage) characteristics for different irradiation values
Lenardic, 2001-2017.

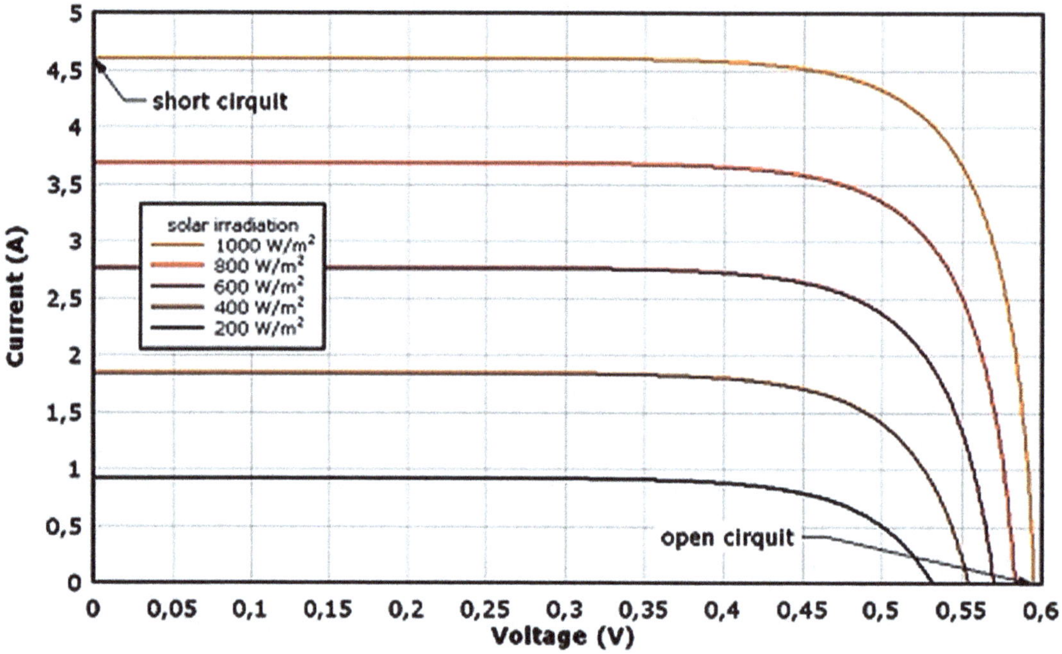

Figure 5. Theoretical Solar cell I-V characteristics, temperature dependence
Lenardic, 2001-2017.

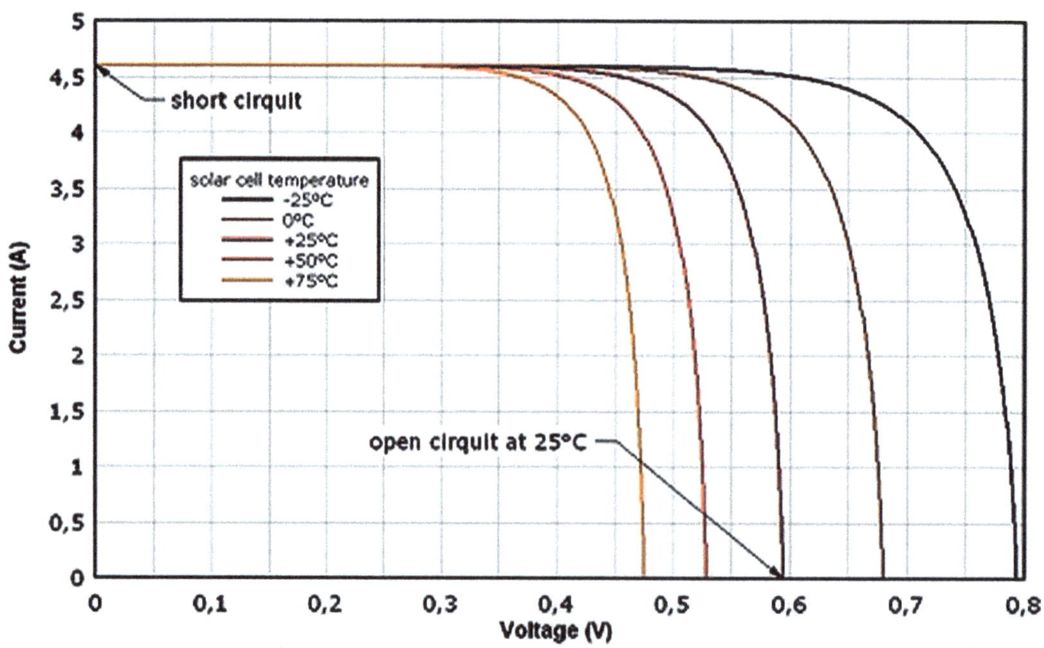

- Solar PV panels – to convert solar energy into electricity,
- Batteries – to store the solar-produced electricity (voltages: 12 V, 24 V, and 48 V),
- A charge controller – to regulate and control the output from the solar panels to prevent the batteries from being overcharged,
- An inverter – to convert direct current (DC) from the solar panels and batteries into 240 V alternating current (AC) for use in the home to power appliances that require AC mains supply, and
- Electrical wires, switches, etc.

In comparison to an off-grid PV system, an on-grid system is additionally equipped with a utility meter for recording the flow of electricity to and from the grid. Depending on the needs, this kind of PV system may or may not contain batteries.

The output power of photovoltaic systems varies with the actual solar irradiance and other conditions, like temperature, shadow, dust. One of the physical quantities given as the technical characteristics of photovoltaic systems is nominal power (peak power). It shall be determined under the following Standard Test Conditions (e.g. IEC 61215-1:2016 Standard): the light intensity is 1000 W m^{-2}, with a spectrum similar to sunlight hitting the earth's surface at latitude 35°N in the summer, the temperature of the cells being 25 °C. The largest PV solar power plant was installed in Zhongwei (China, Ningxia). Its nominal power is 1547 MWp (WIKIPEDIA, 2017e).

Methods of designing photovoltaics systems can be found for example in: (Duffie et al. 2013; Goswami. 2015; Luque et al. 2011; McEvoy et al. 2012).

Solar Collector Systems

A solar collector system absorbs solar radiation energy and stores and distributes heat for particular needs: preparing domestic hot water and heating houses or swimming pools.

The main part of such systems, responsible for collecting solar radiation, is a solar collector. Different classifications are listed below:

- According to the working fluid:
  - Air-based collectors (air is heated in the solar collector),
  - Liquid-based collectors (liquid is heated in the solar collector),
- According to the shape and construction:
  - Unglazed flat-plate collectors,
  - Unglazed perforated plate collectors,
  - Glazed flat-plate collectors,
  - Batch collectors,
  - Evacuated-tube collectors, and
  - Concentrating collectors (parabolic dish, parabolic trough, power towers, stationary concentrating, and solar cookers).

The thermal efficiency of a solar collector is defined as the fraction of available solar energy converted into useful heat during a known period of time. The operational efficiency of the collector depends on operating conditions of the system:

- Ambient temperature,
- Collector temperature, and
- Properties of the absorber and the cover.

The optical efficiency of a solar collector is defined as the rate of optical (short wavelength) energy reaching the absorber divided by the energy coming from the solar resource.

Indicative thermal and optical characteristics of different types of solar collectors are given in Table 3. Thermal losses are proportional to a thermal loss factor. Optical losses are caused by optical properties and are proportional to a conversion factor.

A primitive water heater consists only of a tank where water is stored, warmed up by solar radiation absorbed on its surface and then ready to use, but introducing such a 'system' in a nearly zero-energy building is really problematic.

Generally, the system consists of solar collectors, a storage tank, pipes, and valves. Additionally, in more advanced systems, it has pumps, a pressure relief valve and a controller. If no additional energy is needed to run it (e.g. pumps), the system is called passive; if additional energy to run the system is needed, the system is described as active. If water in a loop is directly used after warming up, it is an open-loop system; if a circulating liquid in the solar collector part is separated from water used for domestic purposes, then such a system is called a closed-loop system.

Basic solar collector systems are shown in Figures 6-9 (Duffie et al., 2013).

The system introduced in Figure 6 represents passive solar water heaters (natural circulation or a thermosiphon). The storage tank located above the collector allows for natural water circulation whenever solar energy warms up water in the collector causing a density difference.

The storage tank can be placed very close to the collector, i.e. on the top of it. That solution is popular in southern European countries. In order to ensure the required water temperature, an auxiliary heater can be used and it can be mounted in the tank or before the tap (a flow water heater).

If forced circulation is needed, a pump is added (Figure 7) and such a system becomes an active one.

Drawbacks of open-loop systems are caused by the flow of chemically-untreated water. Chemical compounds inside the collector and pipes can precipitate, leading to deterioration of heat transfer and flow conditions. Additionally, as temperature rises, solubility of gases in water decreases, resulting in their release. Such conditions are conducive to corrosion. To overcome these problems, a closed-loop system was introduced (Figure 8): the liquid transporting heat from the solar collector to the tank does

Table 3. Indicative thermal and optical characteristics of different types of solar collectors WDD

Type of Collector	Conversion Factor	Thermal Loss Factor, W m^{-2} K^{-1}	Temp. Range, °C
Uncovered absorber	0.82 - 0.97	10 - 30	up to 40
Flat-plate	0.66 - 0.83	2.9 - 5.3	20 - 80
Evacuated-plate	0.81 - 0.83	2.6 - 4.3	20 -120
Evacuated-tube	0.62 - 0.84	0.7 - 2.0	50 - 120
Reservoir collector	about 0.55	about 2.4	20 - 70
Air collector	0.75 - 0.90	8 - 30	20 - 50

Figure 6. Passive water heater (open-loop, one storage tank
Duffie et al., 2013.

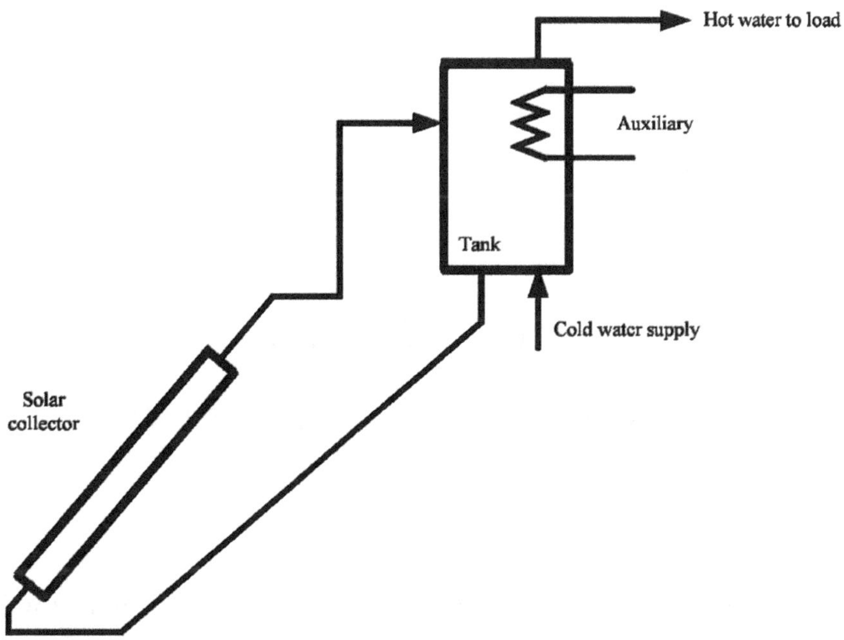

Figure 7. Forced-circulation system (active, open-loop, one storage tank)
Duffie et al., 2013.

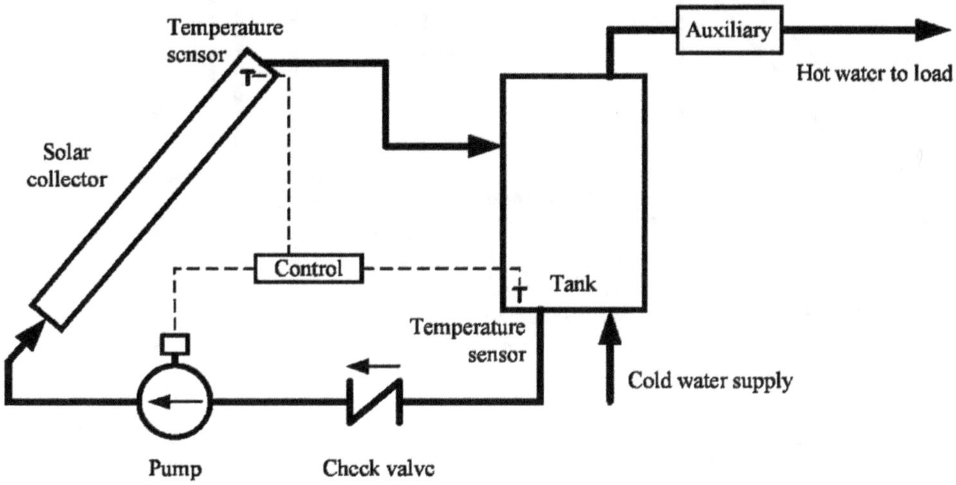

not mix with the heated water (domestic hot water). This kind of system needs electrical energy to run the controller and the pump. If the system is designed to operate when outdoor temperature is below 0 °C, then antifreeze fluid must be used in the solar collector loop. When conditions are insufficient, water is warmed up to a proper temperature by an auxiliary heater.

To decrease the influence of the auxiliary heater on the solar collector loop, a two-tank system can be used (Figure 9).

Figure 8. System with an antifreeze loop and an internal heat exchanger (active, closed-loop, one storage tank)
Duffie et al., 2013.

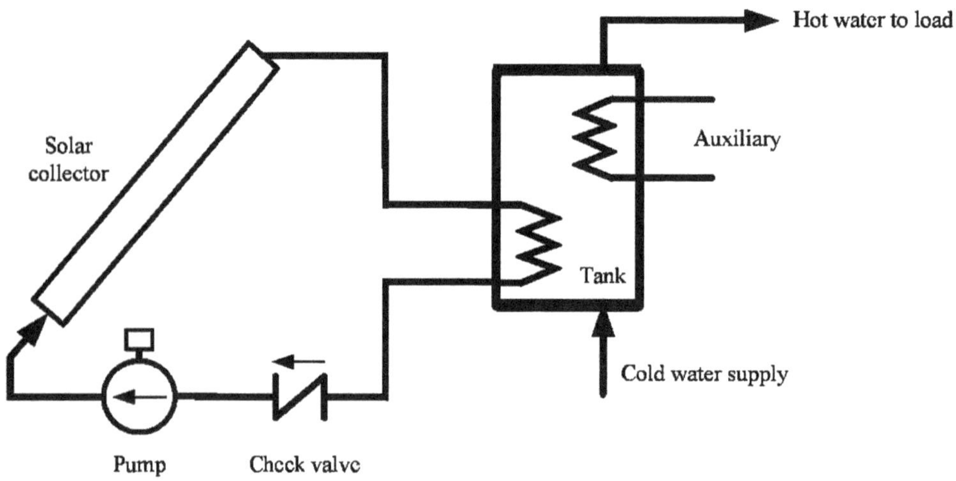

Figure 9. System with an antifreeze loop and an external heat exchanger (active, closed-loop, two storage tanks)
Duffie et al., 2013.

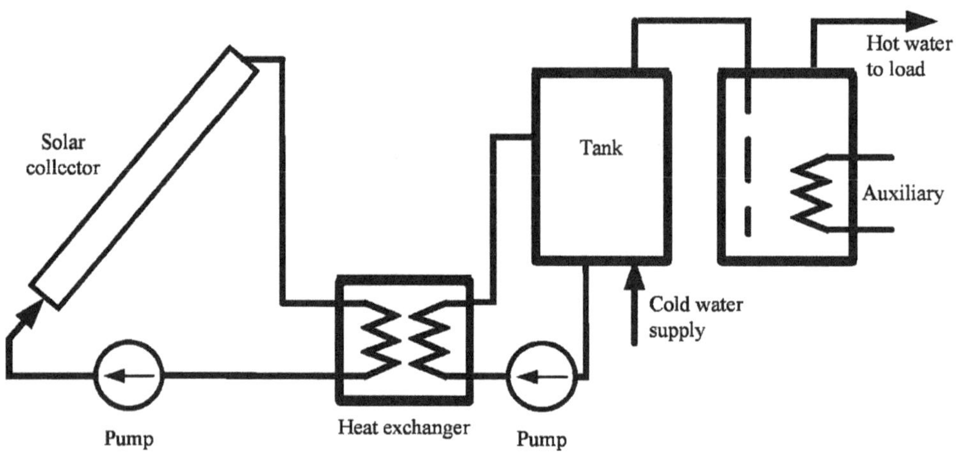

All systems with a pump and an auxiliary heater need a control system. The volume of the storage tank depends on daily needs for hot water or heating needs in a particular period. The area of the solar collector should be designed to match local solar radiation conditions. Oversizing the area of the solar collector causes high water temperatures, while oversizing the water tank results in too low water temperatures; additionally, in the latter case, water in the tank is not fully replaced and some water stays in it for more than one day, which can lead to biological contamination.

As a summary, solar collector systems are compared and described in Table 4.

Table 4. A comparison of solar collector systems

System	Description	Advantages	Disadvantages
Passive water heater (Figure 6)	• Passive • Open-loop • One-tank	• No pollution [1)] • Low investment cost • No running costs [2)]	• Dependence on the availability of solar radiation • No control of water temperature (possible water overheating) • Precipitation of insoluble compounds and danger of corrosion (with chemically-untreated water) • Danger of freezing when temperature drops below 0 °C
Forced-circulation system (Figure 7)	• Active • Open-loop • One-tank	• No pollution [1)] • Small running costs [2)] • Possibility to control water temperature	• Dependence on the availability of solar radiation • Higher investment cost (than in the case above) • Precipitation of insoluble compounds and danger of corrosion (with chemically-untreated water) • Danger of freezing when temperature drops below 0 °C
System with an antifreeze loop and an internal heat exchanger (Figure 8)	• Active • Closed-loop • One-tank	• No pollution [1)] • Small running costs [2)] • Possibility to control water temperature • Usable when temperature drops below 0 °C	• Dependence on the availability of solar radiation • Higher investment cost (than in the case above) • Too high temperatures in the solar collector loop can lead to damage of the collector or parts of the collector loop
System with an antifreeze loop and an external heat exchanger (Figure 9)	• Active • Closed-loop • Two-tanks	• No pollution [1)] • Small running costs [2)] • Possibility to control water temperature • Usable when temperature drops below 0 °C	• Dependence on the availability of solar radiation • The highest investment costs • Too high temperatures in the solar collector loop can lead to damage of the collector or parts of the collector loop

[1)]Pollution depends on the type of the auxiliary heat source.

[2)]In the case of solar heating, no fuel is needed, but total running costs depend on the type of auxiliary heat source.

The largest solar collector plants generating electricity through the use of solar thermal power is located in San Bernardino County (USA, California). Its capacity is 392 MW (WIKIPEDIA, 2017c)

Methods for designing solar collector systems can be found for example in: (Beckman et al. 1977; Duffie et al., 2013; Goswami. 2015).

Wind Power

The principle of using wind power is to convert the kinetic energy of wind into electricity in wind turbines.

As far as construction is concerned, there are three types of wind turbines:

- Horizontal axis wind turbines (HAWT),
- Vertical axis wind turbines (VAWT), and
- Ducted wind turbines (DWT).

The advantages and disadvantages of the types mentioned above are collected in Table 5.

Wind turbines can be free-standing (pole-mounted) or roof-mounted.

In some countries (e.g. Poland), state law regulates the possibility of using this kind of RES.

Power generated in wind turbines is a function of wind speed. There are characteristic speeds of wind (Figure 10):

Table 5. Comparison of basic types of wind turbines (Guzzetta et al. 2007)

Type of Wind Turbine	Advantages	Disadvantages
Horizontal axis wind turbine (HAWT)	• Blades are to the side of the turbine's centre of gravity, helping stability • Ability to wing warping, which gives turbine blades the best angle of attack • Ability to pitch rotor blades during a storm to minimize damage • A tall tower allows access to stronger wind in sites with wind shear • A tall tower allows for placement on uneven land or in offshore locations • Can be sited in forest above the tree-line • Most are self-starting	• Difficult to operate in near ground winds • Difficult to transport • (20% of equipment costs) • Difficult to install (requires tall cranes and skilled operators) • Affects radars in proximity • Difficult maintenance
Vertical axis wind turbine (VAWT)	• Easy to maintain • Lower construction and transportation costs • Not directional • Most effective at mesas, hilltops, ridgelines, and passes	• Blades constantly spinning back into the wind causing drag • Less efficient (than HAWT) • Operates in lower, more turbulent wind • Low starting torque and may require energy to start turning
Ducted wind turbine (DWT)	• Less visual impact on buildings architecture than traditional HAWT or VAWT turbines • Makes use of unused roof space in cities • Allows energy needs to be met on-site, avoiding transmission losses associated with centralized energy generation	• Suitable for urban environments but not households (only effective on urban high-rise buildings) • Uni-directional. Fixed position and is dependent upon wind blowing in the correct direction • Much more research and development is needed. Research in this field is growing as people become more interested in urban wind generation • Research has to be done to determine energy production potential

Figure 10. Characteristics of a wind turbine
WindPower Program, n.d.

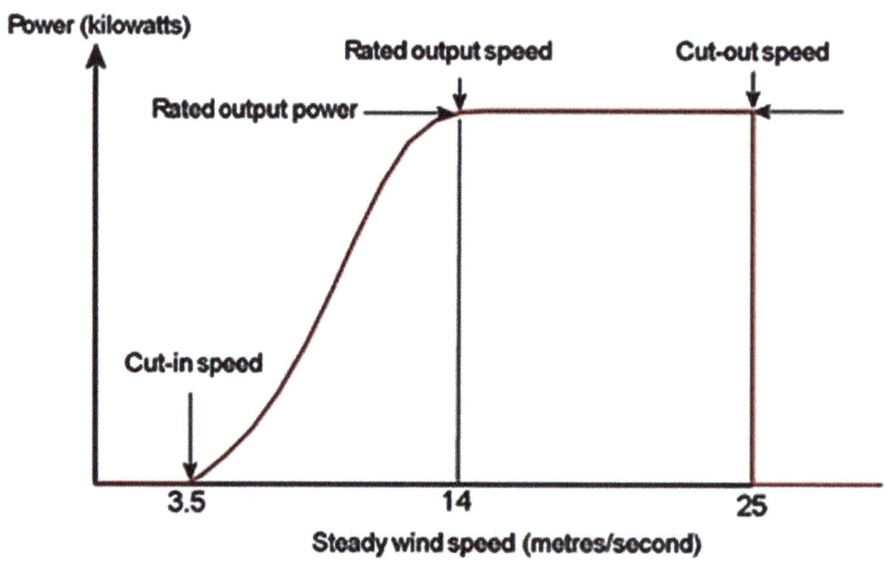

- At very low wind speeds, torque exerted by wind on turbine blades is insufficient to make them rotate.
- Cut-in speed - speed at which the turbine first starts to rotate and generate power.
- Rated output power and rated output wind speed.
- As wind speed rises above the cut-in speed, the level of electrical output power rises rapidly as shown. However, typically somewhere between 12 and 17 meters per second, the power output reaches the limit that the electrical generator can produce. This limit to the generator output is called the rated power output and the wind speed at which it is reached is called the rated output wind speed. At higher wind speeds, the design of the turbine is arranged to limit the power to this maximum level and there is no further rise in the output power. How this is done varies from design to design but typically with large turbines, it is done by adjusting blade angles so as to keep the power at a constant level.
- Cut-out speed - as the speed increases above the rated output wind speed, the forces on the turbine structure continue to rise and, at some point, there is a risk of damage to the rotor. As a result, a braking system is employed to bring the rotor to a standstill. This is called the cut-out speed and is usually around 25 meters per second.

Currently Alta Wind Energy Center (USA, California) is the largest onshore wind farm with current capacity of 1,320 MWp (WIKIPEDIA, 2017f). The largest offshore wind farm is London Array (United Kingdom) with total capacity of 630 MWp (WIKIPEDIA, 2017g).

GEOTHERMAL ENERGY

Geothermal energy is available as heat from within the Earth's crust, usually in the form of hot water or steam. It does not depend on climate conditions. Heat may be contained in trapped vapour, water, or brine or in hot rocks.

This kind of energy can be used directly for heating or to generate electricity. Taking into account temperature, the following types of resources can be distinguished (IEA. Geothermal):

- High-temperature hydrothermal resources (above 180 °C),
- Medium-temperature hydrothermal resources (from 100 °C to 180 °C), and
- Low-temperature hydrothermal resources (below 100 °C).

Depending on the medium and its temperature, different geothermal technologies can be applied (ENERGY.GOV):

- Flash steam plants,
- Dry steam plants, and
- Binary plants.

Electricity and heat produced from geothermal energy are clean, but investment costs of a power station are large. Certainly, the costs depend on:

- The size of the plant,
- Power plant technology,
- Knowledge of the resource,
- Temperature of the resource,
- Depth of the well,
- Chemistry of geothermal water,
- Resource depth and permeability,
- Environmental policies,
- Tax incentives,
- Markets,
- Financing options and costs, and
- Time delays.

For a separate building or for only a few buildings, to avoid high investment costs, the heat stored in the ground up to 150 m is more usable. This layer of ground can be treated as a low-temperature heat storage. This is why the ground is considered to be a low-temperature source for a heat pump which can be used in mechanical ventilation systems for air preheating or precooling.

There are two types of heat exchangers: horizontal (close to the Earth's surface, depth of 1.5 to 2.0 m) and vertical (maximum depth of 150 m).

To design a ground heat exchanger properly, some factors have to be considered, such as:

- Thermal properties of the soil,
- Thermal resistance of the pipe wall,
- Contact resistance at the pipe-soil interface, and
- The effective fluid-pipe wall heat transfer coefficient.

Horizontal Ground Heat Exchanger

A method for simplified dimensioning of a horizontal ground heat exchanger was proposed by IGSPA (International Ground Source Heat Pump Association). In that method, the process of heat exchange between air flowing through the ground heat exchanger and the surrounding ground is described on the assumption that this process is steady and air is incompressible (Bose. 1985).

Vertical Ground Heat Exchanger

To estimate the needed length of a borehole heat exchanger, an approximate value of a specific extraction rate can be used (Table 6). There are some restrictions (Rubik. 2006):

- Operating time at maximum power: shorter than 1800 h per year,
- Depth of the borehole: from 40 m to 100 m, and
- Minimum distance between boreholes:
    - 5 m when the depth is between 40 m and 50 m,
    - 6 m, when the depth is between 50 m and 100 m.

Table 6. Approximate values of specific heat extraction rates of different types of soil (Rubik, 2006)

Soil Type	Thermal Conductivity, $W\ m^{-1}\ K^{-1}$	Heat Extraction per 1 m of the Vertical Ground Heat Exchanger Length, $W\ m^{-1}$
Hard rock	3.0	Max 70
Unconsolidated rock, saturated	2.0	45 – 50
Unconsolidated rock, dry	1.5	Max 25
Gravel, sand, saturated	-	55 - 65
Gravel, sand, dry	-	< 20
Wet clay	-	30 - 40
Limestone (rock)	-	45 - 60
Sandstone	-	55 - 65
Acidic magma rock (e.g. granite)	-	55 - 70
Alkaline magma rock (e.g. basalt)	-	35 - 55
Gneiss	-	60 - 70
Watercourse in sand or gravel	-	80 - 100

HYDROPOWER

Hydropower is power delivered from energy of falling water or fast running water and refers to tidal and wave power as well. It can be converted into mechanical energy but nowadays is usually converted into electricity.

Its applications have the following advantages:

- Environmentally friendly (non-polluting),
- High efficiency of conversion into electricity,
- Zero costs of fuel (only maintenance costs),
- Long operational life, and
- Short time to meet load demand.

The biggest hydroelectric power plant, Three Gorges (China), has a capacity of 22,500 MW, while the second one, in Itaipu (Brazil and Paraguay) has an installed capacity of 14,000 MW (Power.Technology.com). Hydropower can be used on a small or micro-hydroelectric scale to produce enough electricity for a home. Certainly, it is still possible to use hydropower to run mills, sawmills, etc., or simply to do some kind of mechanical work.

Small hydro-electric power plants have a capacity of 1 to 20 MW.

To use hydropower, the availability and quality must be considered in terms of:

- Enough height difference (potential energy), and
- Adequate water flow (kinetic energy).

The use of tidal and wave power is restricted to certain areas, for example next to the sea.

Perhaps it could be interesting that water energy potential can be used for example in a cyclic water pump powered by hydropower – such a device is called a 'hydraulic ram'. Its history dates back to the 'pulsation engine' constructed in 1772 by John Whitehurst (Cheshire, United Kingdom) (Wikipedia. Hydraulic ram).

In some countries (e.g. Poland), state law regulates the possibility of using this kind of RES.

BIOMASS

Biomass is defined in Directive 2009/28/EC as '*the biodegradable fraction of products, waste and residues from biological origin from agriculture (including vegetal and animal substances), forestry and related industries including fisheries and aquaculture, as well as the biodegradable fraction of industrial and municipal waste*'.

Newly harvested biomass can be converted into homologues of natural gas and of liquid and solid fossil fuels and, following various transformation processes (combustion, gasification, pyrolysis), into 'bio-fuels', 'bio-heat', or 'bio-electricity' (Riva et al. 2012).

The biomass chain can be characterised by a negative carbon balance (net removal of $CO_{2eq.}$ from the atmosphere) as well as a positive carbon balance (net addition of $CO_{2eq.}$). The negative carbon balance is achieved when the standing stock of biomass increases or carbon is removed from the carbon cycle via inactive soil carbon, pyrolysis char, or carbon capture and storage (Riva et al. 2012).

The properties of biomass (Table 7), including:

- Lower heating value (LHV), MJ kg^{-1},
- Humidity, %, and
- Ash contents, %,

can vary depending on the location, weather conditions, etc.

Lower heating value (LHV) is the heat amount produced by combustion of one unit of a substance, at atmospheric pressure, under conditions such that all water in the product remains in the form of vapour (McAllister. 2011).

Biomass can be used to produce electricity, heat, and transport fuels.

In order to use biomass on a large scale, one must consider different forms and properties of this fuel, different ways of biomass conversion into energy, and chosen power generation technology (combustion, gasification, pyrolysis, or anaerobic digestion) (IRENA. 2012).

Combustion, applied most often, occurs in a conventional Rankine cycle with biomass being burned in a high-pressure boiler to generate steam, which guaranties net power cycle efficiencies of about 23% to 25%. Additionally, biomass can be co-fired with coal in a coal-fired power plant.

In housing, biomass is usually treated as fuel. In this connection, special furnaces must be used because of different combustion temperatures and relatively high moisture content in biomass, but heating and hot water preparation systems can be conventional. They can be supported by a solar collector system and a heat storage tank.

Table 7. Properties of biomass (Biovalco[1])

Type of Biomass	LHV, MJ kg^{-1} w.b.[2)]	Humidity, %	Ash Contents, %
Wood pellets	16.4	10	0.5 - 1.5
Woodchips – hardwood – pre-dried	12.2	30	0.5 - 1.5
Woodchips – hardwood	8.0	50	0.5 - 1.5
Woodchips – softwood – pre-dried	12.0	30	0.5 - 1.5
Woodchips – softwood	8.0	50	0.5 - 1.5
Bark	8.2	50	5.0 - 8.0
Short Rotation Willow	12.7	25	1.0 - 3.0
Straw (oat)	15.5	10	4.0 - 12.0
Straw (wheat)	14.6	10	4.0 - 12.0
Straw (flax)	15.5	10	4.0 - 12.0
Oat Hulls	15.1	10	2.5 - 5.0

[1)]Adapted by Biovalco from the Handbook of Biomass Combustion and Co-Firing, Phyllis 2.0 database.
[2)]w.b. – wet basis

Hybrid or Multi-Source Systems

To take the advantage of two or more renewable energy sources, several solutions can be proposed. One of them is a hybrid wind and photovoltaic system (Kaldellis. 2010).

By combining photovoltaic and thermal conversion, a photovoltaic thermal hybrid solar collector, also known as a hybrid PV/T, can be made (Kumar et al. 2015). In this solution, to avoid efficiency drops of PV cells due to increased resistance following a rise in temperature, the solar collector is responsible for decreasing temperature. In such a system, heat and power are produced simultaneously.

Another issue is the use of heat pumps in renewable energy systems. In such systems, temperature of the source (ground, air, water) can be raised and then used in heating or hot water systems (Sherratt. 1987).

Economic Analysis

To compare different solutions, especially when risk is an issue, a simple payback period (SPB) can be recommended (Short et al. 1995). The simple payback period is the number of years necessary to recover investment costs of the project in question. It is not recommended when financing and tax features or mutually exclusive alternatives are considered (Short et al. 1995).

The net present value (NPV) is a method of examining costs, i.e. cash outflows, and revenues, i.e. cash inflows (Short et al. 1995).

To compare costs of electricity from different sources, including renewable energy sources, the levelized cost of electricity (LCOE) is applied (Short et al. 1995; IRENA. 2012).

LCOE is defined as a sum of costs over lifetime in relation to the sum of electrical energy produced over a lifetime. The total cost includes: investment, operation, and maintenance costs, and discount rates. The German levelized cost of electricity is shown in Table 8.

Table 8. German LCOE in €/MWh (Kost et al., 2013)

Technology		Low Cost	High Cost
Coal-fired power plants	brown coal	38	53
	hard coal	63	80
CCGT power plants (Combined Cycle Gas Turbine)		75	98
Wind Power	Onshore wind farms	45	107
	Offshore wind farms	119	194
Solar power plants	PV systems	78	142
Biogas power plant		135	250

The comparison of data in the table above shows that costs of electricity from renewable energy sources are relatively high. In terms of costs, only onshore wind farms could compete with coal power plants. Figures in this table clearly indicate the need to improve efficiency, in terms of energy and economics, of all technologies applied in the field of renewable sources.

CASE STUDIES (KODnZEB)

To show what kind of improvement of energy performance can be achieved by using renewable energy sources, two complex examples are given below (KODnZEB). The choice of renewable energy sources was made for two public utility buildings belonging to the Warsaw University of Technology. A description of the buildings can be found in the chapters *Existing buildings – how to meet an nZEB standard?: The architect's perspective* and also in *Improving the energy quality and indoor environmental quality in retrofit buildings*. In order to achieve the standard set in the project, it became necessary to analyse the availability of renewable energy sources. Both examples are based on Polish conditions and cannot be directly transferred to other countries.

Muszelka Dormitory

Prior to the introduction of an RES, the primary energy demand indicator was 76.4 kWh m^{-2} $year^{-1}$. In order to achieve the effect assumed in the project, an RES had to be selected. Due to conservation supervision over the building and specific location in Warsaw, some activities were excluded, such as placing photovoltaic panels or solar collectors on the façade, the use of wind, biomass or geothermal energy, etc.

It was analysed whether it was possible to cover part of heat demand with heat sources with a lower non-renewable primary energy factor than the coefficient 0.68 that characterized the heating system used then.

After the analysis of possible solutions, photovoltaic panels were chosen (Figure 11). Since components of the system had to be installed in such a way that they would not be visible from the ground around the building, maximizing solar energy was possible by placing flat roof collectors or photovoltaic panels. The results of the analysis show that under such conditions the productivity of a PV installation is 162 kWh m^{-2} $year^{-1}$, while that of an installation using solar collectors for DHW preparation is 487 kWh m^{-2} $year^{-1}$. However, given that the non-renewable primary energy factor for electricity is 3.0 and the one for

Figure 11. Computer model of the Muszelka dormitory with a photovoltaic installation and surrounding buildings
Source: KODnZEB internal data.

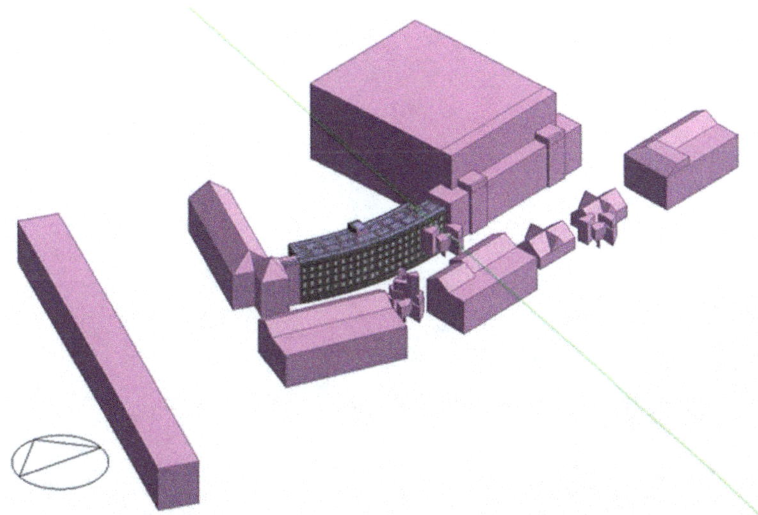

heat is 0.68 (DMID. 2015), the reduction in primary energy demand per unit area is significantly higher with PV panels than with solar collectors. This is why due to a very limited roof surface the idea to use solar collectors instead of photovoltaic panels was abandoned.

A photovoltaic system with a rated power of 79.35 kWp was selected. The panel surface is 374.8 m^2 and supplies 69,953 kWh of electricity annually. The volume of electricity produced is limited by the size of the installation and its inclination to the ground. These features are related to restrictions concerning the roof area and the conservation officer's reservations.

Finally, given the limitations mentioned above, it is proposed to add a heat pump using the air removed from the building as the bottom heat source. Since the value of the heat pump's heating efficiency coefficient depends strictly on the heated fluid temperature, the heat pump is proposed to be used only as an auxiliary source. The purpose of the heat pump in the proposed system is to preheat mains water from 10 °C to a temperature that allows for obtaining the best value of the resultant primary energy indicator of the whole DHW system, taking into account both the value of the heat efficiency factor and its contribution to meeting the heat demand.

After the selection of renewable energy sources, the primary energy factor of 17.9 kWh m^2 $year^{-1}$ is achieved.

Faculty of Building Services, Hydro and Environmental Engineering

Prior to introducing renewable energy sources, the primary energy demand indicator (EP) was 43.1 kWh m^{-2} $year^{-1}$. As in the case of the Muszelka dormitory, it was analysed whether it was possible to cover part of heat demand with heat sources with a lower non-renewable primary energy factor than the coefficient 0.68 that characterized the heating system used then. A starting point for choosing the sources was a schematic diagram of the demand for heating and domestic hot water (Figure 12).

Figure 12. Schematic diagram of the demand for heating (Q_{co}) and domestic hot water (Q_{cwu})
Source: KODnZEB internal data

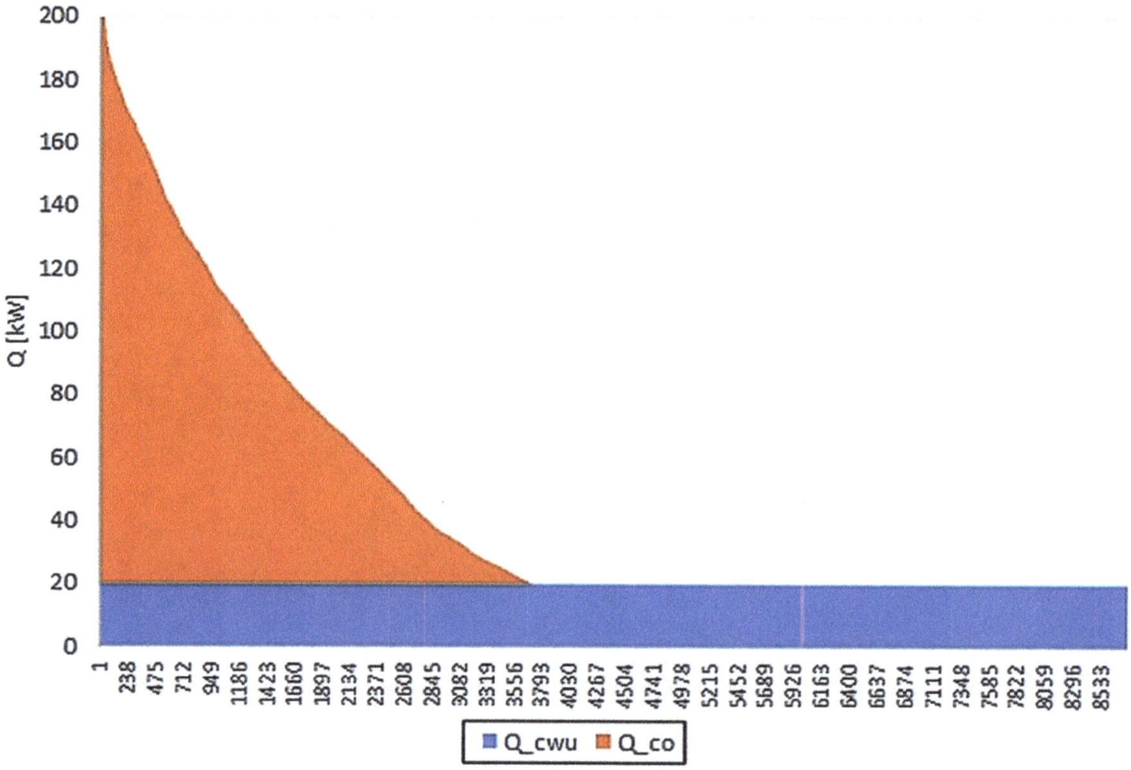

The heat demand was then used to determine what part of the demand could be covered by each renewable source analysed. Two technologies were selected for the analysis: a ground heat pump and a micro-CHP system. Since the value of the heat pump's heating efficiency coefficient depends strictly on the heated fluid temperature, it is proposed to use the heat pump only as an auxiliary source. The purpose of the heat pump in the proposed system is to preheat mains water from 10 °C to a temperature that allows for obtaining the best value of the resultant primary energy indicator for the whole DHW system. The optimal value of the non-renewable coefficient for the DHW system can be obtained by preheating water to 25 °C. In order to reduce the EP index significantly, the use of a combined heat and power generation system based on micro-cogeneration units was proposed. Finally, a cascade system consisting of two identical units of nominal power P_{nom} = 20 kWe was used.

Because the combined sources analysed allow for a better non-renewable coefficient relative to the district heating network non-renewable coefficient, it was necessary to reduce the DHW preheating temperature down to 15 °C. In this case, the non-renewable coefficient is higher. Finally, for the proposed supply system, 6% of heating needs are met by the heat pump, 73.9% via the micro-CHP system, and the remaining 20.1% by heat from the district heating network.

The photovoltaic system was selected taking into account neighbouring buildings. The impact of shading of façades on the efficiency of photovoltaic cells was assessed. The surface of photovoltaic cells installed on the façade and roof of the building is about 600 m^2 and their rated power is 489 kWp (Figures 13 and 14).

Figure 13. A model of the Faculty of Building Services, Hydro and Environmental Engineering with a photovoltaic installation and surrounding buildings
Source: KODnZEB internal data.

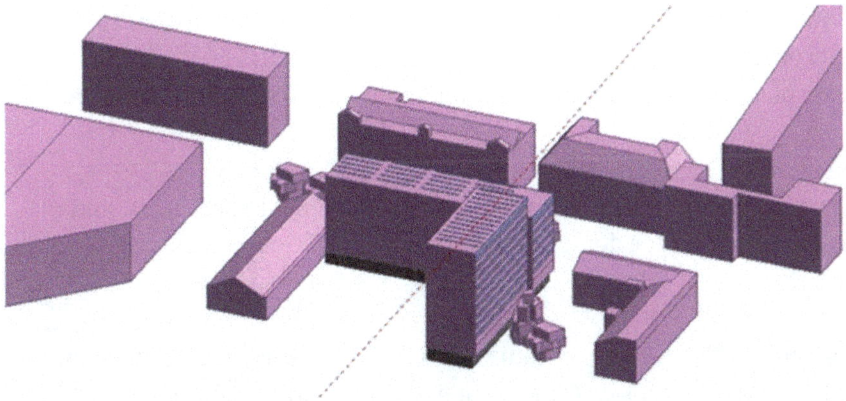

Figure 14. South façade of the Faculty of Building Services, Hydro and Environmental Engineering with photovoltaic cells
Source: KODnZEB internal data.

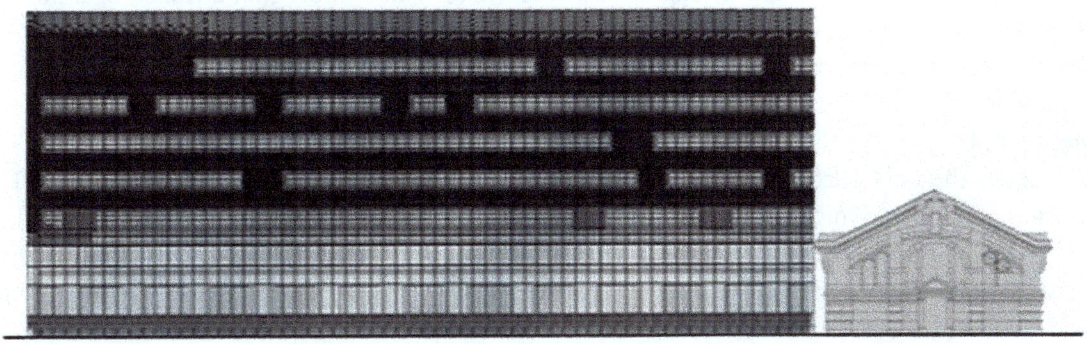

The results for two variants of photovoltaic electricity production are shown in Table 9. In the first case, the influence of neighbouring buildings on the PV installation (without shading) was neglected and in the second one this influence on the PV (shaded) installation was considered.

It can be seen that despite the decision not to place photovoltaic panels below the second storey (possible shading by surrounding trees and buildings), the calculated production in the building model without neighbouring buildings is higher. It follows that local conditions must always be taken into account.

After the selection of renewable energy sources, the primary energy factor is 6.2 kWh m^{-2} $year^{-1}$.

Table 9. Results of photovoltaic electricity production without and with the influence of neighbouring buildings (Source: KODnZEB internal data)

	Electricity Production, kWh Year $^{-1}$	Rated Power, kWp
Building model without neighbouring buildings	185,657	509
Building model with neighbouring buildings	178,139	489

SUMMARY

The availability of renewable energy sources strongly depends on location, and specific data are necessary.

In any case, law regulations, requirements of the spatial development plan and recommendations for conservation supervision (if applicable) must be taken into account.

In most cases, solar energy is the best renewable energy source for use in buildings, as it can be converted into:

- Electricity, which can be consumed directly to run technical appliances, and
- Heat, which can be used for domestic hot water preparation and can also supplement traditional heating.

In the northern hemisphere, the maximum availability of solar radiation occurs in summer, while heating is actually needed in winter. Thus, during summer, solar radiation should be used to prepare hot water. If active solar heating is used in winter (large solar collectors area), this will cause producing excess heat in summer, so it can be used for heating up water in a swimming pool.

Wind power could be individually used on a small scale.

Geothermal energy used by means of vertical or horizontal heat exchangers is attractive, but the cost of such an installation makes it long to pay back (mainly due to digging a trench or a borehole). A horizontal ground heat exchanger requires a relatively large area, but it can be used for preheating ventilation air in winter or precooling it in summer.

A small-scale use of hydropower is limited to certain areas, because availability of this energy source is quite limited.

Biomass is the only RES among those mentioned that causes pollution. Due to its relatively low lower heating value in comparison to that of fossil fuels (e.g. natural gas: about 35 MJ m^{-3}, coal: even up to 35 MJ kg^{-1}), relatively more biomass is needed for heating purposes. Additionally, it can be treated as an auxiliary heat source in transitional periods during the year (spring and autumn), for example in a closed wood-burning stove.

The main advantage for end-users of an RES application is that it is characterised by low running costs (free energy source), but it can be taxed (e.g. electricity production in PV systems in Sweden). Maintenance costs of small and medium-sized installations are usually comparable to equivalent costs of conventional systems. Replacement costs in particular cases can be higher (e.g. batteries in PV systems).

Large-scale applications, e.g. on a regional and state scale, involve high investment costs, but they influence costs of electricity. That electricity is delivered to end-users and also affects cost-effectiveness of local renewable energy systems.

It must be emphasized that solar, wind, hydrothermal, and geothermal energy is clean and safe for local environment.

There are no technical barriers to modernisation of buildings towards the nZEB standard, but the costs of adapting a building and buying and installing necessary equipment may be the problem.

ACKNOWLEDGMENT

I would like to acknowledge the support of all designers from the Warsaw University of Technology and NTNU Trondheim involved in the KODnZEB grant. This research was supported by the *EEA Financial Mechanism 2009-2014 in the framework of the Bilateral Cooperation Fund at the level of the Operational Program PL04 "Saving energy and promoting renewable energy source"* within the framework of the project KODnZEB: "*Design retro-fit nZEB concept for two buildings*" (grant number 315/2015/Wn07/OA-XN-04/D no. SAP 511/00023).

I would like to acknowledge the International Renewable Energy Agency (IRENA) for agreeing to use the publication *Biomass for Power Generation* (IRENA. 2012).

REFERENCES

Akhsassi, M., El Fathi, A., Erraissi, N., Aarich, N., Bennouna, A., Raoufi, M., & Outzourhit, A. (in press). Experimental investigation and modeling of the thermal behavior of a solar PV module. *Solar Energy Materials and Solar Cells.*

Beckman, W. A., Klein, S. A., & Duffie, J. A. (1977). *Solar Heating Design by the f-chart Method.* New York: Wiley-Interscience.

Biovalco. (n.d.). *Biomass Fuels.* Retrieved September 3, 2017, from http://www.biovalco.com/energy/biomass-fuels/

Bose, J. E., Parker, J. D., & McQuiston, F. C. (1985). *Design/data manual for closed-loop ground-coupled heat pump systems.* ASHRE.

Brown, B. (2016). *Shade Losses in PV Systems, and Techniques to Mitigate Them.* Aurora Solar Inc. Retrieved September 3, 2017, from http://blog.aurorasolar.com/shading-losses-for-pv-systems-and-techniques-to-mitigate-them/

Chang, K. K. (2015). Modeling of PV performance without using equivalent circuits. *Solar Energy, 115*, 419–429. doi:10.1016/j.solener.2015.01.025

Cuce, E., Cuce, P. M., Karakas, I. H., & Bali, T. (2017). An accurate model for photovoltaic (PV) modules to determine electrical characteristics and thermodynamic performance parameters. *Energy Conversion and Management, 146*, 205–216. doi:10.1016/j.enconman.2017.05.022

Directive 2001/77/EC of the European Parliament and of the Council of 27 September 2001 on the promotion of electricity produced from renewable energy sources in the internal electricity market. (2001). *Official Journal of the European Communities, 44*(283), 33-40

Directive 2003/54/EC of the European Parliament and of the Council of 26 June 2003 concerning common rules for the internal market in electricity and repealing Directive 96/92/EC. (2003). *Official Journal of the European Union, 46*(176), 37-56.

Directive 2009/28/EC of the European Parliament and of the Council of 23 April 2009 on the promotion of the use of energy from renewable sources and amending and subsequently repealing Directives 2001/77/EC and 2003/30/EC (Text with EEA relevance). (2009). *Official Journal of the European Union, 52*(140), 39-85.

DMID. (2015). *Decree of the Minister of Infrastructure and Development on the methodology for calculating the energy performance of buildings and dwelling or building which is part of the whole technical self-utility. and the preparation and presentation of certificates of energy performance from 27th February 2015.* (in Polish).

Duffie, J. A., & Beckman, W. A. (2013). *Solar engineering of thermal processes*. Hoboken, NJ: John Wiley & Sons. Inc. doi:10.1002/9781118671603

ENERGY.GOV. (n.d.). *Geothermal Basics*. Retrieved September 3, 2017, from https://energy.gov/eere/forge/geothermal-basics

EUROSTAT. (n.d.). *Renewable energy statistics*. Retrieved September 3, 2017, from http://ec.europa.eu/eurostat/statistics-explained/index.php/Renewable_energy_statistics

Fara, L., & Craciunescu, D. (2017). Output Analysis of Stand-Alone PV Systems: Modeling, Simulation and Control. *Energy Procedia, 112*, 595–605. doi:10.1016/j.egypro.2017.03.1125

Goswami, D. Y. (2015). *Principles of Solar Engineering*. Boca Raton, FL: Taylor & Francis Group.

Guzzetta, A., Meyers, G., & Purse, A. (2007). *Types of wind turbines and associated advantages*. Retrieved September 3, 2017, from http://me1065.wikidot.com/types-of-wind-turbines-and-associated-advantages

Huld, T., & Irene Pinedo-Pascua, I. (2012). *Photovoltaic Solar Electricity Potential in European Countries*. PVGIS ©. *European Union*, 2001–2012. Retrieved from http://re.jrc.ec.europa.eu/pvgis/cmaps/eur.htm

IEA. (n.d.a). *Final energy consumption by sector and fuel*. Retrieved September 3, 2017, from https://www.eea.europa.eu/data-and-maps/indicators/final-energy-consumption-by-sector-9/assessment-1

IEA. (n.d.b). *Geothermal energy*. Retrieved September 3, 2017, from http://www.iea.org/topics/renewables/subtopics/geothermal

IEA. (n.d.c). *Unit Converter*. Retrieved September 3, 2017, from http://www.iea.org/statistics/resources/unitconverter/

IEC 61215-1:2016. Terrestrial photovoltaic (PV) modules - Design qualification and type approval - Part 1: Test requirements.

IRENA. (2012). International Renewable Energy Agency. *Renewable Energy Technologies: Cost Analysis Series, 1*(5).

Kaldellis, J. K. (2010). *Stand-Alone and Hybrid Wind Energy Systems - Technology, Energy Storage and Applications*. Woodhead Publishing. doi:10.1533/9781845699628

Kim, K., Kumar, P., Szulejko, J., Adelodun, A., Junaid, M., Uchimiya, M., & Chambers, S. (2017). Toward a better understanding of the impact of mass transit air pollutants on human health. *Chemosphere*, (174): 268–279. PMID:28178609

Klucher, T. M. (1979). Evaluation of Models to Predict Insolation on Tilted Surfaces. *Solar Energy, 23*(2), 111–114. doi:10.1016/0038-092X(79)90110-5

KODnZEB. (n.d.). *Program PL04 Saving energy and promoting renewable energy source within the framework of the project KODnZEB Design retro-fit nZEB concept for two buildings*. Grant number 315/2015/Wn07/OA-XN-04/D no. SAP 511/00023. Retrieved September 3, 2017, from http://kodnzeb.pl/

Kost, Ch., Mayer, J. N., Thomsen, J., Hartmann, N., Senkpiel, Ch., Philipps, S., & Schlegl, T. et al. (2013). *Levelized Cost of Electricity Renewable Energy Technologies*. Fraunhofer Institute For Solar Energy Systems ISE.

Kumar, A., Baredar, P., & Qureshi, U. (2015). Historical and recent development of photovoltaic thermal (PVT) technologies. *Renewable & Sustainable Energy Reviews, 42*, 1428–1436. doi:10.1016/j.rser.2014.11.044

Lenardic, D. (2001-2017). *PVRESOURCES*. Retrieved September 3, 2017, from http://www.pvresources.com/en/introduction/introduction.php

Liu, B. Y. H., & Jordan, R. C. (1963). The Long-term Average Performance of Flat-Plate Solar Energy Collectors. *Solar Energy, 7*(2), 53–74. doi:10.1016/0038-092X(63)90006-9

Luque, A., & Hegedus, S. (2011). *Handbook of Photovoltaic Science and Engineering*. Hoboken, NJ: John Wiley & Sons.

Maleki, S. A. M., Hizam, H., & Gomes, Ch. (2017). Estimation of Hourly. Daily and Monthly Global Solar Radiation on Inclined Surfaces: Models Re-Visited. *Energies, 10*(1), 1–28.

McAllister, S., Chen, J.-Y., & Fernandez-Pello, A. C. (2011). *Fundamentals of Combustion Processes*. Cham: Springer International Publishing AG. doi:10.1007/978-1-4419-7943-8

McEvoy, A., Markvart, T., & Castañer, L. (2012). *Practical Handbook of Photovoltaics - Fundamentals and Applications*. Amsterdam: Elsevier.

NREL. (2017). *Best Research-Cell Efficiencies*. National Renewable Energy Laboratory. Retrieved September 3, 2017, from https://www.nrel.gov/pv/assets/images/efficiency-chart.png

Perez, R., Ineichen, P., Seals, R., Michalsky, J., & Stewart, R. (1990). Modelling Daylight Availability and Irradiance Components from Direct and Global Irradiance. *Solar Energy, 44*(2), 99–108. doi:10.1016/0038-092X(90)90071-J

Power.Technology.com. (n.d.). *The 10 biggest hydroelectric power plants in the world*. Retrieved September 3, 2017, from http://www.power-technology.com/features/feature-the-10-biggest-hydroelectric-power-plants-in-the-world/

Reindl, D. T., Beckman, W. A., & Duffie, J. A. (1990a). Diffuse Fraction Correlations. *Solar Energy, 45*(1), 1–7. doi:10.1016/0038-092X(90)90060-P

Reindl, D. T., Beckman, W. A., & Duffie, J. A. (1990b). Evaluation of Hourly Tilted Surface Radiation Models. *Solar Energy, 45*(1), 9–17. doi:10.1016/0038-092X(90)90061-G

Riva, G., Foppapedretti, E., Carolis, C., Giakoumelos, E., Malamatenios, C., Signanini, P., Giancarlo, C., Di Fazio, M., Gajdoš, J., Ručinský, R. (2012). *Handbook on Renewable Energy Sources*. Retrieved from http://docplayer.net/16926980-Handbook-on-renewable-energy-sources.html

Rubik, M. (2006). *Heat pumps. Guide*. Ośrodek Informacji, Technika instalacyjna w budownictwie. (in Polish)

Sherratt, A. F. C. (1987). *Applications of heat pumps to buildings*. London: Hutchinson.

Short, W., Packey, D. J., & Holt, T. (1995). *A Manual for the Economic Evaluation of Energy Efficiency and Renewable Energy Technologies*. Retrieved from https://www.nrel.gov/docs/legosti/old/5173.pdf

Tiba, S., & Omri, A. (2017). Literature survey on the relationships between energy, environment and economic growth. *Renewable & Sustainable Energy Reviews*, *69*, 1129–1146. doi:10.1016/j.rser.2016.09.113

WDD. (n.d.). *The Worlds of David Darlig*. Retrieved September 3, 2017, from http://www.daviddarling.info/encyclopedia/S/AE_solar_collector.html

Wikipedia. (2017a). *Concentrated Solar Power*. Retrieved September 3, 2017, from https://en.wikipedia.org/wiki/Concentrated_solar_power

Wikipedia. (2017b). *Multi-Junction Solar Cell*. Retrieved September 3, 2017, from https://en.wikipedia.org/wiki/Multi-junction_solar_cell

Wikipedia. (2017c). *List of Solar Thermal Power Stations*. Retrieved September 3, 2017, from https://en.wikipedia.org/wiki/List_of_solar_thermal_power_stations

Wikipedia. (2017d). *List of Cities by Sunshine Duration*. Retrieved September 3, 2017, from https://en.wikipedia.org/wiki/List_of_cities_by_sunshine_duration

Wikipedia. (2017e). *List of Photovoltaic Power Stations*. Retrieved September 3, 2017, from https://en.wikipedia.org/wiki/List_of_photovoltaic_power_stations

Wikipedia. (2017f). *List of Onshore Wind Farms*. Retrieved September 3, 2017, from https://en.wikipedia.org/wiki/List_of_onshore_wind_farms

Wikipedia. (2017g) *List of offshore wind farms*. Retrieved September 3, 2017, from https://en.wikipedia.org/wiki/List_of_offshore_wind_farms

WindPower. (n.d.). *Wind turbine power output variation with steady wind speed*. Retrieved September 3, 2017, from http://www.wind-power-program.com/turbine_characteristics.htm

WSUEEP. (2009). *Solar Electric System Design, Operation and Installation. An Overview for Builders in the Pacific Northwest*. Retrieved September 3, 2017, from www.energy.wsu.edu

KEY TERMS AND DEFINITIONS

Biomass: An RES derived from organic material, mainly from trees and agricultural and urban waste.
Geothermal Energy: An RES, heat stored in the Earth.

Hydropower: An RES, power derived from energy of falling water (mechanics: potential energy) or fast running water (mechanics: kinetic energy).

nZEB: A building characterised by high energy performance, in which energy consumption and production are almost balanced during a year (a nearly zero-energy building).

Renewable Energy Sources (RES): Energy sources that naturally reproduce in the human time scale or are inexhaustible from a human point of view.

Solar Energy: An RES, energy derived from the sun in the form of solar radiation.

Wind Energy: An RES, air flow energy (mechanics: kinetic energy).

Chapter 4
3XE: Efficiency, Ecosphere, Economics

Elzbieta Dagny Rynska
Warsaw University of Technology, Poland

ABSTRACT

For designers proceeding with work in accordance with nZEB, management strategy means that participants of different disciplines have to accept that surrounding environment parameters are in constant interaction with design. This further points out the necessity to understand the correlations taking place between local and existing environments, when buildings should be fitted out with systems working in interaction between the local and general biosphere parameters. Hence, within the construction business, such development means integrated proceedings in all design phases and construction sphere itself, integration of the artificial systems allowing for the buildings function and parameters characteristic to the local surroundings. This condition also applies to nZEB buildings, except that the parameters are more limited to passive and active solutions used for achieving effective energy choices, while maintaining required user comfort parameters and environmental balance. This chapter is dedicated to management procedures for nZEB investments from a Polish perspective.

INTRODUCTION

Pilot projects and design proceeding are very important when having in mind the progress towards achieving nearly zero energy buildings (nZEB). Such projects and most of all understanding of procedures provide relevant examples and practical experience. A lot of different stakeholders want to learn that this type of buildings can be constructed with reasonable cost implications, existing innovative technologies and with full support of designers from various disciplines.

It should be noted that one of the most neuralgic zones within the building investment is design process. In many countries, design process has been treated as a linear system, where the lead designer – usually an Architect – passes his highly aesthetic solutions to be consulted by other disciplines. Representatives of those disciplines also tend to work separately. Therefore, this new strategic management should most of all; include a change within design environment. The main idea is that all participants should understand

DOI: 10.4018/978-1-5225-4105-9.ch004

that technical solutions must be perceived holistically – that effective energy use is not a priority when there is insufficient user ventilation at stake, or low standard of internal use environment. Furthermore, buildings should be perceived as one of many elements of urban context and their influence on natural surroundings and already existing urban space. This outlook must include the impact of existing climatic parameters on their form, functional layout and choice of building materials and technologic systems.

It may be therefore assumed that any issues concerned with the buildings where sustainable development strategy was used, concentrate not just on the ecosphere, but also on people and economic perspectives. Hence, the perceived scope of potential interests is Efficiency of used tools and solutions, in balance with Ecosphere – scoping both humans as well as well as any biotic and abiotic species. Both areas should be joined together with Economic abilities, not always articulated as financial gain, but accepted as a complex phenomenon mirroring human requirements.

Analyses (The Business Case for Green Building, 2013) show that buildings where environmentally friendly solutions have been used do not have to be much more expensive than the standard ones – even though this idea is commonly shared by many. Required financial input may be kept on the same level when from the very start design process is considered as an integrated procedure – for example Performance Based Building Design (PBBD) or dynamically developing Building Information Modeling (BIM).

As any other investments, also building process has to consider various risk factors which have direct influence on the tenant fees and future market value of constructed volume. This has also influence on the return of costs period. Another risk factor is change in legal requirements, which more than often exclude ineffective energy buildings and introduce stricter laws. All decisions undertaken during the early project phases will also have influence on the long term building's value. What follows, holistic approach should be maintained during entire building's life cycle.

Extreme climatic changes and change in local weather patterns can be observed presently on a daily basis, but research is still only touching on the edge of real changes which might follow. Nevertheless they have direct impact on the buildings' function and reaction to negative impacts. New solutions include implementation of complex solutions surpassing sustainable development and allowing for quick redevelopment of destroyed urban tissue or mitigation of natural hazards. In Poland, this path has not as yet received a name – in other more developed countries it is known as resilience.

INTEGRATED DESIGN PROCEDURES

The secondary effects of stricter design process and more energy demanding construction laws are lower design fees and construction costs paid by the investors. This also has direct influence on a better market situation of manufacturers of environmentally friendly building materials and technical systems. What is more, contractors also have better tools to build buildings where these solutions have been introduced. Research proves that additional cost should not exceed 12.5% of total costs (The Business Case for Green Building, 2013), even in case of nearly zero energy buildings (nZEB). In case of those buildings, effective costs measures may include, but not be limited to:

- Acceptation and implementation of environmental solutions at early design stages; strategy must include preliminary cost estimation.

- It is preferable that design team will use integrated design process, allowing connecting energy effective passive envelope solutions, layout of function with regard to required energy needs and possible high efficiency equipment which may be used.
- Such solutions should be preferable instead of investments where environmental solutions are mounted on the rooftops or façades even though the buildings were neither designed nor constructed as energy effective.
- Employment of design teams which have experience and use integrated design process, as well as implement PBBD or BIM management already at early concept stage, through all stages including user participation.
- Utilization of integrated design appears to be the issue not only because of environmental assets, but also of economic feasibility of designed building which will also bring long term financial gains and economically satisfactory costs, due to implementation of durable building materials and equipment.

In most cases sustainable development solutions are not the major issue when calculating the costs in constructed building. Rational balance of costs is a challenge not only for the designers, but also for the representatives of the building sector, who are too often sure that alternative solutions create 10-20% higher costs in comparison to standard buildings (The Business Case for Green Building, 2013). This may be caused by insufficient knowledge, that contemporary expenditures are much lower in comparison to those which date 20 years back. Moreover, it is also characteristic, that demonstration buildings are very often fitted with additional expensive and often irrational solutions. These building also are the ones which appear in publications. They are often fitted with visible over scaled features – photovoltaic panels, wind turbines – which are basically proving alternative technologic solutions – not realistic economy outcomes of implemented features.

So what are the preconditions of the scope and strategy for modernization development?

When preparing analysis of the local characteristics, designer should maintain a holistic management strategy. One of the major conditions is that analysis of urban space should be delimited to an area larger than just the site, and each of existing ecosystems should be checked in relation to all neighborhood ecosystems. Hence, "environment" must be understood within a global context. Analogously, all proposed nZEB solutions should show not just economic parameters, but also other feasibility abilities of actual site.

Complex understanding of environmental construction investment process, with a particular consideration of nZEB solutions, does not mean preservation of the environment from human intervention, but rather creation of relations between the effects of human intervention within the ecosystem and the environment itself. It concentrates on finding of ways to reduce any negative changes due to ineffective energy solutions (Kurnitski, 2011). Efficiency of the nZEB multidisciplinary design solutions depend on the designers' knowledge and abilities. This also has impact on reduction of potential negative influences formed by the construction of the building. Checking procedure should start already during the predesign concept level. The set of issues within the design and construction phase scopes most evident contemporary choice - effective energy solutions – as well as efficient use of building materials, water and soil. Construction process itself should develop with emphasis on complex adaptation solutions adjusting to changing climatic conditions and user requirements, as well as low exploitation costs (Kurnitski, 2011). Additionally market expectations and existing often changing technical conditions place growing demands on the construction sector, especially in the case of effective energy use. More than often investors and contractors have to meet growing tenant requirements which include environmental solutions.

Many existing buildings prove that use of random individual solutions both of passive type as well as technical systems, create negative side effects. For example – higher tightness of the building's envelope causes lower standards than the average accepted ventilation standards, unless we also include adequate window parameters in design process. Hence, the only correct solution is holistic design strategy which should transform typically used linear design process into integrated design while still maintaining aesthetic architectonic values.

When optimizing design process, the designers are obliged to formulate a set of criteria which concern both developers and the users. These may be formulated as follows, but not limited to:

- Highly insulated building's envelope and energy efficiency of chosen installation systems.
- Low environmental influence of chosen building technologies as well as choice of building maintenance procedures which include lower environmental impacts.
- Maintaining high standard user environment.
- The best case scenario prepared with data on durability and efficiency of chosen building materials and installation systems.
- Flexibility of proposed functional layout and chosen building volume – if possible, ability for future change of use should be included.
- Easy facility management of used equipment and effective performance of installation systems.
- Cost efficiency of the building, additionally supported by LCA analysis – both prepared and updated during each stage of design process.
- Acceptation of cultural and social connections within already existing urban context; design should follow possible implementation of user preferences.

Within integrated design traditional phases remain unchanged. The only difference is positioning of the multidisciplinary process in an interactive "loop", where each of the architectonic and technical issues are optimized during consecutive phases of the design process. The key aspect is constant cooperation between representatives of each of the disciplines, interdisciplinary integration of proposed solutions, as well as workshops with the building owner and sometimes final user. Design documents prepared according with Integrated Design Conditions (IDC), allow achieving 30-50% of efficient energy use, while maintaining the same level of construction costs. It has also been practically checked (KodnZEB, 2015-2017), that the best phase to include nZEB solutions – is the concept phase. This condition complies with the expectations included in BREEAM conditions. Hence it should be noted that when using IDC, following issues have to be included:

- Interdisciplinary cooperation starting from design concept level, as only this type of approach that allows for a holistic determination of benchmark parameters and realistic estimation of costs.
- Appointment of a dedicated post of a project manager - a person who will participate in all meetings and will play a management key role within the process of attaining a coordinated set of interdisciplinary drawings and specifications.
- Introduction of final technical parameters defined for each of the disciplines, and as well holistic and economic parameters which will have to fulfill by the building when in use.

- Cooperation with environmental, energy and quantity surveyors preparing initial cost estimation - starting at concept design stage.
- Preparation of main environmental strategy conditions, allowing for a framework where other specialist consultants such as for example daylight consultants may work on the optimization of the daylighting in work areas, other may be consulted on air quality; each of those consultants may be approached to formulate an opinion on each and environmental strategy during each of design phases.

Many of above conditions appear to follow a typical design process. The main difference may be perceived in the fact, that both the client and the end user play the key roles during design process. Usually, complex solutions used in nZEB buildings are subjected mainly to achieve effective energy strategy. Hence, when establishing the initial brief, designers should consider following issues:

- Type of build-in materials and surface finish on the external face of the building's envelope; it is essential to check daylighting options and passive heating solutions of the building's internal areas, as well as options using interactive facades allowing cooperation between local environmental parameters and men made technical systems.
- Choice of heating, ventilation and air condition systems – use of energy effective solutions lowering the energy supply required for the functioning of the systems themselves, use of natural ventilation and securing energy through passive techniques. One of the possible options is hybrid ventilation – an obvious system in most Scandinavian countries.
- Choice of alternative heating source for domestic hot water – used solution mainly depends on the building's function and local climate parameters. In case of low water requirements, it is possible to use solar batteries as the only source of heat, or decide on a point heating systems – to be used "at demand" (Sowa, 2017).
- The choice of the daylighting strategy- level of artificial light system should compensate the existing level of daylight – in total fulfilling demanded level of lux at each working desk (Sowa, 2017). Artificial lights should be switched on only in the areas which are in use (unless the user technology demands otherwise).

Building and finishing materials specified by the designers should have the "eco-friendly" status established through following criteria:

- Low energy use and low primary energy during manufacture and in-build processing of building materials.
- Manufacture of building materials and components from renewable or recycled resources or components.
- Manufacture process allowing use of recycled materials from consumption and industrial waste.
- Use of timber sourced from industrial timber plantations, with certificate proving the place of sourcing.
- Maintaining good standard natural environment parameters – low emission of greenhouse gases;
- Use of materials and processes characterized by low or no emission of dangerous substances which might be emitted during production and in-build process, as well as at the end of technical life.

- Choice of building joints allowing for recycling of elements or dismantling into basic components which can be used in a different location or reused in a new building process.
- Technologies using easy to mount techniques, without need to use additional volatile components.
- Produced in locations placed at a rationally near distances from the site, which in turn has impact on the lower use of vehicular transport, level of emitted energy and emission of greenhouse gases.
- Packaged in materials which can be resourced and reused or recycled

One of the basic tasks placed before the design team is preliminary estimation of construction costs versus foreseen maintenance costs. Design team's main aim is the use of environmental valuable parameters and lowering of potential negative impacts. Hence their tasks should include provision of following macroscale analyses:

- **Local Air Temperature Amplitudes:** Monthly average temperatures for the night and daytime.
- **Daylighting:** Possible screening by other building or landscape features located in surrounding sites; areas shaded due to landscape formations, groups of trees or existing buildings lowering access to the daylight, self-shading (possible effect of building's 3D form).
- **Sunlight:** Landscape of the site and surrounding areas, a case best orientation of the new or transformed building, defining maximum days of sunshine based on the climatic data, angle of sunrays and possibility of natural shading.
- **Wind Parameters:** Provision of a local wind rose for the chosen site, analysis of the level of wind exposition and other characteristic site features, wind turbulence may be used within the building's and site natural ventilation strategy.
- **Preparation of Preliminary Specification of Building Materials:** Research allowing confirmation whether it will be possible to use recycled materials when constructing a new investment or retrofitting, this should include stones, timber and reinforced concrete; in case of already existing buildings undergoing modernization processes, often it is more economic to use existing structure, sometimes in a different configuration than initially (e.g. there is an option to use external glass panels which do not comply with currently expected thermal insulation coefficients, as internal partitions).
- **Analysis of Existing Green Areas:** Geodesic inventory of trees and shrubs, showing their precise location on the site, inventory should also include data disclosing their health status and type of species, this data should be closely connected with the analysis of existing water routes – both surface and underground ones – including the water table levels, water runoff volume and direction.
- **Geodesic Soil Parameters, Including Load Bearing Capacities:** This will have direct impact on the choice of foundations required for the new building, or extension and transformation of existing one, but also on the type of drainage used and choice of on-site greenery.
- Analysis of air quality and acoustic parameters which have direct impact on the ventilation system solutions, especially in areas with high emission of pollutants (Sowa, 2017).
- Environmental analysis checking the level of pollutants found in the soil and underground water systems is required for all investment types; when founding buildings on granite rocks – additional radon analysis and possible other radiation have to be included

- In case of valuable landscape features, it might be required to maintain particular land configuration or endemic plant sites; design team should include these conditions during creation process of building's form and function, so that designed volume will become an integral element of the whole site and immediate surroundings.

In cases when the site is small in comparison to build area, green zones can be designed as winter gardens, "green" terraces or roofs, possibly as internal courtyards often located above underground parking areas. Above all, choice of plants should include these which do not require surplus of water, especially, if neither a rain tank, nor a local purification gray water plant has been foreseen in the design. Use of endemic plants regenerates the outer surfaces of soil, reduces the volume of water runoff outside the site's boundaries. Use of ground covering vegetation also has direct influence on local air cooling phenomenon – this additionally means lower loss of water through evaporation and reduction of local ground temperatures even by 10-15% in comparison with asphalt, gravel or stone slab surface finish. When optimizing the use conditions for the development, high importance is placed on the configuration of land masses. This last feature is often more effective than men made acoustic panels. Small architectonic water features are also good examples of local temperature modifiers. Hence designer's brief prepared for the client should include: analysis of existing site parameters; studies on Master Plan conditions, possibilities and limitations; analysis of the site conditions supporting adequacy of design approach and proposed solutions.

During preliminary feasibility study, design team analyses design solutions which are in accordance with conditions set forth within the Master Study, Local Plan or Conditions for Land Development[1]. Also economically feasible time and financial schedules have to be prepared. Above all, it is required to use the conditions based on the site analysis. Initial Concept Phase deals with volume, form and general functional layout concerning the site and the building, initial cross sections and facades, including:

- In case of existing building – inventory and expert analyses of existing technical status confirming possibility of extension or transformation.
- Definition of the plan's general layout, confirmation of the compatibility of chosen solutions with the design conditions, local planning expectations and legal request.
- Analysis of the buildings functionality, confirmation with the Client's expectations and if required expectations of the final user.
- Choice of the general aesthetic values: structure, building and finishing materials, preliminary estimation of cost.
- Acceptance of conditions set forth during consultation workshop with the representatives of local administration and historic conservation requests – if site is located within historic zone or is included in the register of historic buildings.
- Co-ordination of design works conducted by the engineers of all disciplines.

Documentation must include:

- A report explaining facility management of designed building at the information level corresponding with preliminary design phase;
- All floors layouts and characteristic details; elevations and cross sections; a site plan including specification of hard and soft areas.

- Specification containing data on chosen building and finishing materials.
- Preliminary time schedule with milestones indicating design key dates, as well as dates for the start and finish of each design phase.
- Preliminary construction cost estimation; design fees and author's supervision fees during construction phase.
- Preliminary acceptance of proposed solutions by the consultants and designers of all disciplines.
- Explanation of the scope and impact of proposed technical solutions to the Client, data should include environmental impact and specific conditions.

After conclusion of this phase, all initially accepted solutions become more developed and transformed during consecutive stages. They are re-checked and modified according to need. The building itself becomes more defined in volume and site specific. Decisions concerning type of structural technology and technical systems to be used within the building are made. Initial simulations concerning user comfort requirements and standards are also prepared. All decisions include information on environmental impacts and emission of dangerous substances. Should the Client decide to construct the building according to environmental certification (BREEAM, LEED, WELL, DGNB etc.), certain documents and procedures have to be prepared and followed already at this design phase. It nevertheless should be pointed out that buildings designed and constructed under any of those certifications are not automatically placed within energy efficiency domain. Investors and designers will simply follow very efficient procedure techniques, but the outcomes will depend on the Investors aims.

After Concept Design, Building Permit is usually within the basic scope of services provided by designers. Hence the documents also include various disciplines at different stages of development even if the design process is not considered as integrated. Any major changes in chosen solutions which might take place at later design stages will have to undergo acceptation process by the local authorities according to legal requirements. Regardless of the management type of the design process, produced documents should include all legal and formal requirements, as well as architectonic and technical solutions allowing for the construction of the building in the chosen site. In this phase, the scope of basic required documents and services should include:

- Written technical description of the building's characteristic features, preliminary structural calculations, and evacuation and fire prevention conditions.
- Layouts of all floors in adequate (readable) scale.
- Procedures with all required consultants, expert and specialist opinions.
- Energy certificate data.

Additionally, when using Integrated Design Strategy, designer should present the Client with data containing program analysis, estimated costs and initial bill of quantities. Due to legal conditions, basic environmental solutions should include environmental friendly building materials, energy calculations for the building's volume and calculations containing thermal capacity of building elements. When analyzing most efficient structural solutions, lowest primary energy and life cycle should be assessed. It is often that during the design process, the Client cannot precisely indicate the function of each and every area within the designed building, whereas such knowledge has direct influence on the structural and engineering solutions. In order to maintain flexibility of the design and potential functional change, it may be well justified to design structural elements with higher load bearing capacities than it was calculated.

Execution design can be prepared by the Contractor's architect, but in order to sustain set conditions beside a complete set of drawings, documents should include technical specification of building materials, procedures for execution of each type of works and strategy for the coordination of all disciplines. It would be advisable, that this design phase would be consulted with the initial designing team. Description and conditions of development are integral with the integrated procedure within the design phase including nZEB solutions and choice of foreseen execution procedures, including:

- Building's envelope should be designed with parameters allowing to achieve insulation standards compatible with initially accepted values and adequate technical precision.
- Choice of roofing material and finishing floor surface layers – higher durability of materials will allow for longer use, but at higher costs.
- In specifications concerning designed mechanical systems, the choice of equipment should be based on energy efficiency and high use standards.
- Mechanical ventilation systems should include recuperation units; this is based on a requirement to coordinate all design technical disciplines with the location of culverts and adequate ventilation cross sections connecting all areas with the recuperation steering unit.
- Adequate level of daylight should be supplied to each working area; if possible, designers should propose equipment which emits dispersed light; electric light system is to compensate inadequate level of daylighting.
- Designers should point out possibility to use locally produced building materials – ceramics, concrete blocks, hollow bricks and tiles; this choice allows for a lower environmental impact of the building investment process through lower transport energy (often characteristic when using heavy large scale building elements), it also lowers the cost of transport and indirectly cost of the whole procedure.

In Poland, tender procedure is the most delicate of management tasks. Depending on the chosen investment strategy concerning organization of the building process, data required for the tender procedure is prepared either by the design office, or project manager managing the whole process in close co-operation with the designers. Regardless of the typology, prepared set of documents should be sufficiently complete as to allow the client to start with contractual negotiations with contractors. Design team with the project manager check and verify placed offers, and participate in the tender negotiations. The scope of documents to be prepared and services which should be provided by the design office during the tender procedure includes:

- Preparation and submission of tender documents: contractual conditions, tender specification, existing design documents and other technical and legal information, as well as copies of offer forms.
- Preparation of a report concerning received offers with data on the quality and costs and analysis of all relevant information provided by the tender participants, in case of investment managed according to environmental criteria – estimation whether environmental criteria set for the within the tender specification have been met by the contractor.
- If the client requires environmental solutions, this should be visibly marked within the tender documents, as this will allow for a more realistic pricing and competition between offers.

In case when the general contractor is presenting an offer for the total of foreseen works including turn-key procedures, then the tender data should also include:

- Preferred technology of construction works: durability of the load bearing structure and other building materials, confirmation that the choice of materials will be based on environmental friendly characteristics; this includes technologies when providing finishing works.
- Façade finishing materials: thermal and acoustic parameters of window and door frameworks.
- Finishing wall and floor materials (durable data, foreseen life cycle). Confirmation that all chosen materials comply with the acceptable VOC levels and emission of other toxic substances.
- Environmental construction strategy also includes other works provided by the subcontractors (including building waste management). Requirements should be included in the specification prepared for the contractor by the design team and Project Manager. These conditions should include the need for maintaining a compact site, as well preservation of endemic plant ecosystems. Specific contract conditions supported by schematic drawings should form a set of environmental preservation requirements which must be met by all contracting firms and their subs.

In case of public investments, including modernization works concerning retrofitting of buildings to nZEB standards, it is usual that clients apply for EU funds and this implicates the fact that the contracts are based on FIDIC conditions. FIDIC is a set of procedures and conditions which regulate the obligations and relations between parties – the Client (Investor) and the Contractor (a party providing construction works), as well as Engineer of the Contract who acts as the third party managing the site and adequate standard of executed works.

FIDIC acronym is derived from the name of the organization Federation Internationale des Ingenieurs Conseils – International Federation of Engineers – Consultants, aiming at unification and implementation of schemes and procedures which can be followed in any country and construction site. Wide scope of regulated issues caused the need to provide more than a single scheme, but based on uniform conditions. The choice depends on the type of services required by the Client. Unfortunately in post-soviet countries FIDIC was reintroduced only in the last decade of 20th Century and in some cases the conditions and country specific legal requirements are in contradiction.

Contract for design works should be prepared according to the conditions set forth in the White FIDIC Book. This is especially important in Poland in case of public works, as very often Clients try to change some of the conditions during investment process. Agreement must include all definitions and interpretations defining the scope of services, tasks to be fulfilled by the consultant (designer), tasks to be fulfilled by the Client, a list of the personnel having adequate knowledge to provide required services, responsibilities and guarantees, final dates and strategies how to provide changed scope of services, general conditions and solving of disputes. Designer receives information about the main investment aims which the Client would like to achieve. It is possible to withhold part of the initially contracted tasks concerning design of a new or modernization of an existing building. This allows the investor to lower the design fees and in turn investment costs which should be paid to the contracting parties if the scope remains unchanged.

Contract for construction works signed according to the Red FIDIC Book conditions, contains a typical contract to be signed between the Client and Contractor. It represents a model of contractual conditions for building works and provision of installation systems, volume (buildings) and linear (roads) investments. Client submits complex technical design documents (including execution design), with a preliminary bill of quantities and technical specification to the Contractor. The choice of building materials and technologies is foreseen already at the design stage. Submitting of the documents by the Client means that he is the party bearing the risk for the final investment effects and the Contractor is responsible only for the execution of works according with the provided design. Procedures contained in the Red Book should be used especially for major construction investments such as: buildings, where the design issues are very important to the Client and should remain under his control (e.g. opera or philharmonic buildings, stadiums etc.), or those for which design documentation has been already prepared prior to tender; also linear investments, where such data as land ownership and soil conditions are the major issues. There is a time requirement allowing for the best case choice of the routes, but also the best case choice of technical solutions, provision of economic construction costs and maintenance costs.

This type of conditions is mainly used in cases when technical and environmental conditions, as well costs are the major issue. The Contractor receives all documents and expert opinions, including Building Permit and Permit to use the land as investment site. In this case, it is possible to a certain degree to precisely define required costs, this issue being very important for beneficiaries of EU funds. Provision of full documentation also lowers the risk of not maintaining required completion dates. Contract is of a quantity type, with payments calculated based actually provided works and paid by the Temporary Payment Certificates. Each payment is made according to the on-site measurements, confirmed in a Book of Measurements by all parties involved. It is accepted that the volume of works defined in the bill of quantities, being the base for pricing offer is an estimated number and cannot be used as final and correct when it comes to contractor's remuneration. This type of contract allows introducing additional or exchange works. Any works which could not have been foreseen by the Contractor, appearing during execution of the contract, can be accepted as additional works, to be paid from financial reserves which Client should keep within the budget.

There are two tender procedures within the Red Book, the first one for the Contractor and the other for the Engineer of the Contract, the latter acting as an "Institution" employed by the client for a complex supervision of the site. EC acts in the name of the Client and merges duties of a substitute investor (he organizes and leads the site meetings), as well as coordinator of supervising inspectors (in Poland they should be employed by the Client) and designers. He acts in the name of the Client, but holds the autonomic position and cannot change the conditions contained within the contract signed between the Client and Contractor. In case when adjustments have to be made, he is obliged to lead consultations with each of the parties. It is important to mention that during these procedures often designers are called in to join the Contractual Team. This procedure is possible through an early definition of a lump sum for the author's supervision and placing this fee within the contractual conditions when Engineer of the Contract is signed. In some cases, investors employ their own Engineer of Contract Office, which allows them to use consultants already during the preliminary preparation works for the contract but eliminates Engineer's independence from the client.

Second alternative is Design and Build contract signed in accordance with the Yellow FIDIC Book. In this strategy Contractor not only provides design services but also executes construction works. Design duties as well as construction duties and responsibilities for the correct standard of provided works are placed on the Contractor. It should be mentioned, that based on the Functional and Use Program established by the Client for the investment, Contractor is obliged not only to prepare design, but provide all permissions from local administration offices and expert consultations etc. Within the scope of contractual works, Contractor also has to make choice as to the routes for linear investments, equipment installed in buildings etc. Additional duty is providing in the name of the beneficiary (client) documents allowing the use of the site for construction purposes (permissions from the neighboring sites and permission to start construction works). Contractor is also responsible for any financial compensation to third parties which might be required during the course of contractual works. Presently this type of contract appears to be the most often used in Poland. It allows shortening the time for investment and transfers responsibilities as well provision of all additional opinions to the Contractor. Use of this formula requires preparation of the functional program and a good knowledge of all site location and legal issues. It does not guarantee the level of initially estimated budget (not all issues are evident at the date when the agreement is signed), it does not guarantee the use of technical solutions which allow for low exploitation and maintenance costs. In Polish conditions it might be financially viable to move the responsibility so that execution design will be also provided by the Contractor.

Alternative contract number three is based on the Silver Book conditions, where the Client who is supplying finance requests a "turnkey" investment for a fixed price (a lump sum). These conditions should be used for buildings and engineering structures at "turnkey" standard, where a high level of certainty is required as to the final costs and key dates and Contractor is fully responsible for the design and execution of construction works, Client only assists in some areas.

Within this particular contract Contractor is responsible for the full scope of technical design, he brings in and mounts all equipment and delivers a fully fitted out and made ready for use building. In practice Silver book is used in such investments as production or energy plants, large swimming pools or SPA's. The developer's market is also a good example when the buildings are provided by the contractor on the turn key level to a private or public client. In this case, the price negotiated or accepted during the tender procedure includes the risk of final higher costs but keeping the responsibility on contractor's side, who for a lump sum based on the functional program provides not just the works but also design and fit out, is responsible for the start up of the building and provides training to facility managers. Final agreed price accepted as a lump sum cannot be changed and the payments are made based on the time and work schedule prepared with the submitted tender offer.

In Poland, the cases of using this alternative are rare; hence it is impossible to talk widely on local experience. This formula may be attractive when investing in special use technical buildings such as sewage treatment plants, or water purification stations. The main risk is difficulty in defining costs, and quotation of higher than standard prices in order to omit financial losses or achieve additional gains. The main idea of described contractual procedure is contractor's responsibility to provide duties of the Engineer of the Contract, Designer and Contractor.

Each of presented alternatives has values and defects. In case of retrofit into nZEB standard, the Client's influence on the choice of technical and environmental solutions, as well as estimation of facility management costs is a major issue. Clients also have to be aware of the required construction costs which have to be shown when applying for EU funds. Hence, in case of retrofit buildings (even in case of nZEB ones) it is recommended to use option one – using the Red Book Fidic standard and preparation of design and supervision tasks in accordance with conditions contained in the White Book. Also, due to the environmental character of the planned investment it is recommended to use the contents of Communication from the Commission to the European Parliament, The Council, The European Economic and Social Committee and the Committee of Regions *Public Procurement for better environment* dated 16.07.2008. Within this document Commission has established a number of ecologic criteria concerning public procurement for several groups of products and services (Bogusz A. et al 2012). One of the priority sectors is building industry (scoping such resources as timber, aluminum, steel, glass, as well as building elements such as windows, wall and floor finish, heating and cooling systems, all aspects of facility management and end of building's life, management of the building).

When providing tender conditions for construction or retrofit procedures for buildings, as well as for extension and modernization procedures, client should include such environmental criteria. This criteria point out the level of effective energy use to be reached during construction and use of the building, use of alternative energy sources, environmentally friendly building materials and products, waste and water management as well as many other aspects concerning environmental influence of the construction works. Designers' knowledge and monitoring of build buildings is also a priority. Examples of some of the conditions are shown below:

- Architects must possess adequate knowledge concerning design of environmentally friendly buildings; should have references confirming participation in such investments.
- Contractors should also provide data on their technical abilities (information on qualified personnel employed, with adequate environmental knowledge) allowing implementation of environmental management including:
  - Provision of safety measures for environmental species existing on site and immediate surroundings (especially in case site is environmentally sensitive).
  - Provision of safety measures for lower emission of dangerous waste and other chemical substances, which may unfavorably influence the site.
  - Waste management and reduction of produced waste, maintaining acoustic measures, economy of used transport means
  - Provision of solutions allowing for effective energy and water usage
- Contractors in Poland are also obliged (Bogusz A. et al 2012) to use energy effective transport and site equipment and allow for efficient management of supplies.

In order to maintain use of building materials and products allowing for Client defined environmental criteria:

- Bidders are obliged to state the percentage of materials and elements used during construction will be produced in accordance with standards to achieve marking equivalent to ecologic brand no 1 (in accordance with ISO 14024), or submitting clear information concerning confirmation of those norms through building declaration.
- Bidders are obliged to state which materials and elements used during construction are produced with recycled input.

In Construction phase designers from all disciplines check the standard of executed works with initial design, or in case of preparing execution design by the General Contractor – they verify and confirm detailed solutions with initial concept conditions. Documents and services provided within this phase depend on the scope of the contract with design office and may include following:

- Preparation of documents forming an annex to the contract with the Contractor: including execution design, bill of quantities and technical specification, possibly a list of preferred environmental guarantees.

- Monitoring of the work's standard executed according with the work and time schedule, required building standards and practice, as well as contract conditions.
- Issuing of instructions concerning change in technologies used in construction works, confirmation of appropriateness of proposed changed solutions, including different choice of building and finishing materials.
- Conduct of periodical estimation of the standard of executed works as well as their conformity with initially presented specification.

Construction process should include various parameters: building techniques, quality, time and work schedules, effective control measures, quality of provided works and used building materials, costs, safety precautions on site, and safety precautions against possible negative atmospheric impacts. A lot of attention should be given towards practical use of standards when providing works such as:

- Quality and tightness of the building's envelope layers.
- Provision of insulation layers with special attention paid to solution guarding against local thermic bridges (building's corners etc.).
- Execution of additional waterproofing in the vicinity of door and window openings.
- Provision of insulation, waterproofing and fire resistance measures in the vicinity of wall and slab openings.
- Contractor should manage all construction works with much care taken towards reduction of waste.

Prior to handing over of the building to the usually the design team in the name of the Client checks the building for completeness and correct standard of provided works. Handing over may be connected with preparation of written technical building manuals and as-built drawings by the contractor or his subcontractors. Regardless of the fact whether investment strategy was set for passive or active building's function – presented scheme must be efficient and all systems should undergo start-up trials. Scope of monitoring works will depend on the level of complexity of the building management system. Some buildings will be subject to temperature monitoring during heating and cooling seasons. More than often the systems have an in-build option to be monitored through use of computer software and data being check in other locations. Environmental parameters also undergo through check as to the level of effects achieved during construction. Following tasks can be distinguished in this area:

- Check of unforeseen air infiltration which can have influence on the standard of wall and floor finishing materials.
- Check and analysis of actual level of energy used during heating and cooling seasons; received data should be checked against that submitted by the users with statements showing the level of achieved use comfort.
- Monitoring of temperatures in all areas through standard measuring devices (maximum and minimum indications), or through BMS systems providing constant readings.
- Level of water use is controlled through monthly meter readings and calculation showing daily water usage per single user. Received data should be checked with initial conditions and assumed usage parameters.

Modernization works and fit-out upgrades are the ones with highest environmental influence. Less building materials are needed, but much more design care is required. Prior to the start of works technical state of the building and surrounding structures should be checked. Energy audit with conditions for further development is also a must. Regardless of the effective design of each of the disciplines – all disciplines should form one efficient whole. Lack of cooperation between designers will lead to higher investment cost and waste of construction length. The pre-Concept stage, Client's Brief, as well as designers' experience appear to be the most important aspects within this procedure. Since Polish nZEB buildings are few, it may be said from my private experience, that one of the most important aspect is choice of designers who wish to work with each other in order to reach assumed goals which include not only Economic issues, but also are environmentally friendly and give the final users a high quality building (KodnZEB, 2015-2017).

CONCLUSION

History of a potential construction success starts very early. LCA data should be provided already within the feasibility study. Analysis showing potential effects of chosen design solutions also should be prepared at that stage. When leading a multidisciplinary design process, members of the whole team concentrate on uniform understanding of the aims and tasks defined for each phase of the design. Achieved standard depends on the design quality of each of the technical solutions merged together into a uniform synergic whole.

Decision making by all designers is consciously made and with awareness that a minor change in one discipline may cause a major change in other ones. This attitude can be perceived as additional added value. Energy simulations are prepared already at the design concept stage, whereas the leading designer is responsible for the modeling of final solutions aiming at assumed parameters.

One of the main tasks is giving the building sector a better knowledge – wider education scoping also environmental solutions and understanding the need to a holistic approach to design process and implementation of environmentally friendly solutions.

In environmental construction process, and especially when designing buildings classified as nZEB, the most important phase is the Concept level and initial brief conditions, when most of the key decisions for the whole life cycle of the building including function and maintenance are considered. This especially includes procedures to be used both by the Client and representatives of all design disciplines when considering achievement of proposed aims. A preliminary design specification defining the scope of the services to be provided "for" and "in the name" of the client can be considered as a very helpful document. This document should be prepared in cooperation with the specialists from different disciplines. Scope of tasks should be described in accordance with the designer's best knowledge and client's expectations, accepted standard of the investment and designed internal environmental parameters, time schedule and existing budget. All data and additional requirements initiated during workshops with the Client, including environmental, urban planning and financial effects should be included as part of the project's evolution. In case of integrated design strategy, it is usually the leading designer who is assisting with the final decisions within those initially proposed as potential solutions. Explanations concerning the level of durability characteristic of the chosen building materials, as well as efficiency of designed installation systems, effective energy use strategy – should enable the investor to understand design procedures more deeply.

REFERENCES

Behling & Behling. (1996). *Sol Power: The evolution of Solar Architecture*. Munich: Prestel Publishing.

Bogusz, A., & Polakowski, L. (2012). *Green Public Procurement*. Warsaw: Public Tender Department.

Business Case for Green Building. (2013). *A Review of the Costs and Benefits for Developers, Investors and Occupants*. World Green Building Council.

Communication from the Commission to the European Parliament, The Council, The European Economic and Social Committee and the Committee of Regions. (2008). *Public Procurement for better environment*. Author.

KodnZeB. (2015-2017). *Unpublished grant outcomes and analysis from KodnZEB project, supported by Finance Mechanism EOG 2009-2014, within the framework of cooperation Fund Operation Program PL04 Efficient Energy and promotion of alternative energy sources. Project realized with the cooperation of NTNU Trondheim*. Author.

Kurnitski, J., Saari, A., Vuolle, M., Niemela, J., & Tark, T. (2011). Cost optimal and nearly zero (nZEB) energy performance implementations. *Energy and Building*, *43*(11), 3279–3288. doi:10.1016/j.enbuild.2011.08.033

Sowa, J. (Ed.). (2017). *Budynki o niemal zerowym zużyciu energii*. Oficyna Wydawnicza, PW.

KEY TERMS AND DEFINITIONS

Building Investment Process: Process starting from the initial undertaking by the investor, through all design and construction phases, facility management, and end of technical use (recycle, reuse, and reduce area).

Environmental Management in Building Investments: All issues concerning protection and preservation of the environment during provision of construction services.

Integrated Design Conditions (IDC): Simultaneous design procedure by multidisciplinary design team.

Multidisciplinary Design Process: Design process that includes specialists from different technical disciplines.

Nearly Zero Energy Buildings: There is no one clear definition and understanding varies from country to country; for the purpose of this chapter, buildings that have been retrofitted to a very high energy performance level should be accepted. The low amount of energy that these buildings require comes mostly from renewable sources.

NZEB Retrofit Procedures: All tasks and procedures that have to undertaken in order to achieve a nearly zero energy building based on retrofit scope of design, preliminary feasibility study.

Pre-Concept Stage: Initial idea as to the site, type, and function of the building, usually takes place in investor's firm.

ENDNOTE

1 It must be explained that in Poland there are three types of state planning provisions. Obligatory is the Master Study it contains general information on the function and possible development of sites (usually cities), Master Plan is unfortunately not obligatory therefore only 30-35% of Poland is covered. If the designer deals with a site with no Master Plan he has to submit an application to the local authorities describing basic data concerning proposed development and receive Conditions for Land Development which is a less intricate form of the Master Plan, addressed only to the site and a small surrounding neighborhood area for which the application was made.

Chapter 5
The Concept of Expert System Supporting the Increase of Energy Efficiency in Buildings

Arkadiusz Węglarz
Warsaw University of Technology, Poland

ABSTRACT

This chapter describes the concept of an expert system supporting improvements in a building's energy efficiency. An expert system is a computer program or group of programs that facilitates using knowledge and decision making. The main elements of an expert system are the knowledgebase and a conclusion system. The knowledgebase collects information about a particular area, which was beforehand written in using particular rules. The conclusion system uses the knowledgebase and user input facts to generate conclusions or prove the user's hypothesis. The proposed expert system contains information on building technology, modernization and installation activities, and the values for financial, environmental, and technical indicators characterising these technologies. The user of an expert system defines the problem to be solved using questions. Using the knowledgebase, the system will present the optimal solution or information that the technologies in the existing knowledgebase will be applicable in the case defined by the user.

INTRODUCTION

The improvement of energy efficiency in construction carries with it a lot of challenges and problems: technical, technological, and economic. Solving them requires the involvement of many experts. At the same time access to experts is difficult and costly, which is a problem for many investors, decision makers and designers. An interesting solution for this problem seems to be using expert systems. Described below is a concept for an expert system facilitating the improvement of energy efficiency in buildings. At present such system is being elaborated at the Faculty of Civil Engineering, Warsaw University of Technology, by the author of this chapter.

DOI: 10.4018/978-1-5225-4105-9.ch005

BASIC INFORMATION ABOUT EXPERT SYSTEMS

According to (Mulawka, 1997; Tadeusiewicz, 2012) an expert system is a single piece, or a set of computer software, which facilitates the use of knowledge and decision-making. Expert systems can assist or even replace human experts in a certain field. They can provide the user with advice, recommendations or diagnosis in the problems related to the study field.

The following characteristics distinguish expert systems from conventional systems:

- Explicit representation of knowledge,
- Application to solving specific problems according to the reasoning procedures (inference),
- Ability to explaining the solutions found by the system,
- Knowledge processing mainly concerns symbol processing rather than numerical computations.

Unlike classical software knowledge in the knowledge, database describes the problem domain without providing an exact algorithm for solving a given problem. Knowledge is written according to a specific knowledge representation language, most commonly consisting of: a description of facts, description of the rules used in the inference process, and sometimes meta descriptions of the rules that show a strategy for solving a given problem.

Modern expert systems also use knowledgebase structures such as:

- Frames, which are partially identified with the object approach of the software,
- Decision networks,
- Semantic networks.

A key element of any expert system is it's conclusion system, which uses rules and facts from the knowledgebase and user input to arrive at conclusions. There are two main methods of inference: forward chaining and backwards chaining. It is also possible to use mixed chaining which is a combination of the two mentioned above Forward chaining begins with the facts supplied and engaging rules generating facts which will lead to the conclusion at the end of the process. Backwards chaining begins with a hypothesis (the goal), which the system, by generating appropriate facts, proves, disapproves, or indicates that solving the given problem is impossible.

In cases where multiple rules can be used in one situation the following rules should be applied (Tadeusiewicz, 2012):

- Highest priority,
- Most detailed conditions,
- Last used,
- Last added,
- Containing variable last used,
- Conclusion with highest confidence (in case there are weighted premises, the degree of a premise's truth).

A quality specific to expert systems is the generation of automatic explanations of the problem solving method while working with the user. The most commonly found explanations are:

- How? – system describes how it arrived at the conclusions,
- Why? – system shows the context of the current inference,
- What is – system generates a text explaining the terms in the knowledgebase.

Sometimes systems give explanations by answering the question (Tadeusiewicz R., 2012):

- "Why not?"; in this case the system shows the chain of rules and explains why some rules were rejected.
- "What if?"; in this case the system shows the reasoning and answers assuming a change in fact or rule.

Some claim (Tadeusiewicz, 2012) that a system without an explanation module is not an expert system. Figure 1 shows the general schema for an expert system.

Various IT tools are used in creation of expert systems such as:

- Expert system shells,
- Knowlegde database management programs,
- Expert systems languages e.g. Flops,
- Symbolic (Formal) and algorithmic languages e.g. Visual Basic.

Knowledge bases are created by a knowledge engineer in cooperation with experts. There exists a possibility to create learning expert systems that extend their knowledge database.

Figure 1. Diagram of an expert system
Source: own elaboration based on Rośczak (2003).

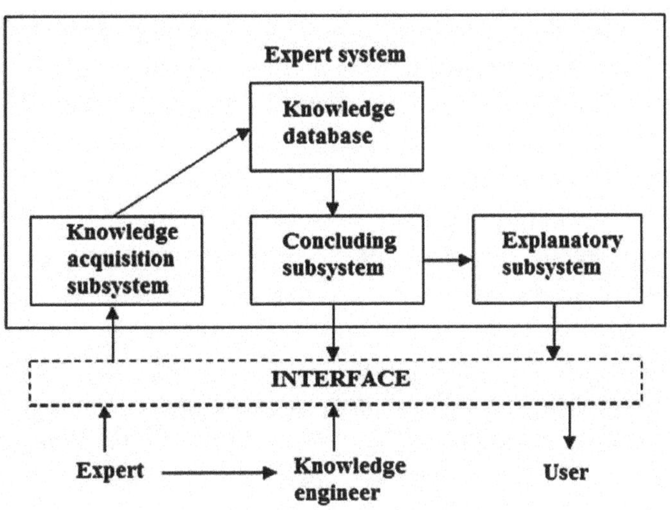

EXISTING EXPERT SYSTEMS IN THE CONSTRUCTION SECTOR

One of the first expert systems in the construction sector was HI-RISE program, which was elaborated to help with the design of tall buildings (Rehak, Fenves, 1984). Another interesting program is called ScaRC and is used to create construction schedules during the construction of repeatable storey floors (Rehak, Fenves, 1984). The expert system called RUMBAu uses the 3D technique to design wall formations in the PERI TRIO system. Also, the ELPSON program is used for the support in decors design in the PERI system. It is characterized by such a level of service simplicity that it can be used by people with very little professional experience. An interesting diagnostic system called REPCON is used during the assessment of buildings structure damage. It also provides hints and proposes methods for repairing them.

The system SPERIL (Fenves, Mahler, Siriam, 1985) is being developed at Purdue University. Its aim is to analyse the building damage which results from hazardous events such as an earthquake or roof overload due to precipitation. It proposes a mathematical model of analysis, using rules of the IF-THAT'S-DIFFERENT type. The inaccuracy of reasoning is modelled by fuzzy logic (between truth and falsity there are many indirect values), and the full program is implemented in C-language.

For the evaluation of a construction investment, the HICOST program (Rehak, Fenves,1984) was implemented as a rule-based program. It consists of a series of subsystems to find the cost of the building's components.

The application called COMVOC proposes support for project planning while operating on estimates of the final value of the construction project as a function of decisions affecting the project. It allows exporting the resulting database to the well-recognized MS environment.

The ACPROM system (Akram, Rahman, Memon, 2014), provides support for the investment implementation phase, estimating the progress of building works based on project and photographic documentation, allowing exporting the resulting databases to MS Project. It allows the integration of executive drawings, digital pictures of the construction site, and digital progress photograph for project monitoring and decision support by the building process managers. As a result of the system automatically interprets the drawings in [*.dwg] format and extracts construction element data so that, comparing the above with digital images, generate information about the percentage progress of the planned work.

The "Value Management for Construction Projects" program supports the process of identifying financial parameters for construction projects. The process of acquiring knowledge into the system consists of an aggregation of knowledge available in the literature on value management in the construction industry by means of a checklist and knowledge available from experts using the submitted forms. It was implemented in C ++ and later verified by experts in the field of value management.

The COMIX program (Akram, Rahman, Memon, 2014), has an advisory role in the design of concrete mix based on New Zealand standards. It is based on rules and framework and helps to choose the type of mix and the type of construction, calculates the correct ratio for the appropriate load and sand mass, aggregate and all concrete components.

The BIDEX program (Akram, Rahman, Memon, 2014), offers support of the process of preparation of offers for a given building objects. It is used by construction engineers when making bid decisions. Built using the expert system shell "Exsys", the program makes a decision in two stages. The first decision concerns the cost of works, materials, and equipment. The second decision concerns the size of the overheads. This last decision includes the type of owner, type of work, the size of work, place of work, market power, and the degree of danger and confidence level of subcontractors.

The platform "Advisory System for Managers" (Akram, Rahman, Memon, 2014), platform is an operational prototype of decision support for planning and management of building projects that can be used by building managers in their day-to-day duties and during the pre-design phase. So far the program has been used to analyse small projects. The main purpose of the application is to systematize the decision-making process. It includes models of cost, raw materials, and administration. The system has mainly an advisory function like: information management, financial management, current and future technical problems on the construction site.

The CONFAULT system is intended to support diagnostic processes related to defects in reinforced concrete construction by identifying subgroups of defects. The knowledge database of the program is divided into modules corresponding to the six basic types of defects. Metadata are used to control and restrict the research. The original approach to confidence allows uncertainty modelling.

At the University of Zielona Góra the "Expert system for the selection of energy source for the building" has been created in the framework of the research task No. 6 "Analysis of technical and operational requirements for buildings with power from centralized heat sources", implemented as a part of the strategic project "Integrated system for reducing the energy consumption of buildings" (Ziembicki, Węglarz, 2013). The concept of the system was based on the integration of many computing tools to create a unified computer system supported by knowledge databases.

The main purpose of the expert system is to analyse the energy of a given building or group of buildings and to make recommendations on energy sources based on CO_2 emissions and investment costs based on the indicators contained in knowledge databases.

ASSUMPTIONS FOR AN EXPERT SYSTEM SUPPORTING THE IMPROVEMENT OF ENERGY EFFICIENCY IN BUILDINGS

The expert system supporting the improvement of energy efficiency of buildings is meant for investors, designers, contractors, building inspection workers, and decision makers responsible for construction on the local level. Its aim is delivering knowledge to the above mentioned people by answering their questions, so that investment, designs, realization decisions which improve efficiency of given building are made.

SOURCES AND METHODS OF ACQUIRING KNOWLEDGE

Most often knowledge for the knowledgebase is usually acquired from experts and specialists in a given field. Another source of knowledge could be academic case studies and expertise. An engineer, most often one who works in IT, usually does acquiring knowledge. Because the process of collecting knowledge is difficult and very work intensive, automatic algorithms collecting knowledge have been created, for example Inductive Algorithm (Cios, Sztandera, 1992; A., Chang, Ku, Chang, 2009). Inductive Algorithm (e.g. ID3 and AQ11) which generate rules for a knowledgebase using training sets. These sets most often contain previously used data describing a problem and the conclusions reached by a person while solving the problem. Based on this data the system learns to solve similar problems by itself. In the case of the proposed expert system knowledge would be collected primarily from trade publications and the practical experiences of experts from the Faculty of Civil Engineering, Warsaw University of Technology and the Polish National Energy Conservation Agency.

KNOWLEDGE BASE FOR THE PROPOSED EXPERT SYSTEM

The knowledge base of the proposed expert system was prepared as a semantic network.

Figure 2 shows a schema of a semantic network.

Figure 2 presents only a part of the semantic network forming knowledge base of the discussed expert system. Rest of the network is build on the similar basis as shown on the figure above, and questions on the network bows as well as respective answers on the network nodes are as follows:

HOW CAN ENERGY EFFICIENCY IN THE CONSTRUCTION SECTOR BE INCREASED?

Energy efficiency in the construction sector can be increased by:

- Building new energy efficient buildings,
- Modernization of existing buildings,
- Energy management in buildings.

What Are Energy Efficient Buildings?

Energy efficient buildings are buildings, which use less energy for, among other, heating, cooling, and preparing domestic hot water. It is assumed that a share of an energy efficient building's energy demand should come from renewable energy sources. There are different levels for energy efficient buildings depending on their energy standard.

Figure 2. Sample semantic network
Source: Own elaboration based on: http://www.rosczak.com/mlp/ai.html.

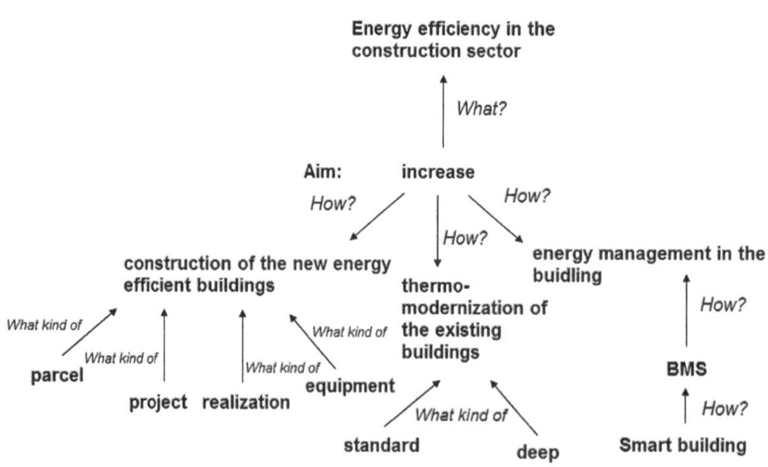

What Is Energy Standard?

Energy standard defines the maximal level of primary, final or useable energy of a building, most often in kWh/m^2/year. It can be defined legally (WT 2021) or defined by the creator of a given technology, e.g. the passive building standard or financial institutions e.g. the NF15 standard. Table 1 shows the classification of energy standards based on the EU criteria of useable energy.

How to Construct an Energy Efficient Buildings?

To construct energy efficient building one has to (Jarzemska., Węglarz, 2009):

- Find a suitable construction plot allowing the southern orientation of a building and using natural the natural terrain conditions,
- Prepare a suitable architectural plan, which will ensure,
  - The best thermal insulation of all outside partitions,
  - The air tightness of all partitions (floors, walls, roof, windows, outside doors),
  - Using mechanical ventilation with heat reclamation,
  - Eliminating thermal bridges,
  - Dividing the building into thermal zones,
  - Using an energy efficient heating system,
  - Using an energy efficient system for preparing domestic hot water,
  - The possibility of passive or active utilization of solar energy,
- Construct the building according to the project and the methods of ensuring appropriate quality of the construction work,
- During furnishing use energy efficient equipment e.g. refrigerator, lighting, water equipment (faucets, W.C., etc.),
- Use the building according to energy management rules.

Table 1. Building energy classification

Symbol of the Energy Standard	Name of the Energy Standard	EU Indicator for Usable Energy EU [kWh/(m^{2}year)]
A++	Near zero	≤10
A+	Passive	up to 15
A	Low energy	form 15 up to 45
B	Energy efficient	from 45 up to 80
C	Average energy efficiency	from 80 up to 100
D	Average energy efficiency (fulfilling current regulation limit)	from 100 up to 150
E	Energy intensive	from 150 up to 250
F	Highly energy intensive	above 250

Source: Association for Sustainable Development

How to Achieve the Best Thermal Insulation of Nontransparent Outside Partitions?

The best thermal insulation of all outside partitions can be achieved by using traditional (styrofoam, mineral wool, polyurethane foam) or new insulation materials such as (Jarzemska M., Węglarz A., 2009):

Intelligent Insulation

The heat conductivity of intelligent insulation changes depending on the outside conditions. They're primarily based on using vacuum insulation panels, which are filled with glass wool or powder and allow transferring the mechanical load on the insulation. Changes in gas pressure in the pores create variable heat transfer coefficients, which adapt to outside conditions.

Transparent Insulation

Transparent insulation combines the characteristics of materials with good optical transmission (light transmission) and thermal insulation. It works by utilizing the greenhouse effect (just as typical glass shields) for collecting solar energy, while at the same time significantly limiting heat loss back to the outside as in regular transparent insulation. Light passes through the transparent insulation and which absorbed by the surface of the walls behind the insulation. Part of the absorbed energy is radiated as heat radiation from the wall at a longer wavelength than the incident rays. Heat radiation can no longer escape outside because the transparent insulation is a "nontransparent" material for those wavelengths. To avoid heat loss through conduction the volume of the transparent material is filled with some kind of insulator, that is air or a noble gas (e.g. argon).

Vacuum Insulation

Using elements of this technology allows the reduction of the insulation layer to a couple of centimetres. In comparison o popularly used materials such as styrofoam or mineral wool, vacuum insulation is characterized by better insulation parameters. These parameters are used by utilizing vacuum, which is a poor conductor for heat energy. In the production of the proposed material, silica or glass fibres with micro pores 0.0001mm are placed in a sealed "package" made out of airtight and watertight multi-layer membranes.

Aerogels

Aerogel is a substance in which 90% of its mass is made up of air. The rest is material creating the aerogel's structure. It is made of silica. Aerogels are currently the lightest solid substances, not much denser than air. Insulation materials based on aerogels are characterized by a very low thermal conductivity coefficient. =0.014 W/(m.K). Their main advantage over other insulation materials is high stretch and press resistance. Modern aerogels are insulation products produced using nanotechnology. Aerogels are used in all cases requiring insulation of the highest parameters and the lowest thickness creating savings in time and area.

Rigid Foil Faced PIR Panels

The shape of the PIR panels is similar to styrofoam panels with shaped edges, so that they can be tongued and grooved together. The difference between them is that rigid polyisocyanurate panels are covered with an aluminium foil and reinforced with kraft type paper on both sides. PIR is currently on of the best insulation materials. Its thermal conductivity coefficient λ is very low, sometimes as low as 0.020W/(m*K). The panels are resistant to moisture and water. Their maximal absorption reaches 9%. They also replace steam insulation. PIR panels can be used to insulate practically everything besides spaces between elements of the structural skeleton.

Spray Foam PIR

Liquid PIR is sprayed under high pressure so that the foam adheres to the surface. It has very good adhesion to most building materials. Sprayed foam has all the advantages of PIR with the added benefit of a homogenous insulation, without any gaps or connections. The spraying technology also ensures rapid installation of the insulation. After work is completed there is minimum waste. The thermal conductivity coefficient λ is between 0.049 and 0.024 W/(mK). Foam of this type is most often sprayed on flat roofs. It increases thermal insulation and replaces water insulation. However the coating created needs to be painted by a special lacquer protecting it from UV light. The foam can be also used to heat skeleton walls, beam ceilings, and the space between roof rafters.

Metallic Mats

Metallic mats are thin (they are up to a dozen mm thick) and are rolled like foil. They're multilayer, there are multiple layers of polyester, metalized foil, or bubble foil between two layers of reinforced aluminium foil. Mats are installed similarly to parainsulationfoil – using fittings with wooden constructions or double-sided aluminium tape to steel constructions. The mat has to be tensed and the joints between rolls and punctures need to be sealed with double-sided aluminium adhesive tape. Their advantage is their low weight and great thermal insulation – the thermal conductivity coefficient λ can reach 0.019 W/(mK). An additional benefit is that one roll of mat, which insulates an area of 60 m^2 replaces 6 m^3 of wool or styrofoam.

Aerogel Mats Mixed With Fibres

Aerogel mats containing fibres provide greater elasticity. They are 5 or 10 mm thick and installed to subsoil either mechanically or mechanically and using an adhesive. They have to stick very close to it. Their thermal conductivity λ is from 0.012 to 0.030 W/(m K).

Glass Foam

The thermal conductivity λ for glass foam is from 0.038 to 0.05 W/(mK). Glass foam provides an extremely rigid insulating layer, resistant to deformation and any damage. Since there is no tendency to absorb water, they can be used to heat the foundations and basements, even those periodically flooded.

Phenolic Panels

These boards are characterized by very good thermal insulation. Their coefficient λ can reach 0.021 W/(mK), are water and moisture resistant. They are easy to cut without causing dust. They are characterized by high mechanical strength. They are hardly inflammable and non-inflammable.

White Wool

White wool is very delicate, and when cutting dusts much less than traditional. It does not cause skin irritation, so you do not need to wear gloves or protective masks to work with it. White wool's λ coefficient is 0.035 W/(mK). White wool mats can be used as a thermal and acoustic insulation in skeletal structures, as a fill between beams, columns, profiles or rafters.

Layered Styrofoam Panels

The construction of such insulation consists of a pair of styrofoam panels separated by an aluminised mat. The mat is made of two layers of aluminium foil laminated with 5 mm polyethylene foam. The styrofoam sandwich has excellent insulation properties. Its heat conductivity coefficient λ is 0.018 W/(m K). This is due to the thermal radiation being reflected from the surface of the aluminized mat. Panels of this kind are mainly used for the warming of external walls.

How to Ensure the Best Energy Efficiency Parameters for Transparent Outside Partitions?

The primary task for transparent partitions is to protect from the external environment, illuminate the rooms with natural light, and to provide visual contact with the natural environment. The number of possible technical solutions for transparent partitions is very large.

From a building's energy efficiency standpoint the following technical solutions are important:

- Using active glass,
- Decreasing heat loss through windows,
- Using architectural elements preventing the building from overheating in the summer.

What Is Optically Active Glass?

There are several types of glazing that have the ability to change their optical properties. The change in properties is due to the effects of varied external factors that allow the classification of active solar glass in the following basic groups:

- **Photo Chromatic Glass:** Change in permeability is a function of the dose of insolation,
- **Thermo Chromatic Glass:** Change in permeability corresponds to the variation in glazing temperature values,

- **Thermo Tropic Glass:** Total solar transmittance decreases as the visible light disperses within a certain range of temperature values,
- **Electro Chromatic Glass, Liquid Crystal and SPD:** Optical properties change under the influence of electric current,
- **Gas Chromatic Glass:** The change in optical properties occurs through the use of a gas mixture inside the system.

Intelligent Electro Chromatic Windows (Using PV)

Electro chromic technology allows you to control the amount of light (heat) entering the building through the windows by dimming or brightening them. In winter, these windows tend to brighten by increasing the amount of heat supplied to the room, thus reducing the need for heat. In the summer they turn dark and blocking the light reduces the amount of heat and therefore reduces the need for room cooling. Lightening or darkening of windows is caused by low voltage chemical reactions. The thin layer of solar cells made of a nickel-magnesium alloy in the windows has the ability to switch from transparent to reflective. The switch can be powered by gas-chromium (hydrogen or oxygen) technology.

How to Reduce Heat Loss Through the Entire Window in the Designed Buildings?

To reduce the heat loss throughout the window, you can use:

- External and internal roller blinds,
- Vertical, horizontal, internal and external blinds,
- Shutters,
- Phase changing window shutters; this concept assumes the use of phase changing materials as components for shutters and shutters installed outside the windows. During the day, casings containing phase changing materials are exposed to solar radiation. Absorption of solar radiation causes the phase changing materials to melt storing heat. At night the shutters are closed and the heat is released into the room.

What Are Phase Changing Materials?

Phase changing materials use the known phenomenon of absorption and heat transfer during phase change. They are characterized by a heat transfer coefficient of $\lambda = 0.05$ W/(mK) and an interesting property of absorbing and heat transfer from and to the environment (rooms). The large heat capacity of these materials is the result of the phase change (melting-solidification) in the range of room temperature changes. Incorporated in different ways in the structure of the building it increases its heat capacity (inertia). The high thermal inertia of the building's structure (heat capacity) contributes to the improvement of its energy efficiency, which is reflected in the reduction in energy consumption necessary to maintain the comfort.

How to Ensure Airtightness of All Partitions?

Airtightness of the external partitions is checked with a special test - Blower Door. This is the name of the device with a fan, set in the input door or window. It pumps out air from inside to a vacuum of 50 Pa. Then the airflow through the leaks is measured.

We can ensure airtightness by:

- Airtight surface materials,
- Airtight connections in critical areas exposed to air movement and material connections.

Therefore it is very important to choose a material, which will ensure appropriate air tightness. Air tightness must be provided by a single membrane. Leakage should not be eliminated by adding additional sealing layers or behind the layer proper (Firląg, 2015).

How Does Mechanical Ventilation Work With Heat Recovery?

The supply and exhaust system requires two units. Through ventilation ducts and diffusers, fresh air from the central unit enters the premises. Separate channels return polluted air to the other unit. Fans force the airflow in the ducts. Mechanical ventilation can work in the system:

- **Centralized:** One ventilation unit for all rooms,
- **Decentralized:** Each room has its own ventilation unit.

Reduction of heat loss is achieved through the use of heat recuperator.

What Is a Heat Recuperator?

A recuperator is a device consisting of a heat exchanger (cross counter current, rotary), two fans - supply and exhaust, and an air filter. Fresh, cold air sucked from the outside passes through a heat exchanger heating up from the exhausted air discharged from the inside of the building to the outside also flowing through the exchanger. Flows are almost silent and take place automatically.

What Should a Designer Planning to Use a Recuperator Know?

The designer planning the place for the recuperator should include:

- Weight of the control panel,
- Provide access to the device so that it can be viewed and possibly repaired,
- Free condensate drain,
- Connection to the electrical system,
- Ambient temperature - do not install the recuperator in rooms where the temperature may be below zero,
- Distribution of rooms at home.

What Are the Conditions of Mechanical Ventilation With Heat Recovery?

The condition of proper, silent operation of ventilation with heat recovery is not only professional design and professional assembly but also proper operation and maintenance.

Dirty, dusty ventilation may cause:

- Contamination of air in the building causing respiratory illnesses and allergies,
- Bad air flow, difficulty in maintaining proper temperature and lower heating costs,
- Higher energy consumption,
- Fire hazard.

It is therefore necessary to systematically maintain both the recuperator and the whole installation, consisting in:

- Cleaning the ventilation ducts using a rigid brush that rotates 600 times per minute - once every 3-6 years,
- Replacement of recuperator air filters - every 2-4 months,
- Check the condensator drainage system - once a year, preferably before the heating season:
- Cleaning the condensate drainage system every 2 years.

What Are the Advantages of Mechanical Ventilation With Heat Recovery?

The benefits of mechanical ventilation with heat recovery include the following:

- Provides air exchange at the desired quantity regardless of atmospheric conditions,
- Allows intensive ventilation by switching the fans to higher gear,
- It allows filtering, heating and cooling fresh air before it enters the premises,
- Enables the use of heat recovery equipment for extracted air, which saves a significant amount of energy.

What Are the Mechanical Ventilation Defects?

The disadvantages of mechanical ventilation with heat recovery are:

- Requires a drive, consumes electricity,
- Mechanical components may fail,
- Does not work during power outages,
- Ducting and ventilation ducts take up a lot of space - they are difficult to install if they were not planned during construction.

What Is a Thermal Bridge?

A thermal (thermal) bridge is a part of the building's outer envelope, with a significantly worse thermal insulation than its neighboring building elements. Thermal bridges in the walls can raise your home's heat demand up to 20%, thereby raising the cost of heating your home. The most common cause of such spots is design and manufacturing errors.

Thermal bridges occur in structural nodes and any combination of external elements made of different materials. These places most often include:

- Ground connections with foundation walls,
- Connections the foundation plate to the outside walls,
- The location of windows and doors,
- Connections between the roof and the wall,
- Balconies,
- Substructures protruding from external walls,
- Support structures such as steel railings, brackets, satellite antennas, and even aluminum rails used in the light-wet method.

How to Avoid Thermal Bridges?

In order to avoid thermal bridges, a simple principle should be applied: the continuity of thermal insulation (laid on the outside of the building partitions) must be maintained at every point.

What Are the Thermal Zones in the Building?

The following thermal zones exist in a residential building:

- 22 ÷24 ° C – bathroom,
- 20 ÷22 ° C - living room, children's room, study room,
- 18 ÷20 ° C - kitchen, bedrooms,
- 16 ÷-18 ° C - corridors, exercise rooms or billiard room,
- 12 ÷15 ° C - utility rooms: storage room, pantry, laundry, veranda,
- 4 ÷8 ° C - garage, storage of tools and garden equipment.

The temperature difference between neighbouring rooms should not exceed 8 ° C, since partition walls can be cheap and relatively thin (12 cm). If the differences were to be larger, the inner walls had to be warmed up or built much thicker.

What Is a Heating System?

The heating system is a set of equipment and components of the installation generating and distributing heat in a building. It consists of a heat source and its distribution system, i.e. internal heating systems (Koc, 2001).

What Are the Types of Heating Systems?

Heating systems in buildings can be divided according to the type of heat source, (Koc, 2001):

- Gas,
- Oil,
- Charcoal,
- Electrical (resistance and heat pump drive),
- Remote (heat from the district heating),
- Renewable sources - biomass, sun, wind.

Due to the way and mechanism of heat distribution in the building we distinguish the following systems of internal heating systems:

- Water,
- Air,
- Electrical (resistance).

What Is an Energy Efficient Heating System?

Energy-efficient heating system is a heating system comprising of an energy efficient heat sources and high-efficiency distribution systems such as:

- **Condensation Boilers:** These are modern heating devices that utilize the phenomenon of heat transfer due to condensation.
- **Pulse Gas Boiler:** The pulse boiler is a heating device adapted to burn gas, in a condensing technology with a pulse combustion system.
- **Heat Pump:** The heat pump receives energy from a low temperature source (bottom source) and transfers it to a higher temperature source (top source), where it is used to heat rooms or heat up hot water.
- **Reversible Air/Air Heat Pump:** The main function of the air / air heat pump is to maintain a constant air temperature in the rooms and control the humidity of the air. Limiting or spreading heat demand creates a significant potential for energy savings. An interesting solution is the of a use of a gas engine in the heat pump. This solution reduces electricity consumption during summer (cooling) and winter (heating). The additional advantage of this system is the ability to recover the waste heat released by the engine hood and exhaust gases, as well as the easy adjustment of the compressor speed by adjusting the gas supply.
- **Heating Nodes:** Where the parameters of the heating medium supplied through the district heating system are changed, according to the needs of indoor installation and the distribution of heat between the heating system and the domestic hot water installation in the case of bifunctional nodes.
- **Infrared Heating:** Involves heating the air with infrared radiation (heat) emitted by preheated heating elements (surfaces of walls, ceilings, and objects in the room).
- **Low Temperature Surface Heating:** Most often used with heating mats or foils placed on building partitions, usually covered by a thin layer of masking.
- **Blower Heating:** Where for heating the air uses a blast furnace with a heat exchanger powered by gas, fuel oil, warm water or equipped with an electric heater. For this purpose you can also use:

condensing gas stove, air-to-air heat pump or solid fuel air heaters. For heating air in single-family homes a fireplace can also be connected.

- **Cogeneration (Micro and Mini Generation):** Cogeneration (micro and mini generation) is the combined generation of heat and power (based on small to medium power plants in the range of 50kW to 3MW). The average efficiency of micro-cogeneration devices is about 85%.

What Are the Ways to Prepare Hot Tap Water?

Hot water can be prepared using:

- Flow heaters,
- Condensate heaters,
- Heating,
- Installations using renewable energy sources.

How Do Flow Heaters Work?

Flow heaters operate only when there is a flow of water, i.e. after tapping or tapping with hot water. Water must also have adequate pressure. The amount of hot water that can be obtained depends on the power of the heater and the temperature of the cold water on the supply.

The most common instantaneous heaters are:

- Gas instantaneous heaters (for natural or liquid gas),
- Electric instantaneous heaters.

What Are the Advantages of Electric Instantaneous Heaters?

The advantages of electric heaters are:

- High energy efficiency,
- The possibility of decentralized production of domestic hot water,
- Low cost of water installation,
- Ability to continuously adjust water temperature.

What Are the Disadvantages of Electric Instantaneous Heaters?

The disadvantages of electric flow heaters include:

- High power consumption,
- Limited to short distances hot water supply,
- Proper water quality in the installation,
- The power consumption occurs directly during the hot water demand period, which often falls into periods when the electricity is expensive - limited ability to control the cost of purchasing energy.

What Is the Passive Use of Solar Radiation Energy?

In passive use of solar energy, the following applies:

- Direct profit system,
- The system of indirect profits,
- Daylight lighting system.

The classic passive way of using solar energy is for examlpe Tromb wall.

Trombe Wall

The Tromb wall construction uses the greenhouse effect obtained by using a window placed in front of a black-painted wall and the air gap between the two elements. The outside wall warms up from the sun while warming up the air inside. The air from the heated wall "emptiness" into the room. The outer transparent coating transmits sunlight and stops the heat. A black brick wall of 11 cm thick acts as an absorber. It heats up during the day, and the stored heat is released with a time delay directly into the free zone between the wall and the insulation. The air in this zone is heated in contact with the heat emitted by the wall. The holes located at the top and bottom of the wall allow the convective flow towards the room. At sunset these openings are closed so that the air does not move in the opposite direction and does not cool the building. The inner insulation layer protects the building against overheating when the air circulation stops due to the closure of the openings.

The passive solar energy system utilizing the Tromba wall model can be modified by replacing the classical brick with building materials containing variable phase components. The validity of such a solution has been confirmed by both theoretical and practical tests. To obtain the same amount of stored heat, the variable-phase materials need less surface area than the massive walls initially used by Trombe and are much lighter.

What Is the Active Use of Solar Energy?

Active use of solar energy is based on the use of devices such as:

Solar collectors - meaning devices that directly convert the energy of solar radiation into heat. Average efficiency of solar collectors is about 70%.

Photovoltaic panels - meaning devices that directly convert the energy of solar radiation into electricity. Average efficiency of solar collectors is: about 15%.

Hybrid Solar Collectors

Hybrid solar collectors are structures that use heat from photovoltaic panels. As photovoltaic heating becomes hot and its temperature rises, the ability to convert solar radiation into electrical energy decreases. In order to counteract this phenomenon hybrid solar photovoltaic panel - solar collector devices are used. This is an effective solution that not only cools the cell, but also gives you the opportunity to make more use of the sun's energy. This design is about 15% efficient in producing electricity and up to

80% efficient in heat production, which is more efficient than the efficiency of a cell or collector operating separately. In total they are able to consume more than 60% of the solar radiation reaching the panel.

WHAT IS A THERMOMODERNIZATION OF A BUILDING?

An thermomodernization of a building is a set of activities on the elements of the structure of the building and its installations, resulting in reduction of the building energy demand for heating, preparation of domestic hot water, cooling and lighting.

The values of the energy performance parameters of buildings defining different types of thermomodernization are presented in Table 2 (Klimczuk K. 2017).

What Thermomodernization Is About

A thermomodernization of the building consists of:

- Modernization of external baffles,
- Modernization of the ventilation system,
- Modernization of the heating system,
- Modernization of the domestic hot water preparation system.

What Is the Modernization of the External Partitions?

Modernization of the external baffles consists in:

- Insulation of external walls,
- Insulation of the roof or flat roof,
- Insulating the floor over unheated basement or warming up the floor on the ground,
- Replacing or upgrading glazed surfaces and exterior doors.

Table 2. Different types of thermomodernization

	Energy Efficiency Modernization According to Existing Legislation	Deep Energy Efficiency Modernization	Deep Energy Efficiency Modernization Using Passive Components
Decreasing final energy demand for heating and preparing domestic hot water	35 ÷50%	min. 75%	min. 70%
Demand for primary energy after modernization	-	>60 kWh/(m²year)	-
Demand for usable energy for heating and ventilation	about 70 ÷80 kWh/(m²year)	-	About 20 kWh/(m²year)

Source: Klimczuk 2017.

What are the Methods of Insulation of External Walls?

Currently the following two methods of thermal insulation of the external walls dominate the market:

- The dry light method consists in insulating the walls with mineral wool or styrofoam panels filled with a wooden or metal grate to which an exterior facing vinyl lining, fibreboard lining, stone facade or profiled sheet is attached.
- The Jointless Insulation System (JIS also referred to as the wet light method) consists of sticking to the wall of the insulating layer (polystyrene board, hard mineral or glass wool board), on which a thin, textured layer is made on a polypropylene or glass fibre mesh. There are several variations and variants of this method. The insulation layer sometimes only glues with special adhesive to the existing external wall, and sometimes also fastened with special dowels.

What Are the Methods Insulating a Roof?

The thermal insulation layer is mainly made of flexible fibrous materials such as mineral wool or glass wool. Not only are they characterized by excellent heat conduction coefficients (lambda = 0.032÷0.045 [W/m.K]), they additionally suppress any external noises and are non-flammable. However, they have to be protected by two additional coatings: exterior winding and vapour barrier from the inside. The total thickness of mineral wool layers in the roof slope should be at least 30 cm (in energy efficient homes).

What Are the Methods of Insulating a Flat Roof?

The choice of insulation method depends on whether we are dealing with ventilated or non-ventilated roofs. In the case of a ventilated ceiling, the best solution is to blow the insulated thermo insulating material into the airspace. In the case of non-ventilated roofs, insulation is applied using traditional methods of laying insulating boards.

What Is the Modernization of the Ventilation System?

Modernization of the ventilation system consists in the use of devices allowing to control the amount of ventilation air (diffusers, exhaust fans).

The most energy efficient in the existing building is constructing an air-exhaust ventilation with heat recovery (recuperation), but usually requires many expensive construction and installation work, which in most cases is not a cost-effective operation.

What Is the Modernization of the Heating System?

There are three main options for a heating system:

1. When Is the Modernization of an Existing Heating System With Replacement of Certain Elements Implemented? Partial modernization is used when there is no planned thorough renovation of the whole building (in particular the insulation of all external partitions) in such a way as to ensure a radical reduction in the heat demand. The modernization of the heat source, including replace-

ment of the fuel or energy carrier, and the installation of systems and devices enabling precise, automatic control of the installation are usually carried out. Therefore, the modernization usually includes the replacement of the heat source, the encapsulation of the installation, the installation of radiator thermostats, the installation of the weather and / or room regulator, and the installation of chimneys resistant to aggressive exhaust gases.

2. When to Complete the Replacement of the Heating System With a New One? Completely replace the heating system with a new one if:
    a. The modernized building has premises of various sources and heating systems,
    b. The installation is in poor technical condition and, due to thermo-modernization, the demand for thermal power has been significantly reduced.
3. When Should I Opt Out of Using an Existing Heating System for Electric Heating? If as a result of the modernization of the building we get the standard of at least passive house it is worth to abandon the central heating system for electric heating.

What Is the Modernization of the Domestic Hot Water Installation?

Basic actions covering the modernization of the domestic hot water installation are:

- Replacement of drain fittings and inefficient wires,
- Execution or repair of thermal insulation of wires,
- Improve the operation of the hot water preparation system,
- Introduction of automatic water temperature control.

What Are the Ways to Reduce Window Heat Loss in Existing Buildings?

Ways to reduce window heat loss are (Koc, 2001)

- Seals,
- Use of additional glass or foil,
- Use shutters and shutters,
- Proper use of curtains and curtains,
- Window replacement,
- Reduce the size of windows (when replacing).

What Is the Scope of Modernization Work in Case of Deep Thermal Modernization?

The scope of the work of deep thermal modernization in a building which aims to achieve a passive standard includes (Firląg, Rucińska, 2007):

What Affects the Amount of Heat Loss in a Building?

- Size of building - heated surface, volume,
- Shape and whether it is one-storey or one-storey,

Table 3. Scope of the work of deep thermal modernization

Type of Operation	Expected Result of Operation
Changes to the architectural project	Increased share of glazed surfaces on the southern facade of the boiling to maximise heat gain from the sun. Decreasing glazed surface on the northern facade.
Improving the thermal insulation of exterior partitions	The value of the thermal heat transfer coefficient should be less than 0.12 $W/(m^2K)$.
Eliminating thermal bridges	Structural details should be modified so that the linear heat transfer coefficient $\Psi \leq 0.1$ W/(mK).
Replacement of window frames	The value of the heat transfer coefficient for whole windows (window pane plus frame) should not exceed 0.8 $W/(m^2K)$ and g glazing factor is not less than 0.5.
Increasing the tightness of the building	Limitation of uncontrolled infiltration of outdoor air, air exchange times at a pressure difference of 50 Pa $n50 \leq 0.6$ 1/ h
Modernization of the ventilation system	Replacement of natural ventilation by mechanical ventilation with exhaust air with recovery efficiency of over 80% and low consumption of electricity (<0.45 $W/(m^3/h)$).
Modernization of the heating installation	Dismantling traditional water heating and replacing it by heating air, using a high efficiency heat source, increasing the quality of regulation
Modernization of the domestic hot water installation	Reduce the consumption of domestic hot water, reduce the losses in the distribution and circulation system, increase the efficiency of the preparation of domestic hot water.
Use of renewable energy sources	Analysis of the legitimacy of the use of renewable energy sources for heating and preparation of ca.
Use of energy-efficient appliances	High-efficiency and energy efficient home appliances and lighting to reduce electricity consumption.

Source: Own elaboration based on Firląg & Rucińska (2007).

- Number and size of windows, glazing area,
- Arrangement of rooms and positioning of windows in relation to the world's sides,
- Materials used to make walls, roofs, floors,
- Thickness of thermal insulation,
- Constructional solutions conducive to the formation of thermal bridges (terraces and balconies),
- Inappropriate ventilation.

What Are the Ways of Managing Energy in Buildings?

There are many ways to manage energy in buildings, but it seems that BMS systems are the most effective. BMS systems, called Building Management Systems (Horyński, M. 2015), are technically advanced solutions that aim to effectively control installations in the building such as electrical, ventilation, heating or cooling installations and adjusting them to changing ambient conditions. The main task of the system is to minimize the operating costs of the building, while enhancing its functionality and security, and ensuring optimal comfort for its users. BMS systems continuously collect, archive and process data related to the state of a particular installation, and control them automatically or semi-automatically. The most advanced building automation systems realize the idea of intelligent buildings.

What Is an Intelligent Building?

The term intelligent building refers to buildings of different uses and sizes, such as office buildings, commercial buildings, industrial facilities, schools, hospitals, residences, private homes, etc., with an integrated management and supervision system. Integrated management system includes many autonomous working automation systems and failure of any one of them cannot disrupt the work of the other.

DATABASE CONTAINING PARAMETERS RELATED TO THE KNOWLEDGE BASE

It is envisaged that the databases containing information on the value of parameters describing projects that increase energy efficiency in construction, e.g. the cost of execution of a given project, the amount of energy saved by the primary energy project, etc. will be linked to the answers placed in the semantic web.

The database fragment is presented in Table 4.

Table 4. Cost of chosen energy efficiency modernization measures

Name	Unit	Cost [zł]	Cost [Euro]
Tile and concrete screed	m^3	14.00	3.28
Cut off the waterproof insulation	m^2	15.00	3.51
Thick concrete mix	m^3	469.00	109.84
Select about 10 cm of sand for export	m^3	45.00	10.54
Reconstruction of lean layer of concrete 10 cm with labour	m^2	30.00	7.03
EMULSION GRINDING OPP 5 L	pc.	9.98	2.34
Reproduction of waterproofing insulation	m^2	20.00	4.68
PAPA ASFALT WOOL SIZE FUNDAMENT SBS SUPER ASSEMBLY 10 M2 (with folds)	m^2	12.80	3.00
XPS board placement	m^2	13.00	3.04
Fibrothermica XPS 10cm frez	m^2	55.94	13.10
STYRODUR AUSTROTHERM XPS TOP 30 7cm	m^2	37.33	8.74
BAUMIT 25 KG CONCRETE TITLE	pc.	7.98	1.87
Made of concrete screed 4cm	m^2	15.00	3.51
WATER HEAT PUMP WATER SILESIA TERM Air SPLIT 13S	set	27 500.00	6 440.28
Installation of the heat pump	pc.	7 000.00	1 639.34
PEREKO KSP DUO boiler 12 kW with feeder	set	6 979.00	1 634.43
Installation of a biomass boiler	pc.	970.00	227.17
Solar Set 3 KSR10-250	set	11 570.00	2 709.60
Solar system installation	pc.	3 000.00	702.58
I Solar JAM6 (L) 290 / PR 290W - 14pcs - for heat pump with assembly	set	20 560.00	4 814.99
I Solar JAM6 (L) 290 / PR 290W - 6pcs - to the collector and gas furnace with assembly	set	9 907.00	2 320.14
MISTRAL PRO 400EC	set	8 870.00	2 077.28
Installation of the recuperator	pc.	6 000.00	1 405.15

Source: Klimczuk K., 2017.

COMMUNICATION BETWEEN THE USER AND THE KNOWLEDGE OF THE EXPERT SYSTEM

It is envisaged that the expert system user will communicate with the knowledge base via an interface mapping the semantic network structure. There are answers to questions on the nets in the nodes of the network. In order to solve a problem that a user has, he has two options:

The first possibility is to direct the user through the semantic web structure and select the questions that the user wants to obtain from the expert response system.

For example:

Question— How to avoid thermal bridges?
System answer: In order to avoid thermal bridges, a simple principle should be applied: the continuity of thermal insulation (laid on the outside of the building partitions) must be maintained at every point.

The second option is to define the facts that describe the problem and to ask the general question in the question-editing pane. The system automatically generates a problem resolution based on forward-looking or generates a response to the problem based on the existing knowledge base.

For example:

The user has defined the following problem:

Fact 1: Detached house with an area of 120 m
Fact 2: Energy efficiency modernization to passive house level, Eu = 15 kWh/(m^2 year)
Fact 3: Current heat consumption, Eu = 167 kWh/(m^2 year)
Fact 4: Maximum budget: 150 000 zł

Question to the system: *How to modernize a building?*

The system response is given in Table 5.

SUMMARY

In the world construction industry, for many years, expert systems are used to support decision-making at every stage of the investment process from design to building. Unfortunately, in Poland, the use of these artificial intelligence tools is still small. This is due to the building tradition and low awareness of the participants in the building process with regard to possibilities of the artificial intelligence applications. The presented expert system concept is an attempt to create a user-friendly tool that is self-learning with a natural language interface that is understandable for the average building and investment industry.

The proposed expert system is intended for the design team i.e. architects, constructors, fitters, managers etc. In a few years' perspective the system should constitute a BIM application and should be used in the process of integrated design. Individual members of the design team master specialist knowledge concerning his or her respective domain, for example metal constructions. In order to communicate with other team members they should possess a part of knowledge from their domains, for example ventilation systems, necessary to design construction in a way that fitters do not have problems with optimal leading of the ventilation ducts. Such knowledge can be delivered by the discussed expert system.

Table 5. Response of the expert system to the question of how to modernize a building to a level of passive standard

Name of the Measure	Detailed Description of the Project	Estimated Cost of the Project
Increased thermal insulation of non-transparent outer partitions	The value of the heat transfer coefficient of the building's external partitions should be less than: • 0.12 W/ (m²K) for external hay, • 0.10 W/(m²K) for the roof, • 0.20 W/(m²K) for the floor above unheated basement	120 m²p.u. 500 zł = 60 000 zł (14 052 Euro)
Elimination of thermal bridges	Structural details should be modified so that a linear heat transfer coefficient of $\Psi \leq 0.01$ W/(mK)	8 000 zł (1 874 Euro)
Replacement of window frames	The value of the heat transfer coefficient for whole windows (window pane plus frame) should not exceed 0.8 W/(m²K) and g glazing factor is not less than 0.5	17 289 zł (4 049 Euro) – The cost of replacing windows on passive windows. Warm three-layer installation of windows around 8 000 zł (1 874 Euro)
Modernization of the ventilation system	Replacement of natural ventilation by mechanical ventilation with exhaust air with efficiency of over 80% heat recovery and low consumption of electricity (≤ 0.45 W/(m³/h))	25 000 zł (5 855 Euro) – Cost of mechanical ventilation installation with heat recovery
Modernization of the heating installation	Abolish traditional water heating and replace it by heating air using an electric heat source	9 000 zł (2 108 Euro) - Cost of electric heating together with disassembly of water heating.
Modernization of the domestic hot water installation	Reduce the consumption of the domestic hot water reduce the losses in the distribution and circulation system, increase the efficiency of the preparation of the domestic hot water. Production of the domestic hot water in 60% in solar collectors installation	5 000. zł (1 171 Euro) - The cost of modernization of the domestic hot water installation 11 500 zł (2 693 Euro) – Cost of installing flat solar collectors
Total cost		143 789 zł (33 674 Euro)

Source: Own elaboration, calculation based on Firląg & Rucińska (2007).

The biggest advantage of expert systems is that they are able to solve technical problems of the qualitative character. This requires broad experience, while expert work is costly. Moreover, a knowledge base of a system can be larger than a knowledge of an individual expert. As it goes for disadvantages of classic expert systems these are, above all, costs of preparation and updating of the knowledge base which parts sometimes are used only once.

The proposed system eventually is to have commercial applications and mass distribution in design offices, that should reduce its implementation costs into the building practice.

REFERENCES

Akram, M., Rahman, I.A., & Memon, I. (2014). A Review on Expert System and its Applications in Civil Engineering. *International Journal of Civil Engineering and Built Environment*, *1*(1).

Chang, W.-J., Ku, C.-C., & Chang, W. (2009). Fuzzy control with passivity synthesis for continuous affine Takagi-Sugeno fuzzy systems. *International Journal of Intelligent Computing and Cybernetics*, 2(2), 386–408. doi:10.1108/17563780910959947

Cios, J. K., & Sztandera, L. M. (1992). Continuous ID3 algorithm with fuzzy entropy measures. *Proc. IEEE Internat. Conf. on Fuzzy Systems*, 469-476. doi:10.1109/FUZZY.1992.258659

Fenves, S. J., Mahler, M. L., & Siriam, D. (1985). Knowledge-based expert systems in civil engineering. *IABSE Periodica*, *4*, 63.

Firląg, S. (2015). How to meet the minimum energy performance requirements of Technical Conditions in year 2021? *Procedia Engineering*, *111*, 202–208. doi:10.1016/j.proeng.2015.07.077

Firląg, S., & Rucińska, J. (2007). Simplified method of designing an air-ground heat exchange. *CESB 2007 PRAGUE International Conference - Central Europe Towards Sustainable Building*.

Horyński, M. (2015). *Energy-efficient automated energy management systems in residential buildings, Wydwanictwo Politechniki*. Lubelskiej.

Jarzemska, M., & Węglarz, A. (2009). *Technical conditions of improving energy efficiency in the construction sector, Department of Civil Engineering*. Warsaw: University of Technology.

Joint work under Praca zbiorowa pod redakcją Dariusza Koca. (2001). *Guide for individual investors on Energy Efficient heating of single family dwellings*. Krajowa Agencja Poszanowania Energii S.A.

Klimczuk, K. (2017). *Analysis of the cost during the lifetime of a building after a deep energy efficiency modernization* (Thesis). Warsaw University of Technology.

Memon, Z. A. (2008). The ACPROM Model: An Expert System for Evaluating the Construction Progress. *Proceedings of First International Conference on Construction in Developing Countries (ICCIDC–I)*.

Mulawka, J. (1997). Expert systems [in Polish]. Warsaw: Scientific-Technical Publishing House.

Rehak, D. R., & Fenves, S. J. (1984). *Expert systems in civil engineering, construction and construction robotics*. Retrieved from http://citeseerx.ist.psu.edu/viewdoc/download?doi=10.1.1.823.781&rep=rep1&type=pdf

Rośczak, P. (2003). *Sztuczna inteligencja*. Retrieved from http://www.rosczak.com/mlp/ai.html

Tadeusiewicz, R. (2012). Computer Decision Support. *Proceedings of XVI Conference of Automation Rytro (CD)*.

Ziembicki, P., & Węglarz, A. (2013). *Expertsystem algorithm for choosing an energy source for a building*. Ciepłownictwo, Ciepłownictwo, Ogrzewanie, Ciepłownictwo, Ogrzewanie, Wentylacja, 5/2013.

KEY TERMS AND DEFINITIONS

Cogeneration: A combined generation of heat and electricity.

Energy Efficiency: A ratio between an output of performance, service, goods or energy, and an input of energy.

Expert System: A single piece or a set of computer software that facilitates the use of knowledge and decision making. Expert systems can assist or even replace human experts in a certain field. They can provide the user with advice, recommendations or diagnosis in the problems related to the study field.

Knowledge Base: A detailed, extensive set of logically connected data regarding certain domain (e.g., construction), along with inference rules formulated by experts from this domain.

Primary Energy: Energy contained in primary energy carriers sourced directly from renewable and non-renewable natural resources.

Semantic Network: A set of interrelated objects connected via various relations. It constitutes a graphic representation of a type of logic, where relations between objects are represented by graph, in which objects are nodes while relations are branches (bows). Nodes and bows (branches) have their names. Weights can be assigned to nodes (or similarly, to branches) that can specify the degree of belief about the fairness of these statements. An inference process can be conducted in the semantic network.

Thermomodernization of a Building: A set of activities on the elements of the structure of the building and its installations, resulting in reduction of the building energy demand for heating, preparation of domestic hot water, cooling, and lighting.

Chapter 6
Energy Simulations as a Tool in Integrated Design Process

Joanna Rucińska
Warsaw University of Technology, Poland

ABSTRACT

In the integrated building modelling, all aspects have the same importance. It means that the architect, construction, and HVAC engineers create a building with interactions. BIM guarantees permanent information about important indicators generated parallel during the whole design process. Sophisticated simulation gives opportunity to observe important factors, for example, thermal comfort, ventilation quality, energy performance of building, operating and investments costs. For effective design process, selection of the appropriate team of people should be done carefully. It is particularly important to have a proper approach to modernisation of buildings. Without careful analysis of the concept and cooperation at various stages, it is not possible to complex retrofit of the building, which is aimed not only at reducing the energy use but also at maintain the desired thermal comfort.

INTRODUCTION

Increasing demands of the energy performance of buildings bring a new approach to the design process. Architects, constructors, HVAC engineers, lighting engineers, energy consultants, contractors and future users must cooperate with each other from the beginning until the end of the process. Communication between them requires understanding each other and new tools that would facilitate exchanging information. Although appropriate modern software can greatly improve the exchange of data, it is also necessary to develop appropriate standards. New building construction and especially modernization to the nZEB standard requires attention to details and accurate design concept development.

BACKGROUND

The Recast of the Directive on the Energy Performance of Buildings (Directive 2010/31/EU) came into force on 9 June 2010. EU member states should until 9 June 2012, publish the relevant laws and ad-

DOI: 10.4018/978-1-5225-4105-9.ch006

ministrative regulations necessary to implement its provisions. All activities aim to improve the energy performance of buildings and its elements. All new public building constructed after 31.12.2018 and all others after 31.12.2020 must be nearly zero energy buildings. Such high requirements are very difficult to meet using standard design methods and standard technical solutions.

In response to market demand the concept of integrated design has been developed. In traditional approach the construction, architects and engineering market is very much focused on the regional and local level. The work of individual specialists is followed in a line. For example, HVAC engineering has no influence on architectural concepts. This approach from the beginning imposes the use of some standard solutions. Other specialists are further forced to adapt the concept. In this solution, it is almost impossible to design a building with very low energy demand for heating, cooling, lighting, domestic hot water and electricity, with good thermal comfort conditions and characterised by comparable to standard ones, construction costs.

The combination of the three objectives mentioned above: low energy consumption, good thermal comfort conditions and low construction costs, represents an integrated design process (ID), sometimes also called IED (Integrated Energy Design). This approach has been developed in many countries independently therefore,, the state of the art for ID in the participating countries is very different. In some countries, GHG emissions or energy embodied in construction materials are taken into account during the assessment (The Research Centre on Zero Emission Buildings 2017).

In response to The Recast of the Directive on the Energy Performance of Buildings in June 2012 a project MaTrID *Market Transformation Towards Nearly Zero Energy Buildings Through Widespread Use of Integrated Energy Design* was lunched. The project was finished in December 2014 and was co-funded by the Intelligent Energy Europe Program of the European Union. The aim of the project was to support the implementation of Nearly Zero Energy Buildings by 2020. Special attention was given to the beginning of the process that is the concept phase. The more workload is placed at the beginning the greater is the possibility of using unconventional solutions. It is also possible to get better final results meaning lower energy use for heating, cooling, lighting, domestic hot water and electricity with lower investment cost while still maintaining the thermal comfort inside the building. Of course, such approach lengthens the design phase but the execution phase itself is more efficient. With the use of right tools, appropriate solutions can be chosen while still meeting all the established criteria.

Widely used tools are simulation programs which allow performing an hourly or even sub-hourly analysis of building variables. Software aimed at analysing energy demand and thermal comfort should already be applied at the conceptual stage. Any change made during the project for example, in the building purpose, its construction, or its external envelope and HVAC systems should be analysed it terms of its impact on energy consumption, thermal comfort conditions and construction costs. In the standard design process, the improvement of the energy standard might turn out to be inefficient, especially if the construction process has already begun. The possibility of achieving good results is decreasing as work progresses, at the same time, the costs increases. For these reasons, integrated design is a right approach when designing buildings with nZEB standard.

Integrated Design

Integrated design is a way to lead the investment process in a way, where the main goal is to improve the project team collaboration (Integrated Design 2012-2014). The members of each project team shall comprise the specialists from all stages of the construction process that are architects, HVAC designers, energy consultants, contractors and future users. The purpose of integrated design process is to reduce investment and operating costs and to better adapt to the future needs of building users (Zimmerman, 2006). One of the method of cooperation form early design stage is described in Planning and Conducting Integrated Design (ID) Charrettes (Todd & Gail, 2013)

The consequence of the traditional approach to building design is primarily a lack of potential benefits of building shape analysis and building location for example:

- Lack of using solar radiation in heating season, this increases the energy demand for heating,
- Lack of including unfavourable orientation or winnows area, which can cause a significant increase in the energy demand for cooling,
- Inadequate selection of heat transfer coefficients for external partitions, in particular transparent partitions, and heat gains transmittance and daylight coefficients of transparent partitions, which often affects the energy use for lighting and cooling system.

The traditional design, process consists of sequentially realized stages. Individual project contractors do not consult their decisions with the other project team members. With such an approach, it is hard to improve anything during design process, because it is very difficult to change previous assumptions and solutions. Sometimes it is even impossible to change anything. Integrated design process requires a change of approach at the very beginning of the design process. It requires from all team members good cooperation and understanding of the decisions made by others contractors. Due to a much larger scope of analysis and, compared to the traditional approach, extended design phase decisions in this process are supported by using advanced tools including simulation software. In early stage of the planning phase, the potential for increasing the energy efficiency of building is the highest. Work load at various stages of design process, impact the energy performance of building, and impact the cost. The workload distribution in relation to the time and building design phase is show on Figure 1.

There are many benefits of implementing integrated design process. The most important ones are described below.

The primary benefit of implementing integrated design process is the ability to achieve low building energy use and low operating costs, with the same or slightly higher investment costs. That is why it is so important to define a goal at the very beginning of the design process.

Experience shows that integrated energy design of buildings raises investment cost around to 5%, but it reduces operating costs by 40-70% (Integrated Design 2012-2014).

Integrated design of building consists of the following stages described below.

- **Selection of the Design Team:** In addition to architects, constructors, HVAC and lighting designers, and contractors, the team project should also include experts from energy management in the building and as well as from environmental protection. All people involved in the design process should be open to suggestions and solutions proposed by other team members. They should also be open for new technologies and non-standard solutions.
- **Analysis of the Boundary Conditions of the Project:** The boundary conditions and limitations should be defined already at the very beginning of the project. Limitations include inter alia building location, shape of the building plot, local spatial development plan, and adaptation of the building to the surrounding architecture. The main assumptions and preferences for the project should be given by the investor for example: the purpose of the building, the architecture

Figure 1. Early design phases offer opportunity for large impact on performance to the lowest costs and disruption
Source: ID Process Guide. www.integrateddesign.eu.

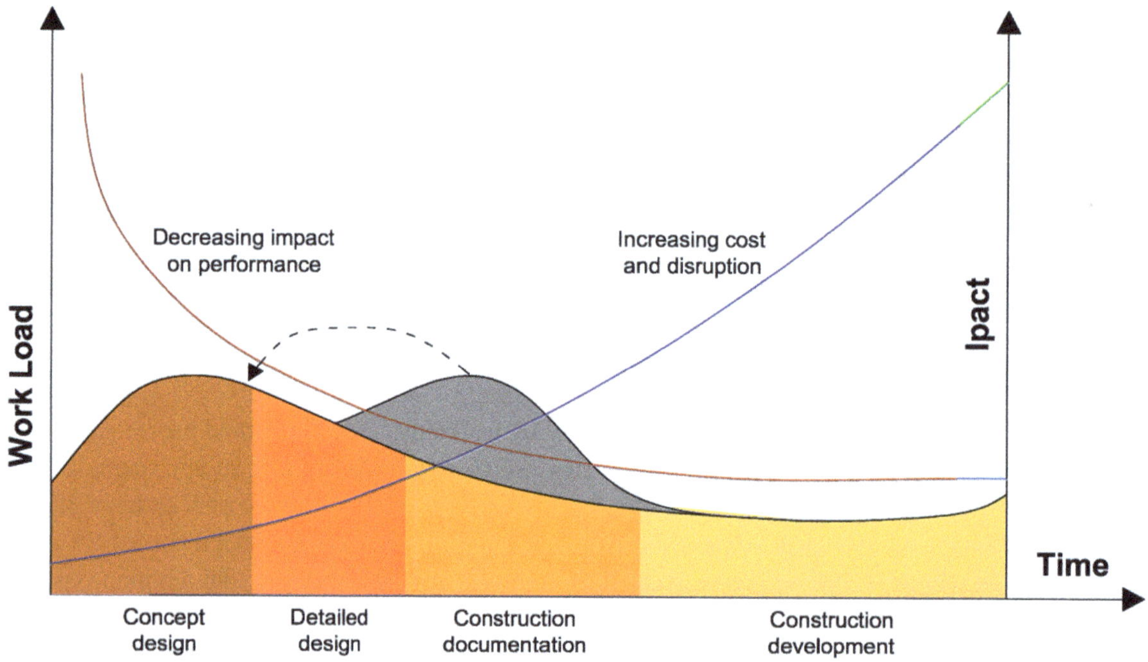

requirements (they may depend on the trend on real estate market), if the building is going to be certificated and in what certification system, what additional requirements building must meet (requirements for the building impact on the environment). The design team must also discuss with the investor, what are his financial possibilities and find out the investor's attitude towards the proposed solutions including energy-efficient strategies.

- **Signing Contracts, Which Motivate to Use Integrated Design:** Contracts signed with participants of project team should in the best way secure the achievement of the target for example energy consumption level and thermal comfort level. Alternative contracting methods encourage to the integrated energy design of buildings. Therefore, at the beginning of the process, the level of energy performance of the building should be determined as well as the punishments or rewards for fulfilling those assumptions. While standard contracting of architects and engineers, generally, incentives for energy efficient design are not used. The salary for the project is usually a percentage of the total project budget or a fixed amount. Such standard approach discourages to additional involvement for improvement of the building energy performance.
- **Development of Quality Assurance Program and Quality Control Plan:** In the integrated design process, it is necessary to define the criteria that the building must fulfil such as thermal comfort, energy demand, natural light, use of renewable energy sources, use of environmentally friendly materials, waste management, transport, water saving, functionality, affordability, and aesthetics. All requirements must be precisely described and their control procedure must be developed at each stage of the investment process. The documents Development of Quality

Assurance Program and Quality Control Plan must be respected and be in force for all participants in the building design process. The Quality Control Plan contains information what exactly will be checked and when and who is responsible for a quality control. A quality report should be generated after each quality control. This report should include any comments that should be known to all participants in the building design process.

- **Project First Meeting:** The purpose of the meeting is to present to all project participants their tasks and project schedule. The main manager should explain the functions performed by each participant and identify the people responsible for each step. Already at the first meeting the decision-making manager must be identified as well as the person who must receive the most important information. It is also necessary to make a schedule of next meetings.
- **Facilitate Close Cooperation Between the Participants in the Building Design Process:** The work should be organized in a way facilitating communication of architects, constructors, HVAC and lighting designers, and contractors participating in the building design process. The first way is setting work stations close together. The second is the organization of working meetings in smaller groups. Communication should be on the same level and the participants should be partners in the project. All persons should clearly and transparently present their ideas and explain why those solutions would be appropriate for the building. All process members should be open to discussion and new solutions. During the design process, all information which could influence energy demand, comfort of users, construction costs and operating costs must be presented to the energy experts. Described in the later part of the chapter analysis used during the construction process might help in making a decision.
- **Motivation and Training of Construction Workers:** The selection of contractors should take into account their previous experience in energy-efficient construction. Contracts should be signed in a way that allows for a control at each stage of the project, not just after the finished work. Construction workers should be motivated and educated in the most important cases such as quality of work and its influence on the final effect, thermal bridges, air permeability and new technologies. Random checks and Quality Tests are recommended during the construction process, not only at the end. It is also suggested to prepare a verification of energy efficiency or the impact on the environment in key moments of the process or in case of unexpected situation. Technical parameters of building components, which are related to energy use in building (efficiency of heat exchanger, SFP factor, efficiency of heating source), must be confirmed by documents, because of their key impact on the energy performance of the building.
- **Preparation of Instructions for Use and Maintenance of the Building:** After finishing the works, the design data should be updated, to provide specific information to the future building manager. Contractors and designers have to recommend to the user or owner of a building a program for monitoring the experimental parts of the systems. Technical department and users should be informed and properly trained how the systems work.

Building Information Modelling

A dozen or so years ago, in response to market demand, a system called BPM (Building Product Models) began to form. These were the first attempts to overcome the double transformation of object information, by two-dimensional plan, up to three-dimensional implementation (3D→2D→3D). The process of such conversion requires very good communication between all design process participants and is very vulnerable to errors related to transitional phases (Bujak, 2013). The ideal seems to be three-dimensional, digital set information about the building, which would reflect all its features, physical and functional details. Such an approach would allow designing architecture, construction and all technical systems in the same time and applying changes with control over existing conflicts and errors. The system that meets these criteria is BIM (Building Information Modelling). It allows crating a three-dimensional model of the building from the beginning design concept, through realization, to the end of the construction process. The elements of the model are given all geometrical and material features, they can be modified, and if any changes are introduced they are immediately available to other specialists. This allows avoiding mistakes and collisions. The virtual model developed in this way is more readable and comprehensible for all participants of design process. Based on the model, a two-dimensional project can be easily generated.

The main purpose of the BIM system is to avoid the problems typical for large projects, such as:

- Exceeding the assumed budget,
- Schedule delays,
- Waste of materials,
- Design errors.

The BIM system allows you to create complex architectural and construction layouts avoiding the long process of the installation coordination. At the same time a three-dimensional model allows investors to track the progress of the design process and to present their expectations. It is also a perfect source of information for contractors that can control the investment schedule and its costs. All participants in the investment process can control all stages of the design and construction work. Any collisions can be quickly noticed and eliminated early enough, so as not to cause changes in the schedule or increase in the investments costs.

Current software for managing the design and certification process improves cooperation between participants in the construction process. Unfortunately, the often appearing problem is the large variety of available software, which is not always compatible with each other. In cases where not all participants of the building design process use the same software it may be a necessary to transfer models between different programs. This causes the risk of errors or even makes it impossible to cooperate. However, if this problem will be solved at the beginning, there will be measurable benefits improving communication between all participants of investment process. It is just necessary to choose the right tool, corresponding to needs of architects, constructors, HVAC and lighting engineers.

Programs allow you to create three-dimensional architectural solids allowing for successive extension of constructions and installations. A full three-dimensional architectural design also gives a better opportunity to sell the project. Investor, often without engineering preparation, is able to fully accept the architect's concept seeing the expected final effect. He can also have some suggestions and ideas, which can be implemented at the design stage. This approach also allows performing economic, ecological and social analysis as early in the process as during the concept stage. This gives the opportunity to analyse the positive and negative effects of the object on the environment (Kaliszuk-Wietecka A. 2017).

ANALYSIS USED IN INTEGRATED DESIGN PROCES

Current requirements and used technical solutions require more and more precise methods of determining energy demand and energy use for heating, cooling, domestic hot water, electricity for existing and designed buildings. In buildings with complicate technical systems commonly used simplified calculation methods often not allow for accurately determine energy needs for all purposes. More and more often, simulation programs are used to calculate energy demand and also its variability in time. Simulation programs allow hourly or sub-hourly results analysis. Energy simulations of existing and designed buildings give information for engineers, architects and researchers about energy demand and demand for energy carriers in each hour in heating, cooling, ventilation, lighting and domestic hot water system.

Simulating of variables in time operational parameters of technical systems allows people involved in the design process of modernized building and new building, make the decision and optimize architectural and installation solutions. This approach reduces the energy and water demand for the designed building.

The analyses used in the integrated design process includes inter alia analysis of building envelope, shading, shading devices and strategies, daylighting, thermal comfort, life cycle costs, life cycle assessment, the use of alternative energy sources, energy analysis, cogeneration and polygeneration energy systems. Some of them are described below.

Each HVAC device can have different usage schedules and settings. Particularly in the case of modernized buildings, the settings and parameters of the selected equipment should be carefully checked.

Sometimes even a little change can improve thermal comfort and reduce energy use for heating, cooling, lighting or auxiliary appliances.

For building certification in systems such as BREEAM, LEED, HQE and others detailed analysis are used to assess the energy performance of buildings and its components (for example evaluation of system work) usually hourly simulation methods.

BREEAM is the world's leading sustainability assessment method for master planning projects, infrastructure and buildings. It addresses a number of lifecycle stages such as New Construction, Refurbishment and In-Use. Globally there are more than 562 000 BREEAM certified developments, and almost 2 265 500 buildings registered for assessment since it was first launched in 1990 (BRE, 2017).

LEED is an internationally recognized green building certification scheme. This program was developed by U.S. The Green Building Council (USGBC) providing building owners and managers with the tools to identify and implement practical solutions for the green design, construction, use and maintenance of buildings (LEED, 2017).

HQE (High Quality of Environment) was created in France for owners, managers, users, developers and investors, as well as for urbanists and local authorities. It is an international standard for managing the quality of the environment in the building industry. The standard is controlled by the Paris-based Association pour la Haute Qualité Environnementale (High Quality of Environment, 2017).

In hourly methods, it is possible to define in the building model the schedules of use of particular zones and accurate schedules of all technical systems. Calculations are based on hourly climate data.

Accurate hourly calculations made for the whole year allow you to check the design assumptions and quality of the performance of each technical system. On the basis of them, apart from energy use for heating, cooling, lighting, domestic hot water and auxiliary appliances, the quality of thermal conditions.

Such calculations also allow the selection of renewable energy sources taking into account the actual needs of the building.

In the next part of the chapter is focused on some of the analyses that should be performed during the buildings modernization process and construction of new buildings. Their execution gives the possibility of a wider analysis of the functioning of the designed HVAC systems. It also allows investigating the impact of components on the energy balance in the building. Thanks to such actions, the nZEB standard can be reached in an easier way.

Examples of analyses described in this chapter are: building shape analysis, building shading analysis, shading devices analysis, thermal comfort analysis, daylighting analysis, life cycle cost analysis and life cycle assessment analysis. The use of simulation methods allow to evaluate the influence of the change of some building elements or properties (building shape, use of additional elements such blinds, different properties of external walls) on the energy demand, thermal comfort of users and costs.

Energy simulations should be used from the beginning of the construction project.

Analysed Building

To show what kind of results, it is possible to receive from analysis used in integrated design, some examples are showed below. The selected building is an office building. The results of the analyses should be performed for each object individually and one example can set out only the general principles. On the other hand, on the basis of one example, it cannot be definitely stated how the modernisation will affect the energy demand, investment or operational costs or thermal comfort of users.

All energy simulations were performed in simulation program DesignBuilder (Deign Builder Software 2017). The analyses described below are recommended for new and modernized buildings. In modernized buildings it is more difficult to reach the nZEB standard, that is why it is so important to analyse the various possibilities. With the use of simulation tools the measures that affect the most the reduction of energy consumption and thermal comfort of users can be identified

Because the type of technical systems used in building affects the energy demand, some fixed values and standard solutions are assumed. Following systems are analysed: heating, cooling, lighting system and also energy use for auxiliary appliances. In this example the demand for hot water was not analysed. The demand for domestic hot water is related to the purpose of the building and the number of users. The more workers, the greater is the energy demand for domestic hot water. Fan coil unit are used for heating and cooling. The design temperature in heating season is 21°C (18°C in the night and non-working days) and in cooling season is 24°C (30°C in the night and non-working days). The building has VAV ventilation system with 75% heat recovery efficiency. The installed lighting power in offices is 10 W/m^2.

In each case building shape factor is given. Building shape factor is defined by ratio A/V_e, where A is the area sum of all external surfaces of a building in contact with ambient air, such as: external walls (including area of windows and doors), roofs, ceilings, floors on the ground or ceilings over unheated basement and ceilings over gates, etc. and V_e is heated or cooled building volume, calculated in accordance with the Polish Standard of the rules for calculation of the volume of buildings (with the volume of heated rooms in the attic or basement and without air space of unheated stairwells, elevator shafts, open cavities, loggias and galleries) (PN-ISO 9836:2015). U-value of external walls is shown in Table 1.

Widows g-value coefficient is 0.35 and light transfer coefficient L_t is 0.55.

Table 1. U-value of walls

Partition Type	U-Value W/(m²K)
External walls	0.25
Roof	0.20
Ground floor	0.25
Windows	1.30
External doors	1.7

Building Shape Analysis

Generally, for energy performance the highest impact is in the concept phase. Therefore it is very important to choose the initial parameters, which affect on observed function the most.

Some guidance for architects are available to predict annual energy performance. Case study presented below has shown that the building shape can have a significant impact on the energy use in building. On Figure 2 the analysed shapes of buildings were presented and in the Table 2 there are characteristic parameters (Area, envelope area, volume, building shape factor) of buildings and results of the energy use.

Architectural design has considerable influence on the need for heating, cooling, electricity and auxiliary appliances. Result shows that building shape has got significant impact on energy use in building. Larger external walls surface results higher energy use for heating as a result of higher heat losses. Larger heat losses through external walls decrease energy use for cooling.

Figure 2. The shapes of building, that were comparison of energy consumption: a)shape 1, b)shape 2, c)shape 3, d)shape 4

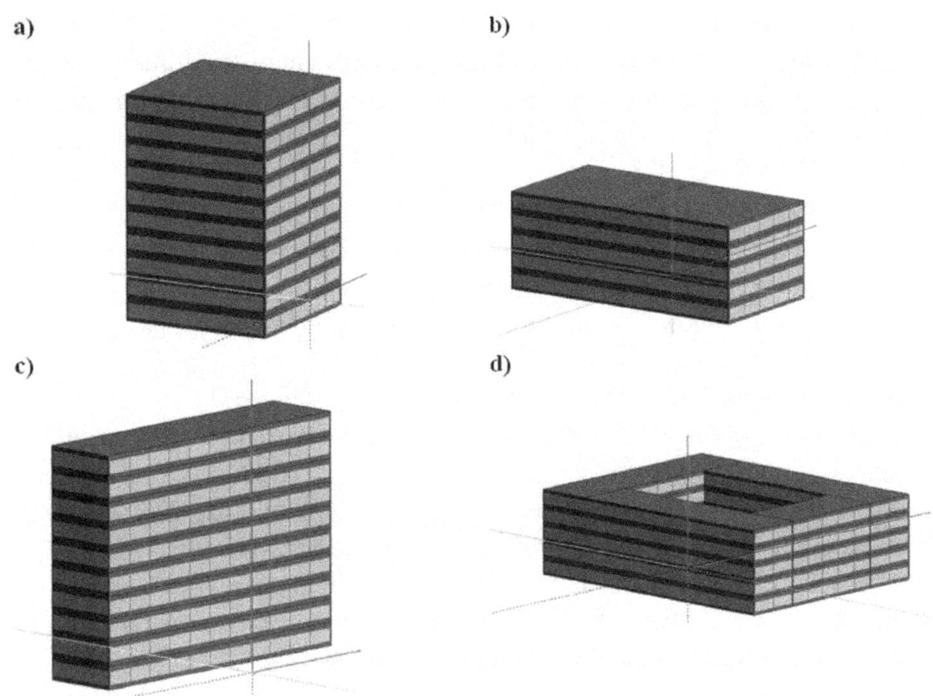

Table 2. Characteristic parameters and energy use in building

Parameter	Shape 1	Shape 2	Shape 3	Shape 4
Usable area [m^2]	8 500	8 585	8 345	14 744
Envelope area [m^2]	7 200	7 650	8 550	15 750
Volume [m^3]	40 500	40 500	40 500	70 875
Shape factor A/V_e [m^2/m^3]	0.18	0.19	0.21	0.22
Heating [kWh/(m^2 year)]	27.00	21.70	33.48	27.20
Cooling [kWh/(m^2 year)]	80.92	98.40	71.09	79.56
Lighting [kWh/(m^2 year)]	69.04	89.03	54.72	69.45
Auxiliary appliances [kWh/(m^2 year)]	21.55	23.17	21.20	22.36

The smaller the depth of the rooms the better use of the daylight and thus less energy demand for lighting system. With the higher energy demand for heating and cooling, the higher the energy consumption of the auxiliary appliances.

To properly evaluate and compare different buildings, the energy demand for all purposes should refer to the total building surface area or the building surface with controlled temperature. Simulations are a good tool to choose the best concept taking into account all needs.

It may seem that this type of analysis is intended for new buildings. However, in the case of deep or complete modernization, it may be necessary not only to change the thermal insulation of external envelope or change the HVAC system, but also the use of the building or its shape. An example of such case is Fredrik Selmers building vei 4, Oslo (FutureBuilt 2010-2020).

Building Shading Analysis

During the building concept preparation, the influence of surrounding environment to the building is not always analysed. In order to carry out an analysis of the impact of shading from other buildings, the energy model of the building and surrounding must be created. Figure 3 presents the model of an office building with surrounding buildings.

The effect of shading from other buildings is changing during the year. In winter, the sun is lower and the shade is larger. The opposite situation is in the summer because sun is then higher and the shade is smaller. Visualization of building shading analysis by surrounding buildings in different months and at different hours of the day is shown in the figures below (Figure 4 – Figure 7).

The main conclusion from the visualization is that the higher density of buildings causes greater impact of surrounding buildings on energy performance of the designed one.

The results of annual energy simulations for the building without and without surrounding buildings are shown in Table 3.

Results for a building model with and without surrounding buildings are different. After taking into account other buildings the energy demand for heating has increased. This is related to the reduction of heat gains from the solar radiation. The energy demand of the lighting system has also increased. This is the result of the reduction of the daylight. Higher internal gains have increased the energy demand for cooling. The energy consumption of auxiliary appliances is higher too.

Figure 3. Analysed building and its surroundings

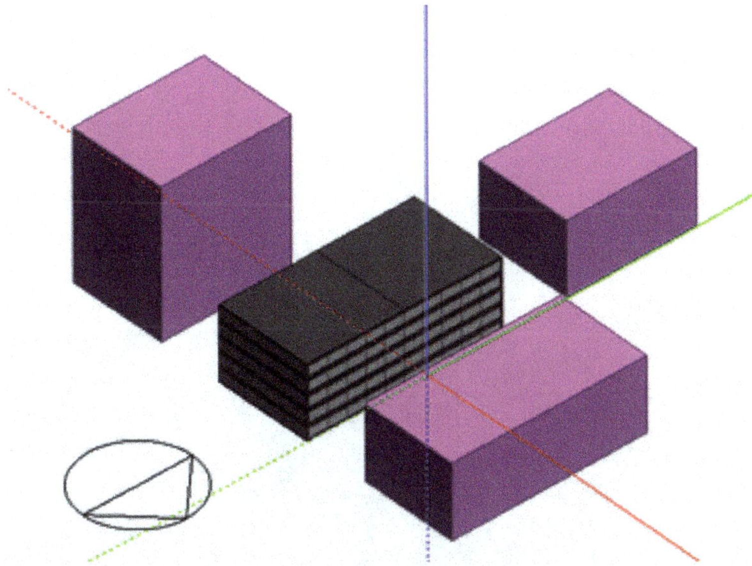

Figure 4. Building shading analysis: 21st of December 9 am

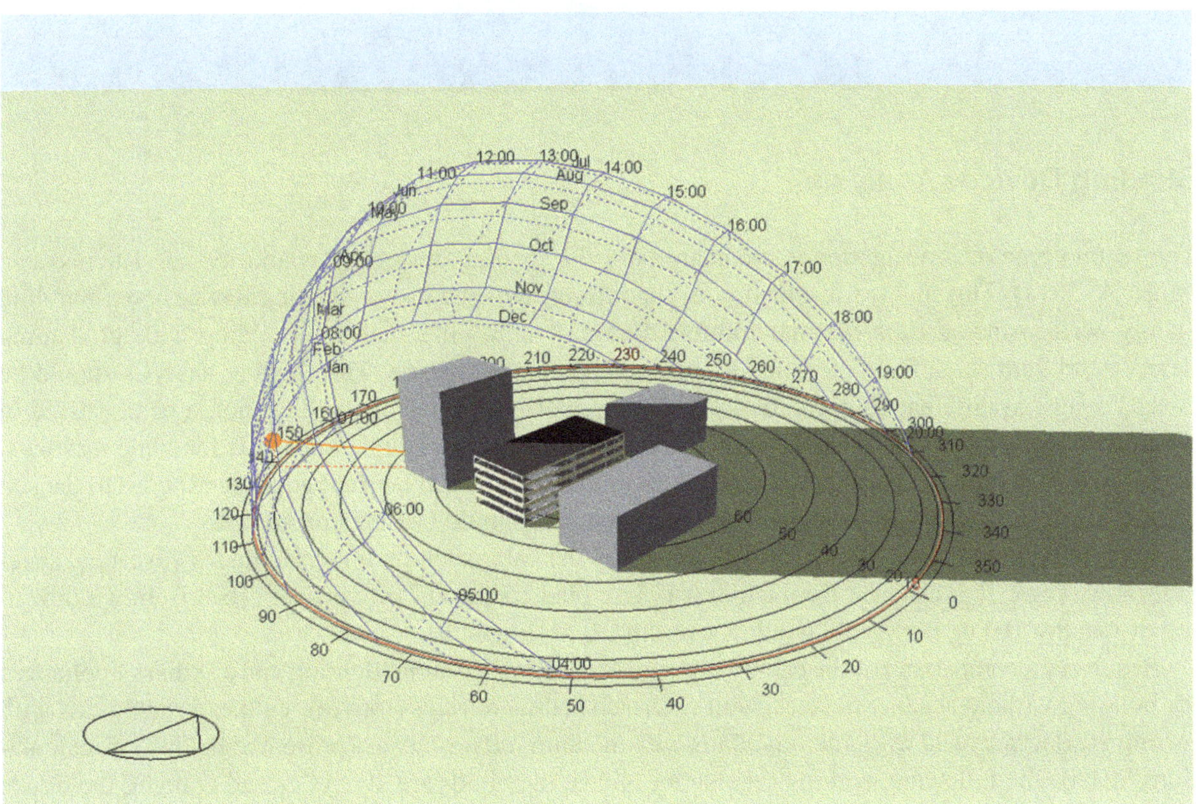

Figure 5. Building shading analysis: 21st of December 3 pm

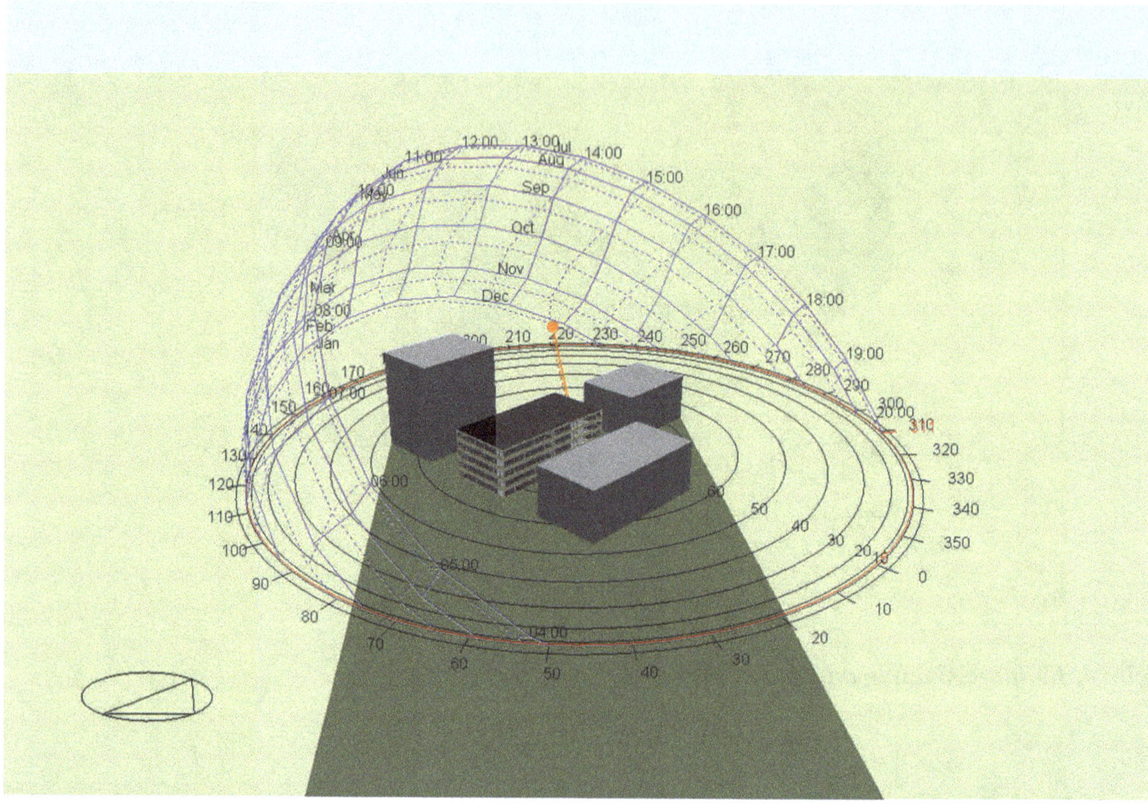

Shading Devices Analysis

The second type of investigation is the analysis of the shading devices (Kwiatkowski J., Rucińska J., Panek A. 2014). This involves evaluation of the influence of the use of shading devices on a building energy performance and on thermal comfort. Figure 8 shows the analysed building without shading devices and with some fixed elements louvres applied on the facade. The shading analysis should be based on several different configurations of shading. Selected types of shading should be discussed in detail with the architect, because they are also a part of the architectural concept. If for some reasons it is impossible to use this solution, in the case of buildings with high internal gains or exposed to the sun or with large glazing surfaces, movable shading elements should be then considered.

In building model applied louvres with the following parameters: number of blades 4; vertical spacing 0.4 m, angle 15°; distance from window 0.3 m; blade depth 0.2 m; vertical offset from window to top of window 0.0 m; horizontal window overlap 0.0 m (Figure 9).

Results of a comparison of the energy demand for heating, cooling, lighting and auxiliary appliances for building without shading devices, with external shading devices (louvre), with external blinds with medium reflectivity and with internal blinds with medium reflectivity are presented below. The calculations include the following working scheme for blinds: the blinds are always closed at night, the blinds are closed during the day when the solar radiation intensity exceeds 200 W/m^2 and cooling in building

Figure 6. Building shading analysis: 21st of June 9 am

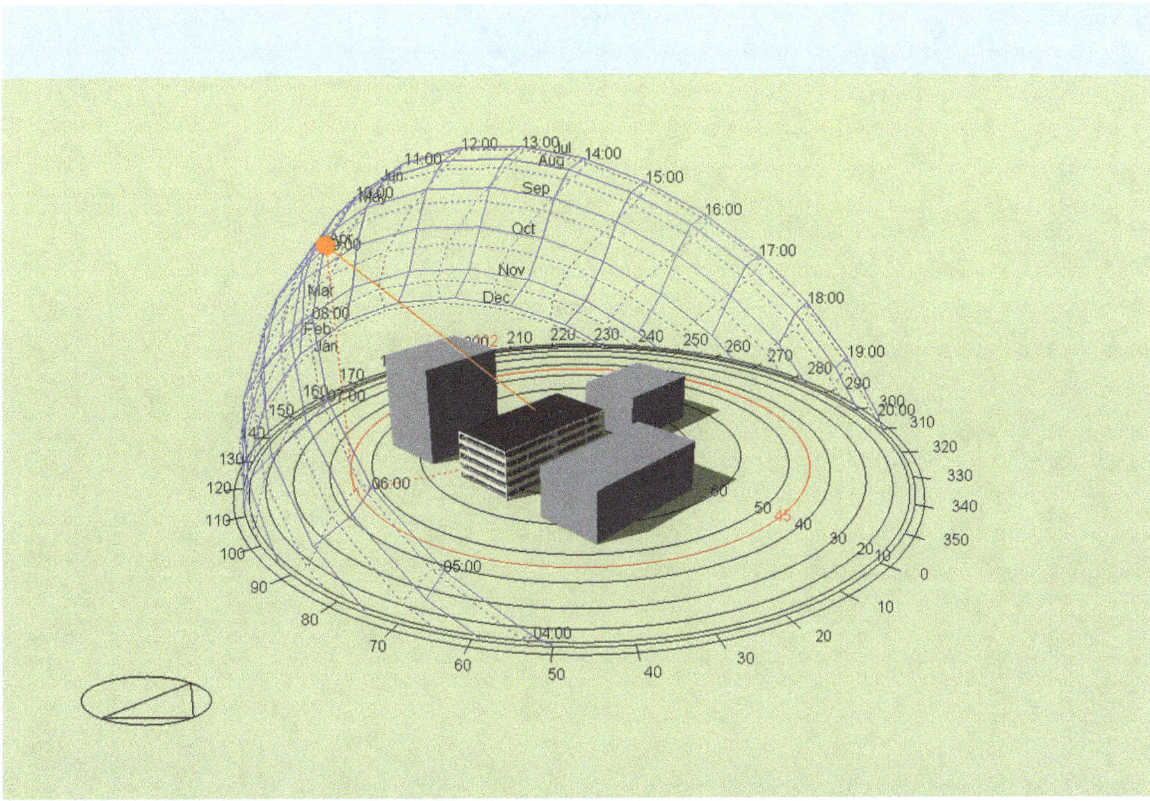

is required. On each facade the blinds work independently of each other. The results of annual energy simulations for the building are shown in Table 4.

The installation of shading devices influences the energy demand for heating, cooling, lighting, and auxiliary appliances. In the analysed example, the use of external shading devices (louvres) has increased the demand for heating energy and reduced the energy demand for cooling which is a result of lower solar heat gains. In addition, a reduction of the energy demand in the lighting system was observed; thanks to the appropriate position more light reaches the interior through reflection from external elements.

Application of external blinds medium reflectivity increases the energy demand for heating and reduces the energy requirement for cooling as well for auxiliary appliances. On the other hand, the use of internal blinds with medium reflectivity results in a reduction in energy demand for heating and lighting and increasing the demand for cooling and electric auxiliary appliances. Energy demand for cooling is higher because of a setting (position) and properties of window blinds. In this configuration of internal blinds, the solar radiation is reflected off the surface of blinds and it reaches the room with greater intensity. These results in increased solar heat gains in office spaces. With such configuration, they do not limit the solar radiation gains. For proper operation, their parameters (for example reflection coefficient) and setting (for example slat angle) should be changed. Energy for auxiliary appliances is higher because of higher energy need for cooling.

Figure 7. Building shading analysis: 21st of June 3 pm

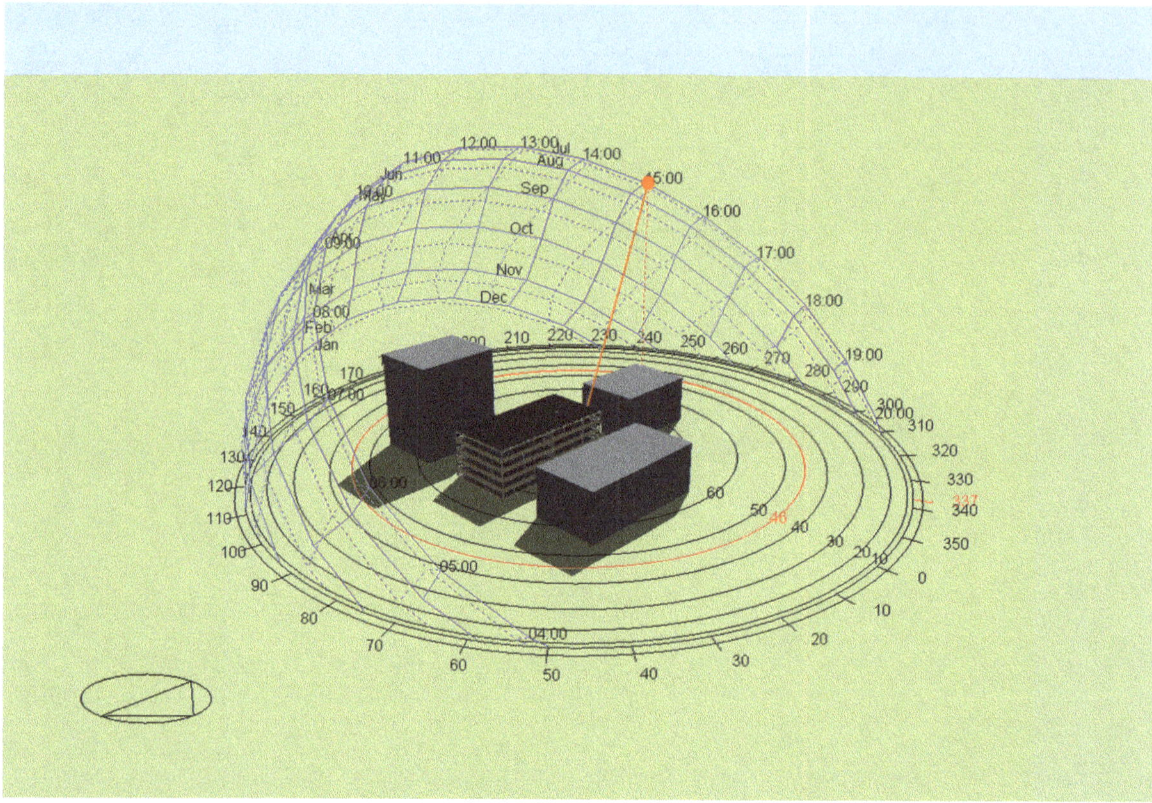

Table 3. Energy use in building: Building shading analysis

Parameter	Building Model Without Surrounding Buildings	Building Model With Surrounding Buildings
Floor area [m^2]	7 650	7 650
Volume [m^3]	40 500	40 500
Shape factor A/V_e [m^2/m^3]	0.18	0.18
Heating [kWh/(m^{2}year)]	21.70	22.41
Cooling [kWh/(m^2 year)]	98.40	102.33
Lighting [kWh/(m^2 year)]	89.03	100.53
Auxiliary appliances [kWh/(m^2 year)]	23.17	23.34

Thermal Comfort Analysis

There are different methods of assessing thermal comfort. Thermal comfort assessment is used together with the assessment of energy demand for heating, cooling, lighting or auxiliary appliances. The full analysis gives better adjustment of design to the user's needs. In case of building modernization often solution will result from analysis of thermal comfort conditions. First the problems occurring in the building

Figure 8. Analysed building: left building without shading devices, right building with louvres

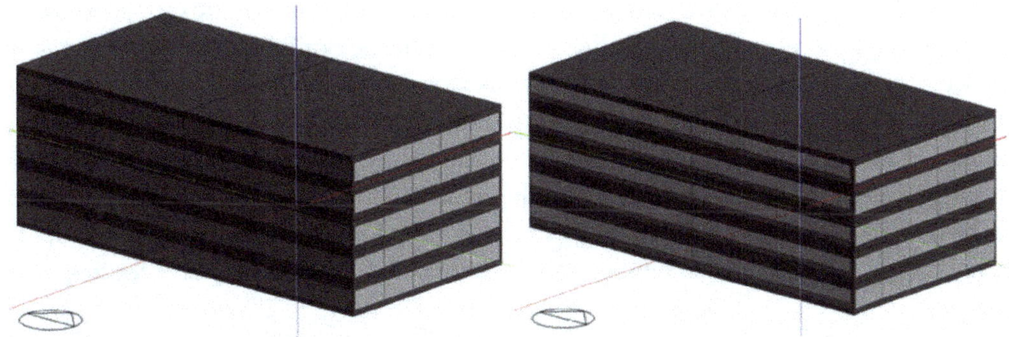

Figure 9. Louvres
Source: DesignBuilder Manual.

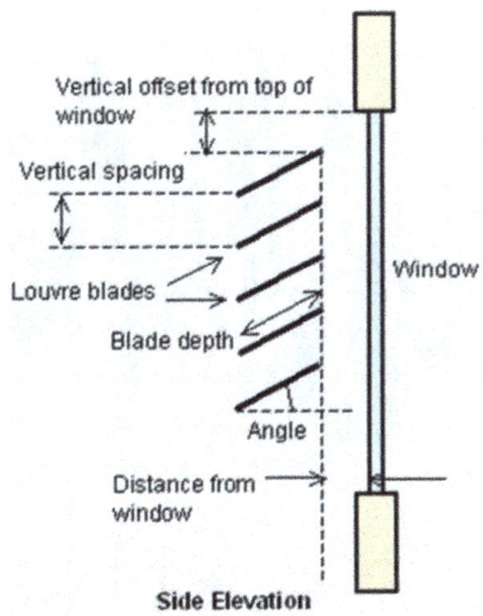

Table 4. Energy use in building: Shading devices analysis

Parameter	Building Model Without Shading Devices	Building Model With Shading Devices		
		Louvre	External Blinds Medium Reflectivity	Internal Blinds Medium Reflectivity
Floor area [m^2]	7 650	7 650	7 650	7 650
Volume [m^3]	40 500	40 500	40 500	40 500
Shape factor A/V_e (m^2/m^3)	0.18	0.18	0.18	0.18
Heating [kWh/(m^2 year)]	21.70	23.47	23.23	20.88
Cooling [kWh/(m^2 year)]	98.40	83.77	95.50	110.94
Lighting [kWh/(m^2 year)]	89.03	66.63	89.03	87.57
Auxiliary appliances [kWh/(m^2 year)]	23.17	21.22	23.06	24.75

should be identified. While choosing the best solution, energy consumption as well as the parameters of thermal comfort should be calculated and evaluated together. Tables and figures below show the results of thermal comfort for presented above shading devices. At the beginning, it was verified whether in the Winter and Summer the temperatures exceeded the assumed design values for the selected zones. The typical design values for temperatures for offices buildings for heating needs are in the range of 20÷21°C and for cooling needs in 24÷26°C. In the investigated analysis following temperatures are assumed: the temperature for heating needs is 21°C and for cooling needs: 24°C. On Figure 10 are shown the values of air temperature and relative humidity during the year in building without shading devices and Table 5 presents the number of hours when temperature is exceeded during heating and cooling season.

Figure 10. Air temperature and relative humidity in analysed building without shading devices

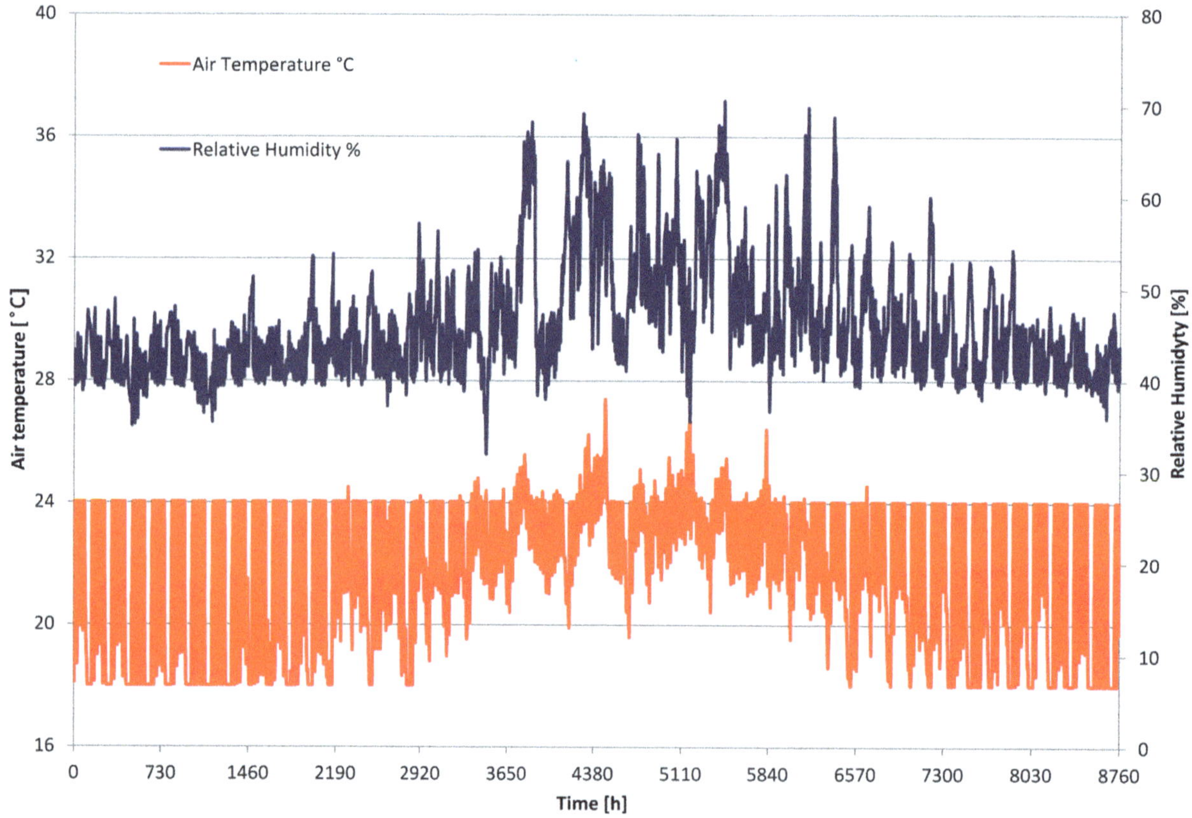

Table 5. Number of hours when temperature is exceeded

Number of Hours When Temperature Is Exceeded	Building Model Without Shading Devices	Building Model With Shading Devices		
		Louvre	External Blinds Medium Reflectivity	Internal Blinds Medium Reflectivity
Heating season (temperature <21°C)	0.0	0.0	0.0	0.0
Cooling season (temperature >24°C)	0.0	0.0	0.0	0.0

After analysis of the data it can be seen that the assumed temperature and relative humidity in heating and cooling season is not exceeded. Temperature is maintained both during and outside working hours. This means that the temperature control system in the building works well. In each variant, keeping the temperature within the assumed range and this affects the energy consumption, that was shown in Table 5.

They are within the assumed range. This means that the heating and cooling system works according to assumptions. For complete information, the thermal comfort according to ISO 7730 (2006) *Ergonomics of the thermal environment - Analytical determination and interpretation of thermal comfort using calculation of the PMV and PPD indices and local thermal comfort criteria* should be calculated (Table 6). Classification of the internal environment for selected week according to EN 15251 *Indoor environmental input parameters for design and assessment of energy performance of buildings addressing indoor air quality, thermal environment, lighting and acoustics* (2007) is shown in Table 7 and 8.

Table 6. Thermal comfort criteria

Comfort Class	PPD [%]	PMV
A	<6	-0.2 < PMV < +0.2
B	<10	-0.5 < PMV < +0.5
C	<15	-0.7 < PMV < +0.7

According to EN ISO 7730 (2006).

Table 7. Percentage share of classes in the analysed period 1^{st}floor: Classification of the internal environment

	Building Model Without Shading Devices			Building Model With Shading Devices								
				Louvre			External Blinds Medium Reflectivity			Internal Blinds Medium Reflectivity		
Comfort Class	A	B	C	A	B	C	A	B	C	A	B	C
Heating Season												
Predicted Mean Vote PMV	31.7	68.3	0.0	30.0	70.0	0.0	33.3	65.0	1.7	26.7	73.3	0.0
Predicted Percentage of Dissatisfied PPD	33.3	66.7	0.0	33.3	66.7	0.0	35.0	63.3	1.7	30.0	70.0	0.0
Relative Humidity	100.0	0.0	100.0	100.0	0.0	0.0	100.0	0.0	0.0	100.0	0.0	0.0
Cooling Season												
Predicted Mean Vote PMV	63.3	33.3	3.3	80.0	18.3	1.7	65.0	30.0	5.0	53.3	45.0	1.7
Predicted Percentage of Dissatisfied PPD	73.3	23.3	3.3	83.3	11.7	5.0	73.3	21.7	5.0	56.7	41.7	1.7
Relative Humidity	11.7	46.7	41.7	11.7	46.7	41.7	11.7	46.7	41.7	13.3	53.3	33.3

According to EN 15251 (2007).

Table 8. Percentage share of classes in the analysed period 4 th floor- Classification of the internal environment

	Building Model Without Shading Devices			Building Model With Shading Devices								
				Louvre			External Blinds Medium Reflectivity			Internal Blinds Medium Reflectivity		
Comfort Class	**A**	**B**	**C**	**A**	**B**	**C**	**A**	**B**	**C**	**A**	**B**	**C**
Heating Season												
Predicted Mean Vote PMV	23.3	73.3	3.3	30.0	65.0	5.0	31.7	66.7	1.7	28.3	70.0	1.7
Predicted Percentage of Dissatisfied PPD	26.7	65.0	8.3	31.7	61.7	6.7	36.7	60.0	3.3	31.7	63.3	5.0
Relative Humidity	100.0	0.0	0.0	100.0	0.0	0.0	100.0	0.0	0.0	100.0	0.0	0.0
Cooling Season												
Predicted Mean Vote PMV	35.0	61.7	3.3	75.0	21.7	3.3	50.0	45.0	3.3	38.3	58.3	3.3
Predicted Percentage of Dissatisfied PPD	38.3	58.3	3.3	83.3	13.3	3.3	58.3	36.7	3.3	43.3	53.3	3.3
Relative Humidity	11.7	46.7	41.7	11.7	48.3	40.0	11.7	46.7	41.7	11.7	46.7	41.7

According to EN 15251 (2007).

In the analysis for the selected summer and winter week, the PMV and PPD was calculated for every hour. The percentage share of comfort classes in the analysed period was then calculated.

Based on obtained results, it can be stated that not in every hour rooms can be classified in A comfort class according to EN 15251 (2007). In the winter period in each case results are very similar. This is due to the high internal heat gains and in those offices it is a bit too warm as for winter conditions. In the Summer, the use of shading devices improves the thermal comfort in offices in case of louvres, external blinds and also internal blinds with medium reflectivity. The best results are in case of using louvres and higher number of hours is in A comfort class. Relative humidity values in Winter fall are within the range because the humidity is controlled, but in the Summer there is no dehumidification. To fulfil requirements of comfort class A of the EN 15251 (2007), following measures can be implemented: better regulation of HVAC systems for example, possibility of temperature regulation by the users, better adjustment to conditions of air velocity.

The best solution determined on the basis of the analysis of energy use for heating, cooling, lighting and auxiliary appliances and thermal comfort is building with louvres. In this solution the summary energy for cooling, lighting and energy appliances represents the lowest value. However, none of those analyses involved economic considerations. Therefore, these analyses should be followed by Life cycle cost analysis or life cycle assessment. These analyses also do not take into account the issues related to daylight.

Daylighting Analysis

An analysis of the use of daylight allows reducing energy needs to internal lighting. Different tools are used in the design concept stages to help determine the availability of daylight. There are specially designed programs for daylight calculations for example Radiance, Dialux, Velux Daylight Visualizer, Daysim and Ecotect. Some simulation program also has a daylighting option for example DesignBuilder, Ida Ice, Velux Energy & Indoor Climate Visualizer. Vasari and Openstudio.

In many simulations software there is a possibility to perform a daylight analysis. In daylight analysis, important parameters are: construction of the external walls in particular transparent elements, floor height, surfaces reflection coefficients. Below on figure 11 is the result of bulging daylight analysis with and without internal partitions.

The assessment of natural light has to be done on the basis of projects and actual arrangements. The most important factor in the analysis is to check the appropriate intensity of natural light in different rooms. For example, too high depth of the office rooms and too small glazed area is not conducive to efficient office work.

It can be seen that the proposed partitions did not significantly change the conditions inside the offices. Especially in the corners of the building in case building with internal partitions the area in which the DF is smaller has increased.

Life Cycle Cost Analysis

LCC analysis involves the economic and technical aspects of the design building in its entire life cycle. Life cycle cost analysis includes not only the investment costs and not only the operating costs. LCC should be a tool widely used to assess the profitability of an investment. There are the following main categories and components of life cycle cost analysis, which should be included in the calculation.

Figure 11. Daylight analysis: left- open space; right- floor with internal partitions

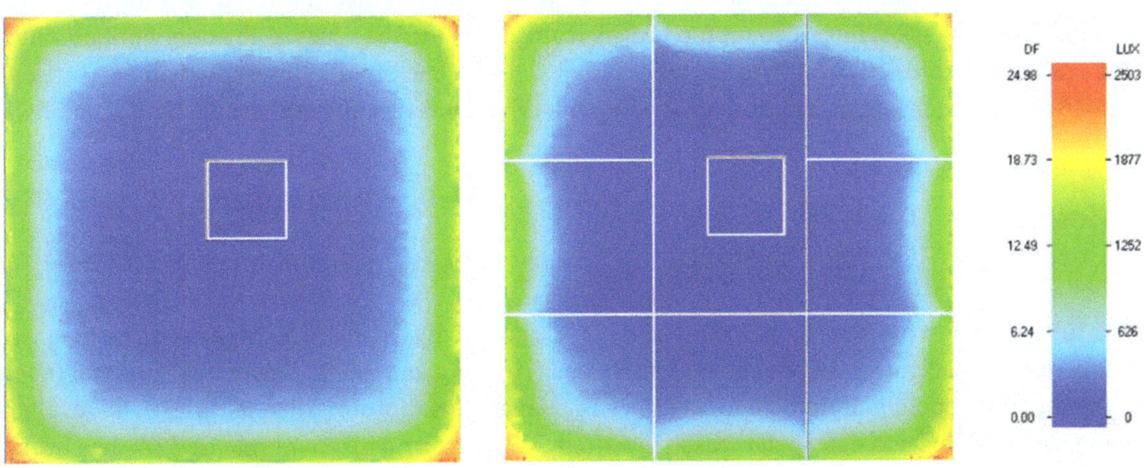

- Investments cost include all costs necessary to construct a building including: preparation costs, project costs, costs of building permits, cost of construction works, costs of materials and installations, transport costs, costs related with energy consumption during construction works, costs involved with waste disposal. Investment costs may also include the cost of buying a building plot.
- Running costs covering maintenance and operation costs, the cost of periodic maintenance such as repair and replacement used or broken components in all technical systems.
- Operating costs for heating, cooling, lighting, domestic hot water etc. associated with consumption of gas, electricity, water, district heating etc. and also sewage, and waste costs.
- Removal costs include for example changing an object's function usually dismantling after the end of its life cycle and materials recycling costs.

Cost data should be based on current and reliable components costs. Results for individual cost categories should be presented separately and also as a sum as total life cycle cost. The total life cycle cost for a building is calculated by summing up the different cost categories taking into account the discount and inflation coefficient. The goal is to express the value of life cycle cost at the beginning of the process (in a zero year).

In order to simplify the analysis, the main elements of the building should be defined such as substructure, construction, walls, roof, partitions, windows, hating system, ventilation system, lighting system, fire protection system etc. Each of these elements must be specified, the construction materials and components used and their quantities and purchase costs. For defined categories usage scenarios should be decribed, including repairs and replacements and associated with them costs. The cost of use should also determine the frequency of replacement of the components and their elements. For each element life time is also determined, for example, for installation elements, it is usually 15 years and for construction elements and envelope elements is 60 years. Estimated life cycle may be longer or shorter than given above calculation period. Maintenance of the building includes cleaning, repairs and materials used in these processes for example: the planned painting and window cleaning should also be included in the life cost. The starting point of calculation should take into account the year in which the calculations are made. The main purpose of this requirement is to take into account current prices and cost levels. Energy costs should take into account not only the cost of fuel but also the fixed fees for its delivery. The calculation also takes into account the profits from generated the renewable energy in building.

An example of calculations was made for external walls for the building presented above. Life cycle cost calculations are made for 60 years of building life time. Two variants of thermal insulation of walls were analysed:

- External walls - U-vale 0.20 W/(m^2K),
- External walls - U-vale 0.25 W/(m^2K).

The following cost has been included in the life cycle cost analysis: investment (materials only insulation material), replacements and repairs, maintenance (cleaning and management), operational (energy needs for heating, cooling, lighting and auxiliary appliances), and of life (dismantling). The results are been presented in real and discounted cash flow. In the LCC study a discount rate of 3.0% has been assumed. The results of analysis are shown in Table 9 and on Figure 12.

Table 9. Results of life cycle analysis for external walls of the building

Parameter	Building U-Vale of External Walls – 0.20 W/(m^{2}K)	Building U-Vale of External Walls – 0.25 W/(m^{2}K)
Heating [kWh/year]	208 546.4	208 589.8
Cooling [kWh/year]	211 209.0	211 218.3
Lighting [kWh/year]	764 387.5	764 387.5
Auxiliary appliances [kWh/year]	198 931.4	198 904.0
Life cycle cost in 60-years real value [€]	14 073 854.2	13 672 037.1
Life cycle cost in 60-years discounted value [€]	6 525 663.3	6 322 307.6

Figure 12. Results of life cycle analysis for external walls of the building

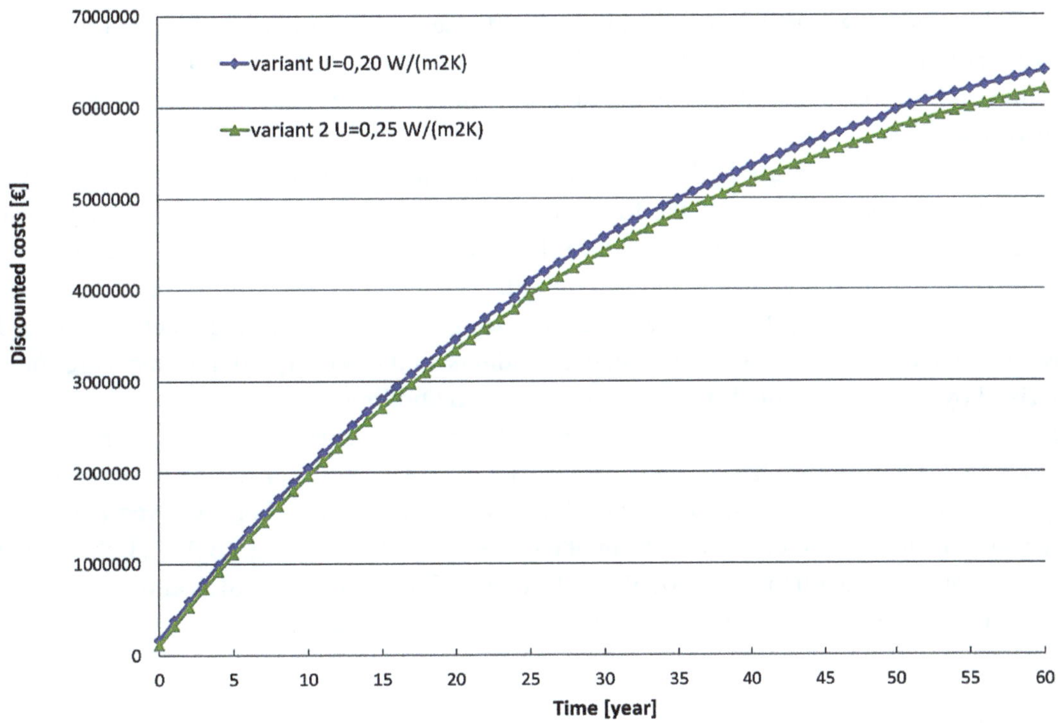

Building with U-vale of external walls 0.25 W/(m^{2}K) has lowest real and discounted life cycle cost in whole life cycle. In this case the changing the heat transfer coefficient of external walls did not significantly affect the energy demand for heating, cooling, lighting and auxiliary appliances.

Life Cycle Assessment Analysis

The Life Cycle Assessment (LCA) takes in to account the environmental impact associated with energy consumption and the associated emissions of pollutants into the atmosphere from extraction of raw material, through all its processing steps and the building maintenance until the end of usage and the waste disposal operation (Komerska A., Kwiatkowski J. Rucińska J. 2015).

Life cycle energy analysis is a detailed method to estimate the total energy inputs, outputs and flows through the life cycle of a building. The boundaries of the system are expanded in order to take into account both the operating energy and the embodied energy of a building.

The operating energy consists of the energy for heating, cooling, ventilation, domestic hot water, lighting, auxiliary appliances and all technical systems.

Embodied energy takes into account the materials manufacturing and their transportation to the construction site, the energy for the construction of the building, the maintenance, repair and replacement of materials and technical systems during the lifetime of the building, the energy for the demolition, the transportation of materials and their end of life management (Chastas P., Theodosiou T., Bikas D. 2016).

In article "Embodied energy in residential buildings-towards the nearly zero energy building: A literature review" (Chastas P., Theodosiou T., Bikas D. 2016) the author presents results of the share of embodied energy in the whole building life cycle for 90 case studies: conventional, passive, low energy and nearly zero energy residential buildings (nZEB). It turned out that the share of embodied energy in conventional buildings ranges between 5% and 36%, in low energy buildings this percentage varies between 10% and 83%.

In another article "Energy use in the life cycle of conventional and low-energy buildings: A review article" 60 buildings were analysed. It was found that the energy use during operation is the largest share during life cycle (Sartori I., Hestnes A.G. 2007). It has also been shown that there is a linear relation between operating and total energy, valid through all the cases despite climate and other contextual differences. It turned out that the demand for energy in standard buildings is higher than in buildings with low energy requirements. On the other hand, buildings with low energy requirement are sometimes characterized by greater embodied energy than in standard buildings.

The largest consumers of energy are buildings. Energy use is over 40% of the energy used in the European Union. On the other hand, a very large amount of energy is consumed in the production of building materials and the elements of the building's technical systems. For these reasons, the use of LCA analysis not only allows for an assessment of the energy need for heating, cooling, lighting domestic hot water and auxiliary appliances, thermal comfort and life cycle cost but also allows for a selective selection of technologies which they are really environmentally friendly.

CONCLUSION

The chapter presents the method of integrated design process. This approach to design allows you to get low energy demand for heating, cooling, lighting, domestic hat water and auxiliary appliances, at the same time it allows to take into account the thermal comfort, costs and emissions. It is necessary to use modern software to enable collaboration between specialists. Buildings energy calculation programs support design decisions, can be used for the design and existing projects. The range of programs is significant so it is important to define the purpose of the analysis at the beginning. However, if not used properly, incomplete information can be generated from the software. Therefore, specialists using them should have a proper knowledge. Applied programs and analyses depend on the complexity of the project. They should be understandable to all process participants. All changes made in the project should be consulted and it is necessary to check their impact on the end result. In the standard design process, making changes to improve energy performance is difficult because there is no direct cooperation between process participants. Integrated design should be accounted as standard not, only in design nZEB buildings.

After the building modernisation to nZEB standard, its energy performance should be verified and evaluated. If the expected level is not reached, the reasons for this condition must be assessed. The results of the modernization together with the instructions for the other contractors should be monitored. Such an approach would allow the development of guidelines for upgrading buildings to the nZEB standard.

ACKNOWLEDGMENT

We would like to acknowledge the support of all designers from Warsaw University of Technology and NTNU Trondheim involved in the KODnZEB grant. This research was supported by the *EEA Financial Mechanism 2009-2014 in the framework of the Bilateral Cooperation Fund at the level of the Operational Program PL04 "Saving energy and promoting renewable energy source" within the framework of the project KODnZEB:* "Design retro-fit nZEB concept for two buildings" [grant number 315/2015/Wn07/OA-XN-04/D no. SAP 511/00023].

REFERENCES

BRE. (2017). *BRE Environmental Assessment Method.* Retrieved from www.breeam.com

Bujak, P. (2013). *Modeling building information in a modern architect's workshop - perspective of application, orginal* [Modelowanie informacji o budynku we współczesnym warsztacie architekta – perspektywy zastosowania] (PhD thesis). Warsaw University of Technology.

Chastas, P., Theodosiou, T., Bikas, D., Kontoleon, K. (2017). Embodied energy and nearly zero energy buildings: A review in residential buildings. *Procedia Environmental Sciences, 38*, 554-561. doi:10.1016/j.proenv.2017.03.123

Directive 2010/31/EU of the European Parliament and of the Council of 19 May 2010 on the energy performance of buildings. (2010). *Official Journal of the European Union, 53*(153), 124-146.

LEED Green Building Certification Scheme. (2017). Retrieved from www.usgbc.org/leed

High Quality Environment. (2017). Retrieved from www.hqegbc.org/accueil

Integrated Design. (2012-2014). Retrieved from www.integrateddesign.eu

International Organization for Standardization. (2006). *Ergonomics of the thermal environment - Analytical determination and interpretation of thermal comfort using calculation of the PMV and PPD indices and local thermal comfort criteria* (Standard No. 7730). Retrieved from https://www.iso.org/obp/ui/#iso:std:iso:7730:ed-3:v1:en

International Organization of Standardization. (2007). *Indoor environmental input parameters for design and assessment of energy performance of buildings addressing indoor air quality, thermal environment, lighting and acoustics* (ISO Standard No. 15251). Retrieved from www.futurebuilt.no

Kaliszuk-Wietecka, A. (2017). *Sustainable Building Selected issues of building physics* [Budownictwo Zrównoważone Wybrane zagadnienia fizyki budowli]. PWN Polish Scientific Publishers.

Komerska, A., Kwiatkowski, J., & Rucińska, J. (2015). Integrated Evaluation of CO2eq Emission and Thermal Dynamic Simulation for Different Facade Solutions for a Typical Office Building. Energy Procedia, 78, 3216-3221.

Kwiatkowski, J., Rucińska, J., & Panek, A. (2014). Analysis of different shading strategies on energy demand and operating cost of office building. *Proceedings from World Sustainable Building Conference SB14*, 5, 512-521.

Sartori, I., & Hestnes, A. G. (2007, March). Energy use in the life cycle of conventional and low-energy buildings: A review artcile. *Energy and Building*, *39*(3), 249–257. doi:10.1016/j.enbuild.2006.07.001

The Research Centre on Zero Emission Buildings. (2017). Retrieved from www.zeb.no

Todd, J. A., & Gail, L. (2013). Planning and Conducting Integrated Design (ID) Charrettes. Whole Building Design Guide. National Institute of Building Sciences.

Zimmerman, A. (2006). *Integrated design process guide*. Ottawa, Canada: Canada Mortgage And Housing Corporation.

KEY TERMS AND DEFINITIONS

BIM: Building information modelling.
Comfort Classes: Comfort classes according to ISO 7730.
HQE: High quality of environment.
HVAC: Heating, cooling, and air conditioning.
ID: Integrated design.
IED: Integrated energy design.
PMV: Predicted mean vote.
PPD: Predicted percentage of dissatisfied.

Chapter 7
Improving the Design of Energy-Efficient Building Retrofitting:
Design Guidelines, Energy Simulations, and Selecting of Technologies

Mari Aino Hukkalainen
VTT Technical Research Centre of Finland Ltd, Finland

Krzysztof Klobut
VTT Technical Research Centre of Finland Ltd, Finland

Tarja Mäkeläinen
VTT Technical Research Centre of Finland Ltd, Finland

Vanda Dimitriou
Loughborough University, UK

Dariusz Leszczyński
E&L Architects, Poland

ABSTRACT

Practical guidelines are presented for improved process for design and retrofitting of energy-efficient buildings, with an aim to integrate buildings better with the neighbourhood energy system, among others through energy matching. The chapter describes the role of energy simulations in an integrated building retrofitting process and how to select technologies for the retrofitting toward nearly zero energy building level. Feasibility of performing a holistic analysis of retrofitting options can be increased through the integration of BIM, well populated, and linked databases and a multi-criteria decision-making approach. Multiple-criteria decision-making methods aid taking into account a number of building energy performance and user-preference-related criteria and the trade-offs between the different criteria for each retrofitting option. The real-life viewpoints and benefits of utilising the developed methods and processes are discussed, especially from the Eastern European view.

DOI: 10.4018/978-1-5225-4105-9.ch007

INTRODUCTION

The building should be considered as a composition, whose functionality is affected by its various elements. The functionality of the building is largely dependent on well-functioning structures that also form a framework for building service systems. In the retrofitting process, replacing a single part, e.g. a window, often affects other structures and systems, so its design must involve understanding the factors affecting the entire building performance. The starting point for planning, also for planning of retrofitting actions, should be the prospective end result in terms of a good indoor climate and energy efficiency. Solutions for retrofitting are always case-specific, and they also include cost-effectiveness analyses.

The primary goal of retrofitting is to repair damaged or obsolete parts but the activity can also aim at improving functionality in all aspects. Since the service life of the repaired structure corresponds to a new one after repair, the target level of the correction needs to be set accordingly (Ojanen et al. 2017). According to Ojanen, requirements may also be other than only regulatory requirements, such as clients' wish to have a higher quality indoor climate and lower energy consumption than before. High quality solutions typically reduce energy consumption and increase living comfort, but the success of the action requires that property is considered as a whole (Ojanen et al. 2017). In this contexts, clients must require the designer, contractor and supervisor to have good expertise on retrofitting. Actions for improving of energy efficiency are undertaken almost exclusively at the same time when some construction or technical systems are repaired or renewed, instead of separate activities.

Choices made during the design of the retrofitting will affect for decades, e.g. the energy demand of the building and other maintenance costs. Better energy efficiency extends the lifespan of the building, improves user comfort and can affect the building's value as a property. An appropriately repaired building consumes less energy, which reduces operating costs. Design is encouraged in the direction of introducing long-term passive measures for saving future heating and cooling energy. When designing a retrofitting project, the order of the design should proceed, as far as possible, toward the minimization of energy demand. This includes, among others, actions to reduce heat losses and the need for cooling, to enhance the efficiency of electricity usage, utilize free (passive) energy sources, control the energy demand, and ultimately select appropriate energy sources (potentially produced on-site).

This chapter presents how these design aims can be achieved by using up to date methodological know-how and the new software tools efficiently employing energy simulations. The chapter is structured as follows. Background describes the current use of energy simulations in retrofitting, and how BIM fosters the paradigm change in the building design processes. Then, the main focus of the chapter is described, followed by describing the design science research methodology. An overview of the developed DESIGN4ENERGY methodology is given for designing energy efficient buildings with holistic approach. Next, developed holistic design guidelines are presented, including to-do lists for designers for the retrofitting phase. These are followed by showing, how energy simulations can be utilised in retrofitting, and how to select the retrofitting options. Finally, future research directions are discussed and conclusions summarise the main findings of the presented research.

BACKGROUND

Energy Simulations in Retrofitting

To improve the energy performance in buildings and achieve the European aims for the Energy Performance of Buildings Directive (EPBD) (European Commission, 2010) and Energy Efficiency Directive (EED) (European Commission, 2012) in retrofitting, a lot of research work has focused on developing appropriate methodologies. Examples include supporting energy audits to provide building energy performance certification for the evaluation of alternative retrofit options to identify the best solutions (European Commission, 2006; Balaras et al., 2004). The decision making process within the retrofitting context requires the incorporation of a number of criteria. The criteria can relate to the energy performance, cost optimality and occupants' well-being to ensure a thorough examination of the limitations and the advantages of the alternatives, informing the decisions reached. Multiple criteria techniques ensure that the trade-offs between the different criteria can be assessed in a quantitative manner. Previous work has focused on using multiple criteria methods for retrofitting (Gero, et al., 1983; Jaggs and Palmer, 2000; Kumbaroğlu and Madlener, 2012; Costa, et al., 2012; and Diakaki, et al., 2010).

One of the main limitations of the retrofit decision making approaches relate to the small number of options explored. Since a small number of retrofit alternatives is explored, decision making is limited and non-holistic. An exhaustive approach in selecting the retrofit alternatives would be more suitable to identify true optimal scenarios for the retrofitting. However, the exploration of a large number of options can be limited by technical barriers. Feasibility of performing a holistic analysis of retrofitting options can be increased through the integration of Building Information Modelling (BIM), well populated and linked databases and a multi-criteria semi-automated decision making approach. BIM increases the speed and the consistency of information exchange, improving the building representation and enabling more accurate modelling results. Currently, the building energy community is focusing on integrating BIM with energy simulation software. GreenBuildingStudio, DesignBuilder IES VE, Energy2 and NREL's OpenStudio are currently some of the BIM compatible energy software (GbXML.org, 2016). Although a number of interoperability issues have been reported (Steel, Drogemuller and Toth, 2012; Cemesova, Hopfe and Mcleod, 2015), using BIM for Building Energy Management (BEM) is a very promising approach for supporting decision making in retrofitting.

In addition to using energy simulation in retrofitting, there also other building and retrofitting evaluation tools exist, such as multi-criteria rating tools from BREAAM and LEED. BREAAM In-Use International is an assessment method for non-residential buildings, assisting property investors, owners and managers to increase sustainability by improving the operational efficiency, including continuing management of the building operations. (BREAAM, 2017). LEED has developed a rating system dedicated to building operations and maintenance issues, called LEED for Existing Buildings: Operations & Maintenance (LEED, 2014). Also many other building evaluation methodologies can be used, complementing each other, such as life cycle analysis (LCA) (as summarised e.g. by Vilches et al. (2017), and cumulative energy analysis. For example, Lee et al. (2015) reviewed 19 retrofit analysis toolkits.

BIM FOSTERS A PARADIGM CHANGE IN THE BUILDING DESIGN PROCESSES

Resent developments in Building Information Modelling (BIM) have enabled a paradigm change in building design processes. BIM can support stakeholder's collaboration at different stages of the building lifecycle, as BIM processes allow stakeholders to insert, extract, update and/or modify information (Alreshidi, Mourshed & Rezqui, 2017).

BIM has been developed and implemented to business processes as data models, modelling principles and solutions based on advanced information management and visualizations. BIM is commonly understood as collaborative delivery system having the use of information and open interoperability between the tools at the core. These key areas to be standardized are classification, data transfer formats and guidelines for the data-flow. BIM has been defined as "a set of interacting policies, processes and technologies generating a methodology to manage the essential building design and project data in digital format throughout the building's life-cycle" (Succar, 2009).

BIM based work has been supported by guidelines, in order to manage the inter-organizational collaboration and data-flow. In early days of BIM implementation, a roundtable session was held in Center for Integrated Facility Engineering (CIFE), with a focus to discuss the barriers to a broader BIM implementation, coordinating strategies to address these barriers, and the process of BIM implementation instead of BIM technology. The early adopters of BIM were "missing software applications that support a BIM approach, missing BIM trained personnel, and missing legal and insurance frameworks that support a collaborative project delivery framework". (Hartmann & Fischer, 2007).

BIM is seen as a larger phenomenon defined as Integrated Design and Delivery solutions (IDDS). IDDS is a systemic innovation that needs changes from multiple organizations and stakeholders to be realized (Taylor and Levitt, 2004). Previously developed collaborative BIM solutions have mainly focused on the technical dimensions and lack often socio-organizational, process, and legal aspects (Alreshidi, Mourshed & Rezqui, 2017).

The benefits of the integrated processes and interoperability of BIM tools are strongly depended on the maturity of BIM uses (BIM functions/BIM applications) and the skills of the stakeholders. Proactive management of all BIM elements is needed on project level in the areas of process, business, people and data, which all are interlinked with each other (Hyvärinen et al, 2013). Alreshidi, Mourshed & Rezqui (2017) conclude, that "achieving a fully integrated and collaborative BIM environment would require governing the collaboration process and data flow, underpinned by Cloud technologies". One of the BIM uses is energy simulations and decision making based on key performance indicators (KPIs).

COLLABORATION AND CO-CREATION, SOCIAL BIM

Many earlier studies emphasize the benefits of BIM for interaction and for collaboration. (Saini & Mhaske 2013; Eadie et al. 2013; Bassanino et al., 2013). One real life example for the need of co-design is a group of designers (e.g. an architect, HVAC designer and an energy expert) developing a solution for an atrium space with conflicting optimum designs from their own perspectives.

Interaction is often based on BIM model uses for merging models of the different disciplines or utilising integrated model for construction. Collaboration can happen in many ways. The visualization of the models helps stakeholders outside of the project team to understand the design solutions and reflect those better. Between design disciplines, the communication evolves towards co-creation as the situational awareness is gained easily with the BIM models. Chief designer can organise co-design sessions, which enable designers to do simultaneous design thinking for better performance. The designers can make informed decisions between different solution alternatives through visualisations of the analyses and simulation results. All these collaboration forms can be supported by information technology to guide the process toward the right way.

Knotworking methodology aims to understand those process phases where collaboration is beneficial (Kerosuo et al., 2013) and define a protocol for collaborative sessions. Groups of people, with relevant tasks and tools, are set to solve a problem or accomplish a task. Knotworking requires collaboration across organizational boundaries and hierarchies. In addition, successful collaboration needs leadership of the people and management of the BIM processes (Hyvärinen et al. 2013).

MAIN FOCUS OF THE CHAPTER

This chapter presents practical guidelines for the design and retrofitting of holistic and energy efficient buildings. The guidelines suggest, how to improve the current design process in order to integrate buildings with the neighbourhood energy system, while considering both foreseeable and unknown future changes. Practical guidance is given through To-do lists for building designers for the retrofitting phase. The presented guidelines are based on a developed DESIGN4ENERGY methodology, which enables studying the building energy performance in a holistic way both at the building and neighbourhood levels throughout the whole building life cycle.

This chapter also shows how energy simulations can be utilised in the building design, and how to select technologies for the building retrofitting toward nearly zero energy building (NZEB) level. Also practical viewpoints of utilising the developed guidelines and tools are discussed. Examples focus on retrofitting activities.

This research was conducted as a part of a large research and innovation project called DESIGN4ENERGY, funded by European Commission. The project focused on developing an integrated and evolutionary design methodology, which targets to highly energy efficient building energy performance throughout the whole life cycle, from planning and design to operation, maintenance and retrofitting.

RESEARCH METHOD

The research presented in this paper followed Design science research methodology (DSRM) (Geerts, 2011). The DSRM is popular among others in engineering and architecture disciplines, as it focuses on finding out how things should be in order to achieve specific goals. DSRM has two important characteristics: 1) relevance (solving of an important problem), and 2) novelty (either addressing an unsolved problem in a unique and innovative way, or solving a problem in a more efficient way). The research steps are structured according to the DSRM approach in Table 1. (Geerts, 2011)

Table 1. Research steps structured according to design science research methodology

DSRM Activities	Description of the Research Activity
Problem identification	Increasing building energy performance requirements need more efficient ways for designing the building and its energy systems.
Define the objectives of a solution	• Improving the energy efficient building design process in a holistic way • Needs to take into account both the building and neighbourhood levels throughout the whole building life cycle • Need to support collaborative working within the building design team and the client • Need to be adaptable to design approaches and methods
Design and development	• Defining of a DESIGN4ENERGY methodology, with adapting three design process approaches: Integrated design and delivery solutions (IDDS), Performance based design (PBD) and Collaborative information service platform. Also methodologies used in the software development have been utilized (waterfall, prototyping and spiral development). • Developing of an improved a building design process, including: • integrated modelling process (BIM) • performance based design • energy simulations, analyses and visualizations • collaborative working environment and related tools • guidelines for the design team members • Developing of holistic design guidelines
Demonstration	• Practical viewpoints • Demonstrating the use of the parts of the improved design process steps (Bassanino M., et al., 2016)
Evaluation	Discussions with the real life users, e.g. architects
Communication	Sharing the design process improvement steps and guidelines etc. among the relevant stakeholders.

DESIGN4ENERGY METHODOLOGY FOR HOLISTIC ENERGY EFFICIENT BUILDING DESIGN

Methodology is defined as a set of methods, process, tools and principals for holistic and energy efficient (EE) design in a project deliverable (Mäkeläinen T., et al., 2016). The developed DESIGN4ENEGY methodology adapts three design process approaches: Integrated design and delivery solutions (IDDS), Performance based design (PBD) and Collaborative information service platform (Mäkeläinen T., et al., 2016; Bassanino M., et al., 2016).

DESIGN4ENERGY building design methodology enables studying the building energy performance in a holistic way both at the building and neighbourhood levels throughout the whole building life cycle. One of the key objectives was to take into account both building and neighbourhood levels, which is done by holistic building energy performance design, including energy matching with neighbourhood energy systems. The key building blocks of the developed design methodology are (Mäkeläinen T., et al., 2016):

- Holistic
- Evolutionary
- Energy efficient oriented
- Adaptable to design approaches and methods
- Serving the decision making process with information, using following key processes:
    - Integrated modelling process (BIM) with EE building simulations and analyses and advanced visualisation,
    - Performance based design process, and

- Collaborative design process (enabled by a virtual workspace, which contains all the tools and the decision support systems, as well as data management, library and visualisation).

The principles of D4E methodology are defined in concrete ways to support design managers and chief architects to apply the methodology in real-life design assignments. The design group searches in collaboration a balanced design solution, which fulfils the targeted key performance indicator (KPI) values. A technical platform supports designers by providing EE building design services. The process maps guide the design team to repeat the design activities to produce models for LOD (Level of model detail) 0 alternatives or for LOD1 and LOD2 models. As the design process is iterative in its nature, it is important to allocate extra time for iterations and pinpoint key iterations for the whole process flow. An intensive collaboration often increases the needed iteration, due to the increasing amount of perspectives, viewpoints and comments from the stakeholders involved. The design process flow has dependences between each domain, and any change in one domain will affect to the others, increasing the needed time. The details of the developed methodology are presented in project deliverable (Mäkeläinen T., et al., 2016).

Using of key performance indicators is one of the central approaches in the DESIGN4ENERGY methodology, because it mitigates problems raising later between conflicting alternatives from different design disciplines (Mäkeläinen T., et al., 2016). An example of this kind of design problem is designing of an atrium space, if an architect would like to have a very spatial space, which would not be effective from the energy efficiency viewpoint. Therefore, all decision making is based on the assessed KPIs, which are set and weighted with the client at the briefing stage. Next, a feasibility study is conducted, aiming to prove if the set target levels are achievable with reasonable design solutions. All these activities aim for developing energy efficient design and retrofitting of a building (Mäkeläinen T., et al., 2016).

The DESIGN4ENERGY methodology relies on common data environment, developed as the collaborative information service platform (Bassanino M., et al., 2016). It supports the process oriented design management, design team collaboration, and information sharing with an automated data exchanges when possible and beneficial. The platform provides collaborative working space for the design team and supports the information management during the process flow of the project.

The process phases of the DESIGN4ENERGY methodology and guidelines have been identified, including activities supported by the platform (main tasks) and other specific activities. Specific activities are options, which can be included to the process based on the characteristics of the project. The process phases are (Mäkeläinen T., et al., 2016):

Phase One: Project Initialisation.

- **Main Task:** Project configuration.
- **Specific Activities:** Define BIM project plan and Model Uses, Visualisation of project objectives.

Phase Two: Need Identification.

- **Main Tasks:** Target setting, Search benchmark data, Collect boundary condition, Feasibility study.
- **Specific Activities:** Define design brief with assessment profile (strategic KPIs), Initial data model, Visualisation of boundary condition (multi-window).

Phase Three: Concept Design (Design and Simulation).

- **Main Tasks:** Explore Technology options, Initial design and Energy simulation, Design Review and select option, Detailed energy performance simulation.
- **Specific Activities:** Evolutionary assessment (Future what-if scenarios).

Phase Four: Concept Design (Energy matching).

- **Main Tasks:** Energy matching, Refine target setting (targets for detailed design), Improve design, Design review.
- **Specific Activities:** Defining brief for detailed design.

Phase Five: Detailed Design.

- **Main Tasks:** Integrated design and simulation, Refine energy matching, Refine target setting (targets for final design), Design review.
- **Specific Activities:** Defining brief for final design.

Phase Six: Final Design.

- **Main Tasks:** Integrated design and simulation, Refine Energy matching, Refine target setting (operational targets), Design review.
- **Specific Activities:** Collecting information for maintenance manual.

Phase Seven: Operational needs.

- **Main Tasks:** Analysis of operational data, Target setting, Benchmark analyses, Generate alternative.
- **Specific Activities:** Facility operational model.

Phase Eight: Retrofit and maintenance design.

- **Main Tasks:** Explore technology options, Initial design and Energy simulation, Design review and select option, Detailed energy performance simulation, Energy matching, Refine target setting, Improve design, Design review.
- **Specific Activities:** Maintenance model.

Design solutions for improving energy efficient through active and passive design have also been identified. Typically, passive design solutions improving energy efficiency are dependent on choices made by architects and structural engineers, while active design solutions improving energy efficiency are done by HVAC designers and energy designers. Passive design solutions are mostly considered in the concept design phase, and active EE building improvement design solutions starting from the concept design and continuing in other design phases. These active and passive design solutions are presented in more detail by Mäkeläinen et al., (2017).

HOLISTIC DESIGN GUIDELINES

The design group develops in collaboration a balanced design solution, which fulfils the targeted KPIs. In practise, this requires iterative design process. An intensive collaboration normally increases the needed iteration, when the amount of perspectives, viewpoints and comments increases gradually with the increasing number of stakeholders involved. The design process flow has dependences between each domain, and any change in the one domain will affect to the others, which increases the needed time. Therefore, it is important to allocate extra time for iterations, and also to identify the key iterations in the whole process flow. The process maps guide the design team to repeat the design activities, and to produce models for LOD (Level of model detail) 0 alternatives or for LOD1 and LOD 2 models. Setting targets for KPIs enable following the impacts of choices throughout the design process (Mäkeläinen et al., 2017).

The developed Design Use cases (UC) for designers are (Mäkeläinen et al., 2017):

Use Case 1.0: Client and key designers identifies needs and targets for the project.
Use Case 1.0.1: Integrated collaborative review in target setting phase.
Use Case 2.1: Designers to consider energy issues at concept design stage.
Use Case 2.2: Designers to consider energy matching at concept design stage.
Use Case 2.2.1: Integrated collaborative design review in concept design stage.
Use Case 3.0: Designers to consider holistic EE building design in detailed design stage.
Use Case 3.0.1: Integrated collaborative design review in detailed design stage.
Use Case 4.0: Designers to consider holistic EE building design in final design stage.
Use Case 4.0.1: Integrated collaborative design review in final design stage.
Use Case 5.0: Defining the need for maintenance/retrofit in retrofit briefing phase.
Use Case 5.1: Retrofit/maintenance modelling in retrofit design phase.

To support the design group in energy efficient design approach, To-do lists have been developed for each specific DESING4ENERGY use case listed above. As this chapter focuses on retrofitting, To-do lists for designers for the retrofitting phase are presented in Table 2 for the use cases 5.0 and Table 3 for the use case 5.1 in the above list.

OVERALL DESIGN ADVISORY

Collaborative design aims to identify and solve design problems affecting other designers' work. In overall, building energy performance and sustainability are composed of many design decisions (Mäkeläinen et al., 2017). When targeting to the best possible building energy design, holistic energy solutions are required. Primarily holistic energy design should minimise the energy demand, without compromising the good indoor environmental quality. Secondarily, it should include optimal use of on-site renewable energy sources, as well as minimising carbon emissions while maintaining profitability of the energy supply (Mäkeläinen et al., 2017).

The developed steps for the design thinking are presented in Mäkeläinen et al., (2017). At first, the target values need to be set for the energy demand, on-site renewable energy use, carbon emissions and energy costs in the target setting phase. Then, accordingly, the potential and available energy design options are identified, and the energy solutions concept for the building is drafted in the concept design phase. In practise, this includes firstly minimising the energy demand with passive design solutions and giving priority to local and on-site renewable energy sources, such as solar energy, heat pumps, etc.). Profitability of the energy solution should be considered against the targets, and planning on how to manage the energy demand and the selling of excess energy produced on-site (if available). Often, the priority goal is to minimise the energy costs, and in this, it is expected that in the future demand response and matching of energy demand and supply need to be part of the building energy solution. When making detailed design decisions at the detailed design phase, the overall energy concept should be kept in mind. The overall design advisory is completed with checklists. The design thinking steps considered are listed in Table 4 (Mäkeläinen et al., 2017).

Table 2. To-do list for designers in use case 5.0: Defining the need for maintenance/retrofit

ID	Use Case 5.0: Defining the Need for Maintenance/Retrofit
Phase	Retrofit Design
Phase consists of	• Collecting initial data of the building consisting of monitored EE building performance data. • Analyses of retrofit concept and making a decision between: 1) retrofit, 2) extra maintenance, or 3) change of behaviour/change of purpose of use supported by an analysis tool. • Designing retrofit alternative based on client's needs and performance target set with help of conducted analyses.
Description	• In this use-case, based on calculated and monitored energy efficiency data of existing building, the experts (representing the domains of facility manager, client, architect and energy expert) analyses potential retrofit concept solutions based on needs. • The client prepares a list with objectives and performance targets for the design group to start preliminary retrofit concept design. • Based on the list the facility manager suggests and architect sketches potential architectural alternatives for the retrofit to be assessed and reviewed. • Outcome from review is client decision on which design concept alternative to proceed with.
Primary Actors	Facility Manager, Client, Architect
Pre-conditions	• Monitored and pre-analysed energy performance data available • Geometrical initial data model is available • Design group with competence in modelling, BIM based processes and performance based design.
Post-conditions	• Generation of operation and maintenance data for deep analyse • Design alternatives validated by KPIs for next steps of retrofit design >> hard gate
To-do list for facility manager	• Collect monitored data about the building operation • Study Benchmark data • Run Preliminary data analyses • Prepare the specification of requirement for facility manager • Set project objectives (level of energy savings etc.) and performance targets • Create and describe potential retrofit alternatives from maintainability perspective • Run analyses of retrofit concept: retrofit, extra maintenance, change of behaviour (change of purpose of use) • Prepare the report, present the results
To-do list for client	• Review the report from facility manager • Specification of requirements (KPIs and their target levels) • Present the brief for architect and other designers
To-do list for architect	• Review the design brief and report from facility manager. • Review FM model (as monitored and inventory model (as inspected) and other information. • Update the BIM model (as–designed). • Create and describe retrofit alternatives based on usability of spaces and technical improvements. • Analyse maintainability together with facility manager. • Create potential EE building solutions for each alternative. • Produce a report and upload alternatives for HVAC designer for energy simulations and for energy expert for energy matching analysis. - wait for results from holistic energy matching analysis. • Review Energy experts report. • Assess KPIs for each alternative. • Perform the alternative selection (matching KPI levels with target values). • Prepare the selected alternative for further design or actions.
To-do list for energy expert	• Review reports of Architects and Facility manager. • Check neighbourhood implications and potentialities for each drafted retrofit alternatives and EE building solution. • Analyse holistic energy matching for each alternative. • Produce report for Architect to assess KPIs.

Mäkeläinen et al., 2017.

Table 3. To-do list for designers in use case 5.1: Retrofit/maintenance modelling

ID	Use Case 5.1: Retrofit/Maintenance Modelling
Phase	Retrofit Design
Phase consists of	Design with developing BIM model from LOD2 to LOD4 with help of analyses and collaborative reviews and final performance assessment to support client's decision making.
Description	• In this use-case, based on preliminary concept design alternatives the client produces design brief with KPIs. • Design work begins with concept design models (LOD2) towards more detailed design (LOD4). Simulations and analyses are conducted based on BIM plan, especially analyses of neighbourhood implications in energy supply, adaptability analyses and choosing building material for retrofit/maintenance and analysing energy demand and energy matching. • Design with developing BIM model from LOD2 to LOD4 with help of analyses and collaborative reviews and final performance assessment to support client's decision making.
Primary Actors	Chief designer, Architect, FM, Energy expert, HVAC, Structural designer, Client, User
Pre-conditions	• The selected design alternative is agreed for further retrofit design. • Comments from FM are available. • Architectural concept model is available. • KPI assessment reports of the selected alternative available. • Report of the made decisions and their arguments is available.
Post-conditions	Final solution of the retrofit design combining holistic neighbourhood level energy matching and low energy demand ready and agreed. >> **hard gate**
To-do list for the client	• Produce retrofit or maintenance brief. Define KPIs • Upload the brief to D4E portal. • Studying assessment results of the concept alternatives and choosing one alternative for detailed design. • Studying assessment repost of the detailed design • Decision making for the final alternative
To-do list for architects or chief designer	• Access as designed (LOD2) concept models from previous phase. Search from D4E project information platform. • Develop the architectural design using D4E component library and using performance indicators as search criteria • Review neighbourhood implications • Review adaptability (evolutionary aspect) • Improve design (LOD2) • Prepare concept design alternatives for client review • Prepare the BIM models for coordination • Facilitate clash checking review using preferred indicators of each design discipline. • Compile the KPIs from the analyses results according to the client's preferences and select the best holistic EE building design solution. • Review (design group internal) the KPIs agree with the proposed solution to the client. Use the checklists to facilitate the review. • Prepare the presentation to client, including results from analyses and final KPI assessment. • Presentation to client and client's approval on final design models to be used for procurement • Prepare detailed design documentations and the integrated model for call for tenders.
To-do list for the design team	• Develop HVAC, Structural and electricity model (and other design models) using D4E component library. • Prepare all design models for a quantity take off and detailed cost calculation and CO_2 analyses • Prepare analytical model for energy analyses and for energy matching • Perform the analyses and generate reports from analyses results. • NOTE: Each designer has a design responsibility of solutions covering their own substance area. Other designers do not have competence nor authority to creating solutions on another design area.

ENERGY SIMULATIONS IN RETROFITTING

For successful energy efficient retrofitting one needs to take into account criteria such as the cost of the retrofit technologies, the thermal comfort for the building's occupants, current policies and regulations, building specific information and the human factor. According to Ma et al. (2012) there are five phases

Table 4. Energy efficiency related design thinking steps, based on the performance based design principles

Performance Based Design Principles	
Target setting	Setting of target values for the energy demand, renewable energy use, carbon emissions, and energy costs based on a feasibility study. Retrofit design: before target setting: run the energy study of the existing building with monitored data about energy performance
Design management	Specify and document technical specification, which fulfils strategic KPIs.
Energy Efficiency Design Advisory, Steps 1 and 2A-B-C	
1. Minimize energy demand	• **Concept Design:** Minimise energy demand with passive EE building design solutions • **Detailed Design:** Minimise energy demand with active EE building design solutions
2A. Optimize the distribution of renewable energy sources fulfilling energy demand	• **Concept Design:** Give priority to solar thermal production (prosumers) and PV (if there is surplus), when design energy mix. Check with Spatial design solution and facade concept. • **Detailed Design:** Select and size equipment and assess solution against KPIs: OER (on site energy ratio) and cost.
2B. Minimize carbon emissions	• **Concept Design:** Consider district heating with generation by renewables (biomass, pellets, etc.) Give priority to solar thermal and PV. • **Detailed Design:** e.g. material choices
2C. Energy costs, energy management, excess energy selling	• **Concept Design:** Optimize the use of RES potential • **Detailed Design:** Minimize the cost of heating and DHW by optimizing (matching) the operation of the entire neighbourhood. An example of a business model principle: Energy matching is usually profitable if energy can be produced and used in same system without selling it to the energy operator of municipality (when feed-in tariff not used).

Mäkeläinen et al., 2017.

for successful implementation of a retrofit process, from a pre-retrofit survey to identify the operational needs for the specific building to a post-occupancy evaluation and assessment of the energy savings from the implemented retrofit solution. To ensure that the retrofitting targets can be met it is important to be able to assess the potential impact of the proposed retrofit solutions very early on and as accurately as possible at the design stage through building energy performance simulation.

To be able to perform accurate building energy analysis, consistency across different building related disciplines can be enabled using Building Information Modelling (BIM). BIM enables a collaborative way of working and can be used to better inform decision-making processes from the early design stage through to the building operation and maintenance stage. Automated and rapid analyses of the low energy strategies of buildings are enabled through BIM (Hyun, Marjanovi-Halburd and Raslan, 2015). To use BIM to perform energy analysis to inform retrofit decision making, the he first step is the BIM model development (e.g. Autodesk's REVIT). Software enabling the use of BIM to proceed with energy analysis (e.g. OpenStudio and DesignBuilder) can be used to facilitate the data transfer process. However, there are limitations related to the transparency and accuracy of the data exchange, with missing data frequently being reported and resulting in inadequate building representation. To cope with some of the missing data a non-energy expert user relies on default settings regarding thermal characteristics of the building envelope, building systems and the occupants' behaviour. Limitations in the data transfer process result in interoperability challenges between BIM and BEM (Miller et al., 2014). A number of interoperability issues between BIM and energy analysis tools have been identified in previous research work (Steel, Drogemuller and Toth, 2012; Cemesova, Hopfe and Mcleod, 2015). Transparency of data exchange at each step is essential in overcoming these limitations. Such examples exist and include a semi and fully-automated interface for energy analysis, enabling a varying degree of user involvement in the selection of the energy analysis parameter values (Ahn et al., 2014).

The main prerequisite for the retrofit analysis using modelling is the development of a building model, which is representative of the real-life building. This model can then be used to implement and assess different retrofit scenarios. To achieve a representative model, the use of actual measurements from the real-life building to inform the model development could help to increase the accuracy of the results reported and eventually of the predicted savings from the implementation of retrofit changes. An automated process to obtain a number of retrofit alternatives and be able to compare their impact on the building's thermal performance whilst taking into consideration the client's needs can help to achieve a more exhaustive search to identify an appropriate retrofit option. At the same time decision making support should enable the user to explore the available options and the results obtained from each alternative to ensure that the selected retrofit action is appealing to the client on a variety of criteria (see Figure 1).

RETROFITTING DESIGN: SELECTING RETROFITTING OPTIONS

There is large variety of technologies available for retrofitting different types of building, which include the building construction components, the HVAC systems and renewable energy technologies. Figure 2 shows how the end-user (the energy expert, engineer or architect) can use a BIM model and input from a technology database to choose the components to be used for retrofitting and to generate retrofit alternatives of all possible combinations of the selected components.

Figure 1. Current practice from BIM to BEM for the retrofit scenario and the two key improvements when using the proposed methodology. Autodesk's REVIT and EnergyPlus are used as example BIM and BEM software respectively.

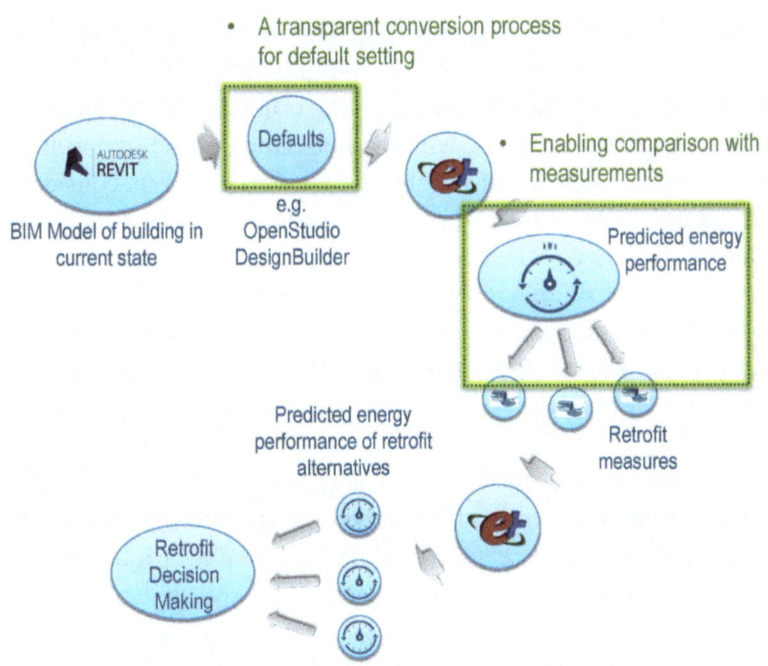

Figure 2. Proposed workflow of data model for energy efficiency in existing buildings

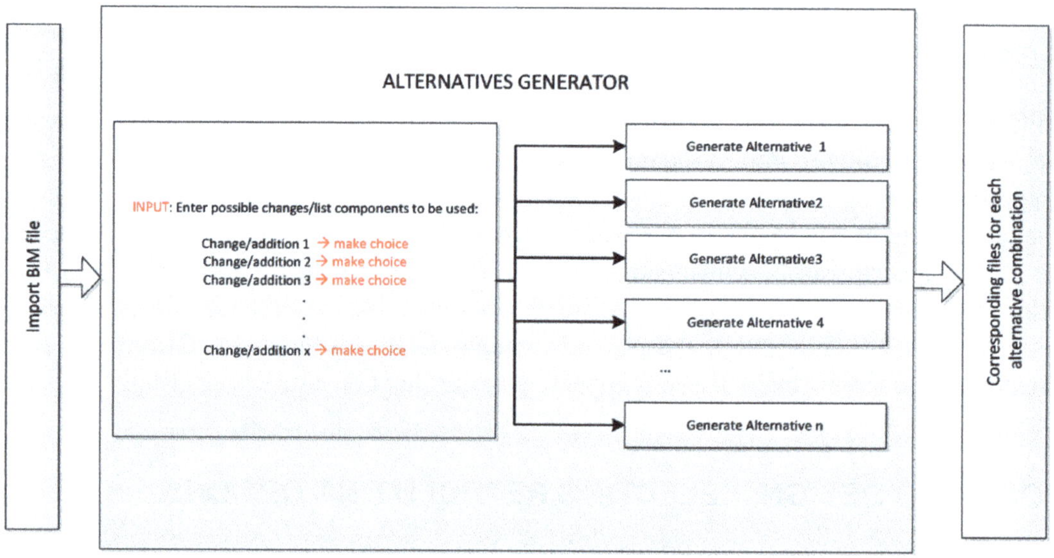

The selection of appropriate retrofit alternative actions relies on the availability of energy analysis results and requires the integration of user needs considerations. The task of identifying appropriate retrofit actions from a virtually infinite number of available designs requires a sophisticated approach. The requirement to take into account a number of building energy performance and user preference related criteria suggests that Multiple Criteria Decision Making (MCDM) methods are needed. MCDM uses multiple indicators to calculate a single meaningful index that reflects how well each retrofit alternative performs for the criteria under consideration. The criteria can relate to energy performance, cost, CO_2 emissions, thermal comfort and indoor air quality. The resulting indexes are directly comparable and the best performing retrofit options are easily identifiable. The user's preferences are accounted for in the calculation process of the indexes using user-defined weights for each of the criteria for selection. The weighting values are used to specify the importance of the criteria. The higher the weight assigned to a criterion the more important the criterion is, with the magnitude of the importance being signified by the relative difference between the different weights across all the criteria considered.

Figure 3 shows an example of the architecture of a decision support tool using energy analysis results to inform decision making processes. The energy analysis results for all retrofit alternatives generated by an options generator are the first input to the tool. Benchmark values for the corresponding building type and location, available through a comprehensive benchmark database, can be used to filter out some of the retrofit alternatives that deviate significantly from the best-case building performance scenarios. The user-defined weights reflecting the user preferences as explained previously serve as additional inputs to the decision making tool. The core of the tool uses a MCDM approach to rank the options based on the selected performance criteria. The main output of such a tool is the ranking of the available retrofit options. To allow for further investigation of the retrofit options available, the Key Performance Indicators (KPIs) for each of the criteria can be used. A graphical user interface can be used to enable the visualisation of the KPI results with appropriate schematic representations and metrics to facilitate more in-depth analyses of the highest ranked, best performing retrofit alternatives.

Figure 3. Example of system architecture for retrofit decision making

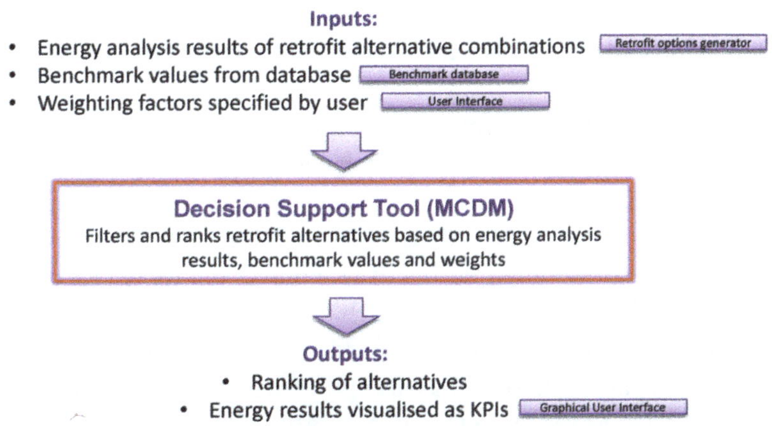

PRACTICAL VIEWPOINT TO THE RETROFITTING DESIGN AND DEVELOPED TOOLS

The idea of an innovative approach to building design considering energy efficiency is being constantly put to test during day-to-day practice at every architectural office. While popular and widely discussed in the mass media, the "green design" is usually not precisely described technically nor economically, until it becomes a postulated way of proceeding in real case scenario – the real architectural project.

In the current harsh economy times, the everyday architectural practice does not encourage designers to compete much in the field of energy efficiency. This is especially true in the East European countries, e.g. in Poland. While being a rising economy, with a rapid growth of GDP, this does not directly convert into much interest among the investors to take on the ambitious projects, designed with energy efficiency in mind. There are very few examples of building designs attempting to implement design standards or tools for energy optimisation in the local Polish market. Some rare cases of individual houses or small scale office buildings being designed and build by "environmental aficionados", but at present, this is regarded a niche market in Poland. The obvious reason is the popular consideration that innovative technologies for energy efficiency are very expensive and not profitable.

Architects involved in the process of energy efficient design are responsible to optimize their projects in order to meet the design criteria they were confronted with, set by their clients. Nowadays, most of the clients are interested in economically savvy designs that save their wallets rather than world's climate – this is a very serious argument that designers should keep in mind. And while they may not be able to prove that paying more for energy efficiency makes the building cheaper to build, they should at least try to give their clients a chance to understand both the opportunities and trade-offs of investing in innovative solutions.

There are numerous ways of preparing the economic analysis of the investment, starting from the basic tools and price catalogues up to very advanced methods of financial assessments of the building. The major characteristic of those traditional methods, however, is that they come in play at the end of design – when there is usually too late for changing major cost factors. This is where such novel ideas like DESIGN4ENERGY show their general strength – the ability to aid the architects and investors in the very beginning of the design process.

During first phases of design, the architects and investors hardly discuss exact figures and prices of detailed solutions. At this phase, the very general strategic decisions are usually taken, e.g. overall shape and mass of the building, general list of materials and technologies to be used, the size and the number of windows and openings, etc.

The decision making process should involve the comparison of different design solutions and their economic consequences. Depending on the desired level of accuracy, such preliminary economic analysis may or may not consider exact cost information. The benefit of innovative design approach of DESIGN4ENERGY and alike is that such tools allow for working out the statistical correspondences between several crucial factors of building design. The comparison tools based on building database gathered by DESIGN4ENERGY, are able to calculate basic dependencies between major building elements and investment options. Another benefit embedded in such novelty tools is utilizing the BIM technology. At every level of analysis, the user has the option to link his real design with the calculations, e.g. by uploading the detailed BIM model to DESIGN4ENERGY platform.

FUTURE RESEARCH DIRECTIONS

This chapter presented the developed high level guidelines. As a next step for these would be to develop detailed guidelines for different stakeholders, as well as manuals for related tools. The development of practical guidelines raised some ideas for re-engineering the energy efficient building design process. Among others, it could be beneficial if the feasibility study would contain energy simulations and the definitions of neighbourhood energy matching concept and related KPIs, such as on-site energy ratio) both at the building and neighbourhood level, e.g. as a part of the urban (energy) plan. The building design and energy simulations could reflect accordingly the neighbourhood level optimal energy concept and its energy matching targets. And as a continuation for this development, the feasibility study could be brought to a part of the briefing phase. In practice this would mean, that a new pre-concept phase would be added to the design process, and there energy designers would define the main elements of the building energy concept, possibility together with the architect. Currently, there is no phase before the concept design (making of architectural model), and therefore there is a need to find the most potential energy solution by running several energy simulations to study the energy performance and its KPIs, e.g. onsite energy ratio.

BIM based working methods, new technology and methodologies can be implemented in retrofit projects. Utilizing of the full potential of the BIM based integrated work would require improved BIM working processes, model uses, and data-flow processes. There are also areas of new competency demands. The needed skills could be trained in real life building projects in a form of piloting, with active support of the clients, especially public bodies. With action research methods, the lessons learned in piloting cases can be documented and disseminated to reinforce the adaptation of BIM. BIM strategy and guidelines, as well as definitions of the use cases to be tested, is at the core of piloting, with the assessment of the success of the piloting, in order to ensure the scalability. One further development area is to define detailed manuals for the use of BIM tools in collaborative process, and BIM guidelines for retrofitting, with viewpoints of each stakeholder. Furthermore, for the retrofitting cases, there is a need for more detailed studies in large pilot projects, to confirm that the technologies are solid.

Organizational theories and more specific methodologies can help to understand the inter-organisational relationship on project level business based on BIM collaboration (Mignone et al., 2016). The issue has been pointed out in the early socio-technical research and development of BIM, also emphasizing a need to development contractual agreements as a part of the procurement (Kazi & Koivuniemi, 2006; Miettinen et al., 2013).

CONCLUSION

The work presented here has its roots in the development of guidelines for holistic design of optimal energy efficient buildings matching with neighbourhood energy systems. This means that the energy demand and on-site energy production at the building are integrated to the local energy systems in an optimal way. In general, the processes are established for designing of buildings or design of the district (as part of city planning). During the design integration and solution iteration, it is important to have the neighbourhood aspect clearly in mind. The same principles apply also to the retrofitting activities addressed in this chapter.

The principles of DESIGN4ENERGY methodology are defined in concrete ways, which can support design managers and chief architects to apply the methodology in real-life design assignments. As the main principle, the design solutions are based on studies of the local operation environment (local boundary elements) and alternatives have been analysed taken into account neighbourhood energy efficiency potentialities, especially the use of renewable energy sources in the energy supply mix. Holistic energy efficient design means design in connection to building's surrounding and the local energy systems in district.

Feasibility of performing a holistic analysis of retrofitting options can be increased through the integration of BIM, well populated and linked databases and a multi-criteria semi-automated decision making approach. Multiple Criteria Decision Making (MCDM) methods are needed to take into account a number of building energy performance and user preference related criteria and the trade-offs between the different criteria for each alternative retrofitting option.

In the real life projects, and especially in the East European countries, innovative technologies for improving energy efficiency are typically regarded as very expensive and not profitable. Designers have a challenge to show the benefits and trade-offs of investing in innovative energy efficient solutions. Architects and investors are aided to consider the novel ideas at the very beginning of the design process through the DESIGN4ENERGY methodology with KPI assessments, guidelines and related tools. The benefit of innovative design approach of DESIGN4ENERGY, and alike developments, is that such tools allow for working out the statistical correspondences between several crucial factors of building design and support comparing of basic dependencies between major building elements and investment options.

ACKNOWLEDGMENT

This research is funded by the European Commission under contract FP7-2013-NMP-ENV-EeB through the Design4Energy project (Grant agreement no: 609380). All project partners are acknowledged.

REFERENCES

Ahn, K. U., Kim, Y. J., Park, C. S., Kim, I., & Lee, K. (2014). BIM interface for full vs. semi-automated building energy simulation. *Energy and Building*, *68*, 671–678. doi:10.1016/j.enbuild.2013.08.063

Alreshidi, E., Mourshed, M., & Rezgui, Y. (2017). Factors for effective BIM governance. *Journal of Building Engineering*, *10*, 89–101. doi:10.1016/j.jobe.2017.02.006

Balaras, C.A. (2004). Decision support software for sustainable building refurbishment. *Transactions-American Society of Heating Refrigeration and Air Conditioning Engineers, 110*(1), 592-601.

Bassanino, M., Fernando, T., & Wu, K.-C. (2013). Can virtual workspaces enhance team communication and collaboration in design review meetings? *Architectural Engineering and Design Management*, *10*(3-4), 200–217. doi:10.1080/17452007.2013.775102

Bassanino, M., Fernando, T., Wu, K.-C., Ghazimirsaeid, G., Klobut, K., Mäkeläinen, T., & Hukkalainen, M. (2016). A collaborative environment for energy-efficient buildings within the context of a neighborhood. In *Proceedings, eWork and eBusiness in Architecture, Engineering and Construction*. CRC Press.

BREAAM In-Use International. (2017). Retrieved 22.08.2017 from http://www.breeam.com/in-use

Cemesova, A., Hopfe, C. J., & Mcleod, R. S. (2015). PassivBIM: Enhancing interoperability between BIM and low energy design software. *Automation in Construction*, *57*, 17–32. doi:10.1016/j.autcon.2015.04.014

Costa, A., Keane, M. M., Raftery, P., & O'Donnell, J. (2012). Key factors methodology — A novel support to the decision making process of the building energy manager in defining optimal operation strategies. *Energy and Building*, *49*, 158–163. doi:10.1016/j.enbuild.2012.02.001

Diakaki, C., Grigoroudis, E., Kabelis, N., Kolokotsa, D., Kalaitzakis, K., & Stavrakakis, G. (2010). A multi-objective decision model for the improvement of energy efficiency in buildings. *Energy*, *35*(12), 5483–5496. doi:10.1016/j.energy.2010.05.012

Eadie, R., Browne, M., Odeyinka, H., McKeown, C., & McNiff, S. (2013). BIM implementation throughout the UK construction project lifecycle: An analysis. *Automation in Construction*, *36*, 145–151. doi:10.1016/j.autcon.2013.09.001

European Commission. (2006). *EC Directorate general for energy and transport. Intelligence energy – Europe Programme. Multiplying success in buildings: 21 innovative projects for an energy-intelligent Europe. Energy performance assessment for existing non-residential buildings*. Brussels: EPA-NR.

European Commission. (2010). *Directive 2010/31/EU of the European Parliament and of the Council of 19 May 2010 on the energy performance of buildings*. Retrieved on 30th May 2017 at http://eur-lex.europa.eu/legal-content/EN/ALL/?uri=CELEX:32010L0031

European Commission. (2012). *Directive 2012/27/EU of the European Parliament and of the Council of 25 October 2012 on energy efficiency*. Retrieved on 30th May 2017 at: http://eur-lex.europa.eu/legal-content/EN/TXT/?qid=1399375464230&uri=CELEX%3A32012L0027

GbXML.org. (2016). *Software*. Retrieved on 16th March 2016 at http://gbxml.org/software.php

Geerts, G. (2011). A design science research methodology and its application to accounting information systems research. *International Journal of Accounting Information Systems, 12*(2), 142–151. doi:10.1016/j.accinf.2011.02.004

Gero, J. S., D'Cruz, N., & Radford, A. D. (1983). Energy in context: A multicriteria model for building design. *Building and Environment, 18*(3), 99–107. doi:10.1016/0360-1323(83)90001-X

Hartmann, T., & Fischer, M. (2007). Applications of BIM and Hurdles for Widespread Adoption of BIM Event Report. AISC-ACCL eConstruction Roundtable, New Orleans, LA.

Hyun, S., Marjanovic-Halburd, L., & Raslan, R. (2015). Investigation Into Informational Compatibility Of Building Information Modelling And Building Performance Analysis Software Solutions. *WIT Transactions on the Built Environment, 149*, 543–553. doi:10.2495/BIM150441

Hyvärinen, J., Mäkeläinen, T., Peura, J., & Rönty, J. (2013). Strategies, Guidelines and Project Level Leadership as Methods for IDDS Practices in Transition. *Proceedings of the 19th CIB World Building Congress 2013: Construction and Society.*

Jaggs, M., & Palmer, J. (2000). Energy performance indoor environmental quality retrofit—a European diagnosis and decision making method for building refurbishment. *Energy and Buildings, 31*(2), 97-101.

Kazi, S., & Koivuniemi, A. (2006). *Sharing through Social Interaction: The Case of YIT Construction Ltd. Real-Life Knowledge Management, Lessons from the Field.* KnowledgeBoard in collaboration with VTT – Technical Research Centre of Finland.

Kerosuo, H., Mäki, T., & Korpela, J. (2013). Knotworking - A novel BIM-based collaboration practice in building design projects. *Proceedings of the 5th International Conference on Construction Engineering and Project Management ICCEPM*, 9-11.

Kumbaroğlu, G., & Madlener, R. (2012). Evaluation of economically optimal retrofit investment options for energy savings in build. *Energy and Building, 49*, 327–334. doi:10.1016/j.enbuild.2012.02.022

Lee, S. H., Hong, T., Piette, M. A., & Taylor-Lange, S. C. (2015). Energy retrofit analysis toolkits for commercial buildings: A review. *Energy, 89*, 1087-1100.

LEED. (2014). *LEED 2009 for Existing Buildings Operations and Maintenance*. USGBC U.S. Green Building Council. Retrieved on 22.08.2017 at: https://www.usgbc.org/sites/default/files/LEED%20 2009%20RS_EBOM_04.02.2014_current.pdf

Ma, Z., Cooper, P., Daly, D., & Ledo, L. (2012). Existing building retrofits: Methodology and state-of-the-art. *Energy and Building, 55*(0), 889–902. doi:10.1016/j.enbuild.2012.08.018

Mäkeläinen, T. et al. (2016). *Deliverable 2.2a Visionary scenarios of holistic energy efficiency design.* Design4Energy project.

Mäkeläinen, T. et al. (2017). *Deliverable 2.3: Guidelines for optimal building design: Holistic design of buildings matching with neighbourhood energy systems.* Design4Energy project.

Miettinen, R., & Paavola, S. (2014). Beyond the BIM utopia: Approaches to the development and implementation of building information modeling. *Automation in Construction, 43*, 84–91. doi:10.1016/j.autcon.2014.03.009

Mignone, G., Hosseini, M. R., Chileshe, N., & Arashpour, M. (2016). Enhancing collaboration in BIM-based construction networks through organisational discontinuity theory: A case study of the new Royal Adelaide Hospital. *Architectural Engineering and Design Management, 12*(5), 333–352. doi:10.1080/17452007.2016.1169987

Miller, C., Thomas, D., Irigoyen, S. D., Hersberger, C., Nagy, Z., Rossi, D., & Schlueter, A. (2014). BIM-extracted EnergyPlus model calibration for retrofit analysis of a historically listed building in Switzerland. Proceedings of SimBuild, 2014.

Ojanen, T., Nykänen, E., & Hemmilä, K. (2017). Guidebook for Structural Energy Efficiency in Renovation [Rakenteellinen energiatehokkuus korjausrakentamisessa – Opas]. Ministry of the Environment, the Confederation of Finnish Construction Industries RT and the Federation of Finnish Woodworking Industries.

Saini, V., & Mhaske, S. (2013). BIM an emerging technology in AEC industry for time optimization. *International Journal of Structural and Civil Engineering Research, 2*(4), 196–200.

Steel, J., Drogemuller, R., & Toth, B. (2012). Model interoperability in building information modelling. *Software & Systems Modeling, 11*(1), 99–109. doi:10.1007/s10270-010-0178-4

Succar, B. (2009). Building information modelling framework: A research and delivery foundation for industry stakeholders. *Automation in Construction, 18*(3), 357–375. doi:10.1016/j.autcon.2008.10.003

Taylor, J., & Levitt, R. (2004). Understanding and managing systemic innovation in project-based industries. In D. Slevin, D. Cleland, & J. Pinto (Eds.), *Innovations: Project Management Research* (pp. 83–99). Newton Square, PA: Project Management Institute.

Vilches, A., Garcia-Martinez, A., & Sanchez-Montañes, B. (2017). Life cycle assessment (LCA) of building refurbishment: A literature review. *Energy and Building, 135*, 286–301. doi:10.1016/j.enbuild.2016.11.042

KEY TERMS AND DEFINITIONS

BIM: Building information modelling.

DESIGN4ENERGY: A large research and innovation project funded by European Commission. Focused on developing an integrated evolutionary design methodology targeting to highly energy efficient building energy performance throughout the whole life cycle, from planning and design to operation, maintenance and retrofitting.

DSRM: Design science research methodology.

EE: Energy efficient.

HVAC: Heating, cooling, and air conditioning.
KPI: Key performance indicator.
LOD: Level of model detail.
MCDM: Multiple-criteria decision making.

Chapter 8
Improving the Energy Quality and Indoor Environmental Quality in Retrofit Buildings

Joanna Rucińska
Warsaw University of Technology, Poland

ABSTRACT

Energy requirements are variously specified in different countries. In many cases, different technology is applied. For this reason, improving the energy quality and indoor environmental quality in retrofit buildings requires a detailed study case. Modification of the building structure and technical systems is not sufficient. Therefore, it is necessary to change the behavior of inhabitants and create uniform zones in terms of usage. Preparation of the modernization of building requires elaboration of a detailed concept. Taking all aspects into account is not easy; that is why, in modernized buildings, the design phase often requires more work, time, and nonstandard solutions.

INTRODUCTION

The energy quality of existing building is very important from many reasons. The most noticeable is considered operating costs. In this chapter thermal comfort models and problems in modernisation of buildings are described. In the complex modernization process the project team and the goal should be selected properly. The architects and engineers must be open for new nonstandard solutions and ready to change their concept in case of problems with the implementation of the earlier assumptions. Apart from the aspects related to energy consumption for heating, cooling, domestic hot water, lighting, and auxiliary appliances in building, the project team should not forget about the occupants needs. The building without users does not consume any energy at all. Is it possible to consider such solution as ideal one? No, because the purpose of modernization should be to reduce the energy demand with the improvement of thermal comfort. There are several methods for assessing thermal comfort in buildings. In a modernized building, the most important problems should be first identified. Then, with the use of the available tools, right solutions can be applied to improve thermal comfort at the same time with

DOI: 10.4018/978-1-5225-4105-9.ch008

a considerable reduction of energy use in the building. Sometimes, however, after the building retrofit the energy use might not be at the previously set level. Often the reasons for these condition (especially) in residential buildings are behaviors and habits of the users. That is why it is extremely important to explain to the users how the systems work and which behaviors affect energy demand and how. If any actions are not accepted by users, the introduction of the proposed solutions should be reconsidered.

BACKGROUND

The modernisation process of energy efficient buildings is complicated due to a large number of energy performance parameters of various building components (Sowa J. (red.), 2017). Often the first step in the process of building modernization is to define the energy performance indicator which fulfils the assumed standard. The definition of nZEB was introduced with the Energy Performance of Buildings Directive (EPBD recast, 2010/31/EC). According to Article 9 Member States are required to set a detailed national nZEB definition. However, in many countries, the standard is mainly related to the energy requirements of a building or building components (BPIE 2015).

An extremely important aspect is also the thermal comfort of occupants. Depending on the needs, the assessment of the thermal comfort in building might be performed with the use of, one of the comfort models described in the standards

NON-TECHNICAL BARRIERS IN RETROFIT BUILDINGS

There are many non-technical barriers that have to be overcome in order to achieve energy efficient retrofitting. The first one is skepticism to deep and overall modernization of the building including modernisation of external walls, installation system and also user behaviour. Next one is the owners' or users' lack of knowledge about new technologies and about additional benefits coming from increasing building energy performance such as possible improvement of the thermal comfort. (SQARE, KODnZEB)

Before modernisation the building owner should thus increase the knowledge about current user and technical employees' problems of building (European Parliament's Committee on Industry, Research and Energy (ITRE 2015).

The approach to modernization will be different for one-family houses, residential buildings and office buildings. In single-family buildings, in most cases, users know which measures affect the energy demand. Therefore, owners of single-family homes are more interested to make improvements.

In multifamily houses the modernization process will bring effects when tenants will be involved in the retrofitting process at its early stage. Another aspect is to increase their knowledge about energy saving measures and improvements of the thermal comfort of their flats. In office buildings, users should know how their behaviour affects energy use in building. For this purpose, appropriate training and rules should be provided.

The regulation barriers could also be related to the national legal requirements, such as lack of requirements on energy efficiency for existing buildings or the preference of only standard modernization measures (EU 2050).

Another group is financial barriers. The most common is high costs of realization of a complex retrofitting and the lack of suitable support programs for such investments.

Problems may also occur at the beginning stage of the modernization process. First of them could be lack of innovative design team approach. Another barrier may be lack of technical data on the structure of the building or the restrictions resulting from its construction.

Thermal Comfort

Thermal comfort is the condition of mind that expresses satisfaction with the thermal environment and is assessed by subjective evaluation (ANSI/ASHRAE Standard 55). Maintaining this standard of thermal comfort for occupants of buildings or other enclosures is one of the important goals of HVAC (heating, ventilation, and air conditioning) design engineers.

Improving the energy quality and indoor environmental quality in retrofit buildings requires detailed case study. Modification of the building structure and technical systems only is not sufficient.

Building after modernization should use less energy while still improving the thermal comfort of users. Occupants should feel comfortable. Thermal comfort is difficult to be measured because users may have different needs and preferences. Moreover, it depends on the air temperature, air humidity, radiant temperature, air velocity, metabolic rate and clothing insulation.

Thermal comfort is the occupants' satisfaction with the surrounding thermal conditions and it must be taken under consideration when designing or modernization a structure that will be used by occupants or employees.

Until recently in the Polish climate the main aspect in energy analysis was to reduce the energy requirement for heating. Therefore, buildings are usually designed to increase the use of sunshine during the heating season. This may, however, lead to a situation in which the thermal comfort will be maintained in winter conditions, but during summer the rooms might be overheated.

In order to avoid overheating in some buildings, shading system should be used and if is not stuffiest the cooling system must be additionally added during modernisation

Energy Quality in Retrofit Buildings

Improving the energy performance of an existing building requires a design team. It should consist of architects, constructors, HVAC engineers and energy experts. The first standard operation is to increase the thermal insulation of the external walls and replacement of the windows. These measures, properly made, should also improve the air-tightness of the building. Lower air-tightness causes usually the need to adjust or replace an existing ventilation system. Properly modernised ventilation system usually reduces the heat losses and improves thermal comfort of users. After reducing the heat loss or/and solar heat gains it is necessary to make improvement or replacement of the heating and cooling system. In many cases, it is also necessary to comprehensively upgrade or replace the lighting system. Proper operation of all HVAC systems and lighting system is guaranteed by an individually designed control system. It is important to measure, monitor and control the energy use in building after the retrofit process as well as after some time of building operation.

There are many examples of renovation of different kind of buildings presented in many articles. Paper (Almeida, Ferreira, Rodrigues, Bragança, 2014) presents the results of modernisation of two residential buildings of that emerged from the analysis and identification of the most cost-effective packages of renovation measures needed to adapt existing buildings to zero energy balance and comparing them with those resulting from the calculation of cost-optimal levels. Another article presents the aims of a method regarding the selection of technical solutions for thermal and energy rehabilitation and modernization of buildings (Almeida, Ferreira, Rodrigues, Bragança, 2014). However, they are not concerned about aspects of users' thermal comfort.

THERMAL COMFORT ASSESSMENT IN RETROFIT BUILDINGS

In order to improve the thermal comfort in existing buildings, thermal comfort should be itself assessed. Thermal environmental parameters can be divided into two groups, first those that can be measured directly and secondly those which must be calculated from results of other measurements. The first group primarily includes: air temperature, wet-bulb temperature, dew-point temperature, water vapour pressure, total atmospheric pressure, relative humidity, humidity ratio, air velocity and mean radiant temperature (it can be also calculated). The second group of parameters includes mean radiant temperature, plane radiant temperature and radiant temperature asymmetry.

The mean radiant temperature is defined as the uniform temperature of an imaginary enclosure in which the radiant heat transfer from the human body is equal to the radiant heat transfer in the actual non-uniform enclosure (EN ISO 7726, 1998).

The plane radiant temperature is the uniform temperature of an enclosure in which the incident radiant flux on one side of a small plane element is the same as that in the actual environment. The plane radiant temperature describes thermal radiation in one direction, and its value thus depends on the direction (Korsgaard, 1949; ASHRAE, 2013).

The radiant temperature asymmetry is the difference between the plane radiant temperatures of the opposite sides of a small plane element. This parameter describes the asymmetry of the radiant environment and is especially important in comfort conditions. Because it is defined with respect to plane element, its value depends on the plane's orientation. Orientation of plane element may be specified in some situations (e.g. floor to ceiling asymmetry) but in some it is impossible to determine it (ASHRAE, 2013).

Factors in Human Comfort

There are six main parameters taken into account during evaluating the thermal comfort of rooms with a cooling system. Thermal comfort parameters include the following:

- **Metabolic Rate (met):** The energy generated from the human body,
- **Clothing Insulation (clo):** The amount of thermal insulation the employee is wearing,
- **Air Temperature (°C):** Temperature of the air in the room,
- **Radiant Temperature (°C):** The weighted average of all the temperatures from surfaces surrounding the employee,
- **Air Velocity (m/s):** The speed of air moving across the employee,
- **Relative Humidity (%):** Relative humidity is the ratio between the actual amount of water vapor in the air and the maximum amount of water vapor that the air can hold at that air temperature.

Factors dependent on individual preferences of the person are metabolic rate and clothing insulation. Factors that are controlled by HVAC systems and depend also on, for example, the structure of the building (wall thermal insulation, windows U-value coefficient) are: air temperature, radiant temperature, relative humidity and air velocity.

Thermal comfort is calculated as a heat transfer energy balance. Heat transfer through radiation, convection, and conduction are balanced against the occupant's metabolic rate. The heat transfer occurs between the environment and the human body, which has an area of 19 ft2 (1.81 m2). If the heat leaving the occupant is greater than the heat entering the occupant, the thermal perception is "cold." If the heat entering the occupant is greater than the heat leaving the occupant, the thermal perception is "warm" or "hot."

A method of describing thermal comfort was developed by Ole Fanger and is referred to as Predicted Mean Vote (PMV) and Predicted Percentage of Dissatisfied (PPD) (Fanger, P.O. 1970).

Analysis of Thermal Comfort

Analysis of thermal comfort in building should include the following elements:

- Determination of the thermal comfort parameters that were assumed during building designing process.
- Selection of representative rooms and measurements of the thermal comfort conditions.
- Assessment of thermal comfort issues and checking the compliance with design conditions.
- Investigation and determination of the most important problems.
- Introduction of changes in the HVAC installation or planning comprehensive modernisation to reduce energy consumption and improve thermal comfort in the building.

Depending on the building type and the criteria, an appropriate tool should be selected. To evaluate the hourly variability of for example air temperature, humidity and air velocity the dynamic thermal modeling is used. However, sometimes is same cases a simplified tool may be used for calculating a specific parameter. In case of thorough modernisation it is worth to use more complex tools.

From the beginning of the work, the purpose and scope of modernization is very important. That means that the goal should be always clearly defined and known to all design process participants. Carrying out a number of iterations is normal practice to help to find the best solution. Unfortunately some measures that reduce building energy demand affect indoor comfort conditions reducing its standard. For instance, application of external shading devices will reduce overheating but also reduce daylight level and increase energy use for artificial lighting.

Basic factors that need to be determined while modeling are: air temperature, air speed, air humidity, walls, flor and ceiling surface temperatures.

For dynamic simulation of building, it is necessary to define following input data: building geometry and zoning, construction of internal and external partitions (with definition of characteristic parameters of materials), shading devices, climatic data, internal gains (people appliances and processes), occupancy, airtightness level, ventilation rate, heating and cooling setpoints. In addition to defining the building partitions and HVAC systems to the thermal comfort model, it is needed to enter: users clothing (clothing insulation), activity level and metabolic rate.

Measurements

Before modeling variants, thermal comfort in the existing state should be assessed and main problems identified. The first step is to divide the building into zones corresponding to their use. The choice of representative spaces cannot be incidental. Measurements of comfort parameters should be carried out in spaces: exposed to overheating, where temperature in heating season is not maintained, rooms with large glazing, rooms on different building floors and with different external wall orientation, rooms with different people density and different internal gains. Measurements should be done in different periods.

It is also important to properly place the sensor in the room.

The sensors should not be placed below or over devices emitting heat gains and also can not be exposed to direct solar radiation. Sensors should be placed on the internal walls in places where they can be naturally ventilated by the air flow.

Poorly or not properly performed measurements can provide misleading information about the conditions in rooms.

THERMAL COMFORT ASSESSMENT MODELS

According to the Ashrae 55 there are three basic types of standards:

- Temperature out of range,
- Heat balance,
- Adaptive thermal comfort (ANSI/ASHRAE Standard 55).

Temperature out of range – is most appropriate for buildings without heating or cooling systems or intentionally designed undersized systems. In this case, the number of hours out of defined range, for example number of hours during the year when temperature in the room is over 26°C, is usually determined.

Heat balance – is most appropriate for buildings operating with controlled heating, cooling or air conditioning systems.

Adaptive thermal comfort – is most appropriate for all buildings during mid-season conditions when heating or cooling system is not used or for buildings without heating or cooling system. The adaptive model is based on the idea that people are able to adjust to the conditions by, for example, adjusting the clothing insulation.

Three basic standards are commonly used: EN 15251 (2006), EN 7730 (2005) and Ashrae 55.

Predicted Mean Vote

PMV coefficient (predicted mean vote) allows evaluating thermal comfort on a 7-step scale with range ±3. Determining the sensation of the thermal conditions in scale is given in the Table below. This thermal scale was originally developed by Fanger and later adopted by an ISO standard. The original data was collected by subjecting a large number of people (reputedly many thousands of Israeli soldiers) to different conditions within a climate chamber and having them select a position on the scale the best describing their comfort sensation. A mathematical model of the relationship between all the environmental and physiological factors considered was then derived from the data. The result relates the size thermal comfort factors to each other through heat balance principles and produces the following sensation scale (Table 1).

The recommended acceptable PMV range for thermal comfort from ASHRAE 55 is between -0.5 and +0.5 for an interior space.

According to EN ISO 7730, the PMV is calculated with equation:

$$\begin{aligned} PMV = & \left[0.303 \cdot \exp(-0.036 \cdot M) + 0.028\right] \cdot (M - W) - 3.05 \cdot 10^{-3} \\ & \cdot \left[5733 - 6.99\left(M - W\right) - p_a\right] - 0.42 \cdot \left[\left(M - W\right) - 58.15\right] - 1.7 \cdot 10^{-5} \\ & \cdot M \cdot \left(5867 - p_a\right) - 0.0014 \cdot M \cdot \left(34 - t_a\right) - 3.96 \cdot 10^{-8} \\ & \cdot f_{cl}\left[\left(t_{cl} - 273\right)^4 - \left(\bar{t}_r - 273\right)^4\right] - f_{cl} \cdot h_c \cdot \left(t_{cl} - t_a\right) \end{aligned}$$

$$\begin{aligned} t_{cl} = & \; 35.7 - 0.028 \cdot \left(M - W\right) - I_{cl} \\ & \cdot \left\{3.96 \cdot 10^{-8} \cdot f_{cl} \cdot \left[\left(t_{cl} - 273\right)^4 - \left(\bar{t}_r - 273\right)^4\right] + f_{cl} \cdot h_c \cdot (t_{cl} - t_a)\right\}, \end{aligned}$$

$$h_c = \begin{cases} 2.38 \cdot \left|t_{cl} - t_a\right|^{0,25} & for \quad 2.38 \cdot \left|t_{cl} - t_a\right|^{0.25} > 12.1 \cdot \sqrt{\nu_{ar}} \\ 12.1 \cdot \sqrt{\nu_{ar}} & for \quad 2.38 \cdot \left|t_{cl} - t_a\right|^{0.25} < 12.1 \cdot \sqrt{\nu_{ar}} \end{cases},$$

$$f_{cl} = \begin{cases} 1.00 + 1.29 \cdot l_{cl} & for \quad l_{cl} \leq 0.078 \dfrac{m^2 \cdot K}{W} \\ 1.05 + 0.645 \cdot l_{cl} & for \quad l_{cl} > 0.078 \dfrac{m^2 \cdot K}{W} \end{cases},$$

where:

e = The Euler's number (2.718),
M = The metabolic rate, [W/m^2],

Table 1. Predicted mean vote sensation scale

Value	Sensation
-3	cold
-2	cool
-1	slightly cool
0	neutral
+1	slightly warm
+2	Warm
+3	hot

W = The external work, [W/m^2],
I_{cl} = The clothing insulation, [(m^{2}K)/W],
f_{cl} = The clothing Surface area factor, [-],
t_a = The air temperature, [°C],
$\bar{t}_r$ = The mean radiant temperature, [°C],
v_{ar} = The relative air velocity, [m/s],
p_a = The water vapour partial pressure, [Pa],
h_c = The convective heat transfer coefficient, [W/(m^{2}K)],
t_{cl} = The clothing surface temperature, [°C].

PMV comfort indicator values can be specified in different combinations of metabolic rate, clothing insulation, air temperature, mean radiant temperature, relative air velocity and air relative humidity. Equations describing the clothing surface temperature and convection heat transfer coefficient can be solved iteratively.

Clothing insulation describes the heat flow in time through 1m^2 of material at temperature difference on both sides of the clothes of 1 K. The unit of thermal insulation is clo (1 clo = 0.155 (m^2·°C)/W).

The value of 1 clo means the thermal insulation of clothing necessary to maintain the thermal balance between the human body, in a sitting position, and the environment with defined parameters: air velocity 0.1 m/s, air and partitions temperature 21°C, relative humidity lower than 50%. Methods for calculating the thermal insulation of clothing and thermal insulation for typical combinations of garments described in the EN 9920:2009 *Ergonomics of the thermal environment — Estimation of thermal insulation and water vapour resistance of a clothing ensemble* and EN ISO 15831:2006 *Clothing - Physiological Effects - Measurement Of Thermal Insulation By Means Of A Thermal Manikin* (EN ISO 9920:2009, EN ISO 15831). The values of thermal insulation for typical combinations of garments are shown in Table 2.

The metabolism rate depends on the type of performed work, the unit of metabolic rate is met = 58.2 W/m^2. Table 3 shows the value of metabolic rate depending on the type of performed activity.

Table 2. Thermal insulation for typical combinations of garments

Combinations of Garments	**Clothing Insulation I_{cl}**	
	(m^{2}K)/W	**clo**
Without clothes	0	0
Panties, T-shirt, shorts, light socks, sandals	0.050	0.3
Underpants, shirt with short sleeves, light trousers, light socks, shoes	0.080	0.5
Underwear, shirt, trousers, socks, shoes	0.110	0.7
Panties, shirt, trousers, jacket, socks, shoes	0.155	0.7
Underwear with long sleeves and legs, shirt, trousers, V-neck sweater, jacket, socks, shoes	0.200	1.3
Underwear with short sleeves and legs, shirt, trousers, jacket, heavy quilted outer jacket and overalls, socks, shoes, cap, gloves	0.220	1.4

Table 3. Metabolic rates

Activity	Metabolic Rate	
	W/(m²)	met
Reclining	46	0.8
Seated, relaxed	58	1.0
Sedentary activity (office, dwelling, school, laboratory)	70	1.2
Standing, light activity (shopping, laboratory, light industry)	93	1.6
Standing, medium activity (shop assistant, domestic work, machine work)	116	2.0
Walking on level ground: 2 km/h 3 km/h 4 km/h 5 km/h	 110 140 165 200	 1.9 2.4 2.8 3.4

Predicted Percentage of Dissatisfied

Predicted Percentage of Dissatisfied (PPD) predicts the percentage of occupants that will be dissatisfied with the thermal conditions. PPD is a function of PMV, given that as PMV moves further from 0, or neutral, PPD increases. The maximum number of people dissatisfied with their comfort conditions is 100%. The case where all employees in the same time are satisfied is not actually possible. The recommended PPD range for thermal comfort according to ASHRAE 55 is less than 10% persons dissatisfied for an interior space with cooling system.

The value of this parameter is based on the value of the PMV, according to the equation:

$$PPD = 100 - 95 \cdot \exp\left(-0.03353 \cdot PMV^4 - 0.2179 \cdot PMV^2\right).$$

According to ISO 7730: 2006, people who are unsatisfied with the thermal conditions in a room are those for whom the predicted average thermal comfort rate is hot (+3), warm (+2), slightly warm (+1), neutral (0), slightly cool (-1), cool (-2), and cold (-3). Acceptable are conditions for those where the predicted average thermal comfort rate is included in range $-1<PMV<+1$ (slightly cool, neutral, slightly warm).

Local Discomfort

PMV and PPD indicators refer to the body's overall thermal sensation, but discomfort may also be caused by local sources of unwanted heating, cooling or air movement occurring for a particular part of the body (ISO/FDIS 7730:2005)0. Such case is referred to as local discomfort. Local discomfort can be caused by: too high air velocity (DR – air draft), too high vertical air temperature difference (too high temperature difference between temperature on the height of head and feet), too high or too low floor temperature or too much radiation temperature asymmetry.

For these reasons, the maximum allowable values of local discomfort factors are also taken into account in the classification for rooms comfort classes.

Requirements based on internal temperature and humidity are described in PN EN ISO 15251 (EN ISO 15251). The standard specifies the minimum temperature values for the winter period and the maximum summer temperature values for each of the three room categories.

In office buildings, humidity regulation is typically used to improve thermal comfort. Too much humidity can cause microbial growth, while too low air humidity (below 15-20%) causes dryness and can lead to dry mucous membranes. This can cause irritation, for example of the eyes or respiratory tract. Air conditioning processes related to humidity control require the use the additional equipment and therefore lead to an increase of energy consumption in building.

Asymmetric thermal radiation in old modernized building may be caused for example by low surface window temperature (high U-value of window), low surface wall, roof or floor temperature (high U-value of external partition, uninsulated partitions). It can also occur if heating or cooling devices were improperly designed or are incorrectly used.

Draft is an undesired local cooling of the human body caused by air flow. In modernized building draft may be caused for example by leaky windows or improperly working ventilation system. If people feel draft then they often demand higher temperature in the room. During the heating season it increases energy use for heating and in the cooling season it decreases energy use for cooling.

Since feet have direct contact with the floor, discomfort of the feet is often caused by too high or too low floor temperature. Usually the low temperature is due to low thermal insulation of the floor above unheated basement, on the ground or above other kind unheated space. If users feel cold discomfort in their feet then they often demand higher temperature in the room during heating season which increases energy use for heating.

Adaptive Comfort EN ISO 15251

Concept of adaptive comfort is based on the idea that people adjust to external conditions during different times of the year (EN ISO 15251)

Annex A.2. EN ISO 15251 also describes the rules for calculation the maximum room temperature (in the summer period) if the building is not equipped with an air conditioning system (with mechanical cooling system) (EN ISO 15251). Although the analyses should be provided mainly for office buildings, they can also be extrapolated for other types of buildings.

In the case of residential buildings, by appropriate behaviour and adjustment of the clothing insulation, users have more adaptability to the indoor conditions. However, if in there is no data, it is possible to use the rules for rooms, in which the occupant has only limited possibilities to adjust the clothing insulations to current conditions. An important feature in buildings not equipped with a mechanical cooling system, is a possibility of window opening. In such rooms windows are a basic tool for users to improve in the summer the indoor climate.

For summer period in buildings without mechanical cooling the upper operative temperature limit for room category I should be calculated from following equation (EN ISO 15251):

$$\theta_{i\max} = 0.33 \cdot \theta_{rm} + 18.8 + 2, \left[°C \right],$$

and lower operative temperature limit:

$$\theta_{i\,\max} = 0.33 \cdot \theta_{rm} + 18.8 - 2, \left[{}^{\circ}C \right],$$

For room category II upper and lower operative temperature limit:

$$\theta_{i\,\max} = 0.33 \cdot \theta_{rm} + 18.8 + 3, \left[{}^{\circ}C \right],$$

$$\theta_{i\,\max} = 0.33 \cdot \theta_{rm} + 18.8 - 3, \left[{}^{\circ}C \right],$$

For room category III upper and lower operative temperature limit

$$\theta_{i\,\max} = 0.33 \cdot \theta_{rm} + 18.8 + 4, \left[{}^{\circ}C \right],$$

$$\theta_{i\,\max} = 0.33 \cdot \theta_{rm} + 18.8 - 4, \left[{}^{\circ}C \right],$$

where

θ_i = Upper and lower operative temperature in the room
θ_{rm} = Running mean outdoor air temperature calculated according to equation:

$$\theta_{rm} = \left(1 - \alpha\right)\theta_{ed-1} + \alpha\theta_{rm-1}$$

where

θ_{ed-1} = Exponentially weighted running mean of the preceding day's daily mean outdoor temperature
θ_{rm-1} = Exponentially weighted running mean temperature on day
α = Constant value of range 0 ÷ 1, it is recommended to take 0.8 according to (EN ISO 15251)

Design values for the indoor operative temperature for buildings without mechanical cooling systems are on Figure 1. Values calculated according to this procedure, should be used for designing a solar passive protection solutions to avoid overheating for example: calculating maximum area of glazing and its orientation, designing shading devices, proper selection of thermal properties of external walls. When it is not possible to obtain an acceptable air temperature despite application of aforementioned solutions, it is necessary to use a mechanical cooling system. Maintaining comfort conditions during periods with increased heat gains can also be achieved by the increase of the air flow velocity.

Figure 1. Design values for the indoor operative temperature for buildings without mechanical cooling systems

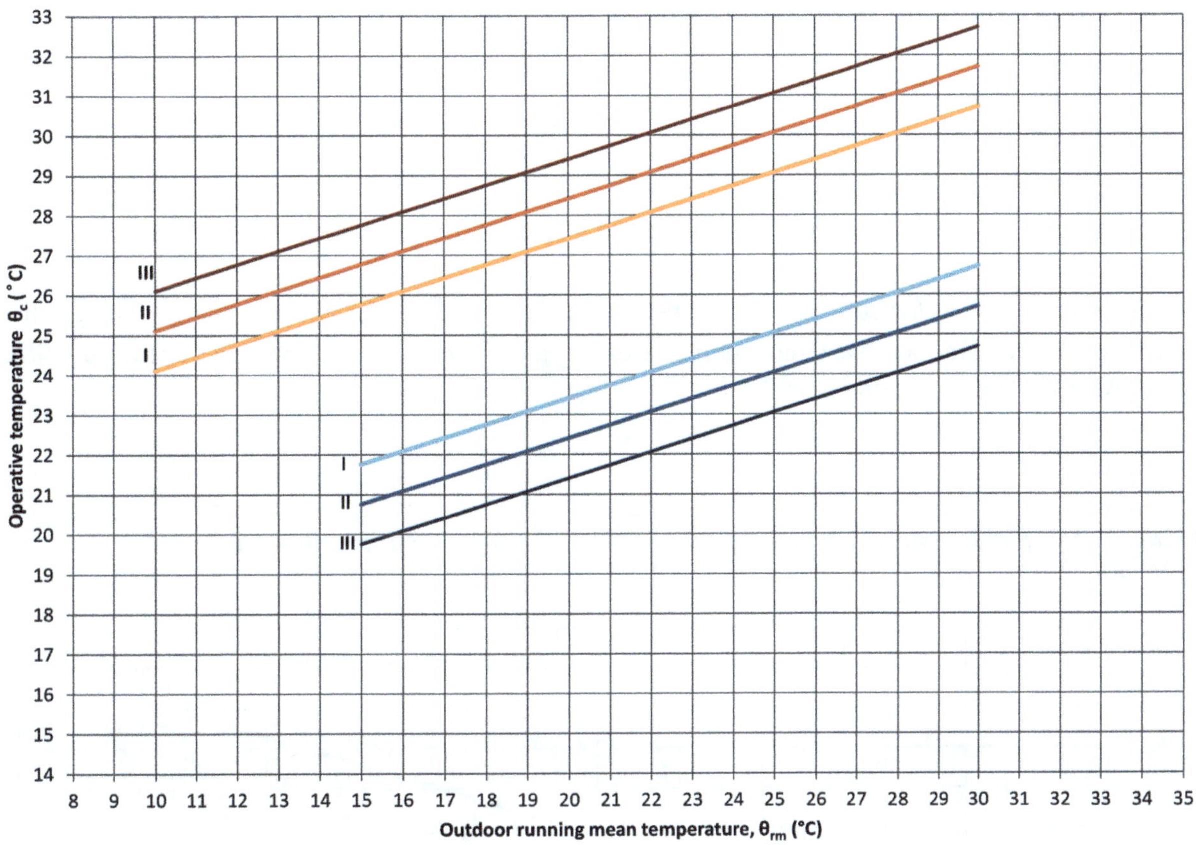

Adaptive Comfort ASHRAE-55 2010

Numerous researchers have conducted field studies worldwide in which they measure thermal comfort in naturally ventilated buildings. This extensive database show, that occupants of naturally ventilated buildings accept and even prefer a wider range of temperatures than their counterparts in sealed, air conditioned buildings because their preferred temperature depends on outdoor conditions (de Dear, Brager, 1998). The results of the study were included in ASHRAE 55-2004. However, the ASHRAE-55 2010 standard has introduced the prevailing mean outdoor temperature as the variable input for the adaptive model. The adaptive comfort equations are provided by ASHRAE

$$\theta_{comf\ op} = 0.31 \cdot \theta_{out\ mm} + 17.8\left[°C\right],$$

where:

$\theta_{comf\ op}$ = Indoor comfort operative temperature,

$\theta_{out\ mm}$ = Monthly mean outdoor air temperature,

The acceptable comfort temperature is defined as:

$$\theta_{accept\ op} = 0.31 \cdot \theta_{out\ mm} + 17.8 \pm \theta_{limit} \left[{}^{\circ}C\right],$$

θ_{limit} = The range of acceptable temperatures for 80% and 90% of occupant being satisfied .

The limits are: θ_{limit} for 80% of occupant being satisfied 3.5 K and for 90% of occupant being satisfied 3.5 K. Acceptable operative temperature ranges for naturally conditioned spaces according to ANSI/ASHRAE Standard 55 are on Figure 2.

Figure 2. Acceptable operative temperature ranges for naturally conditioned spaces
ANSI/ASHRAE Standard 55.

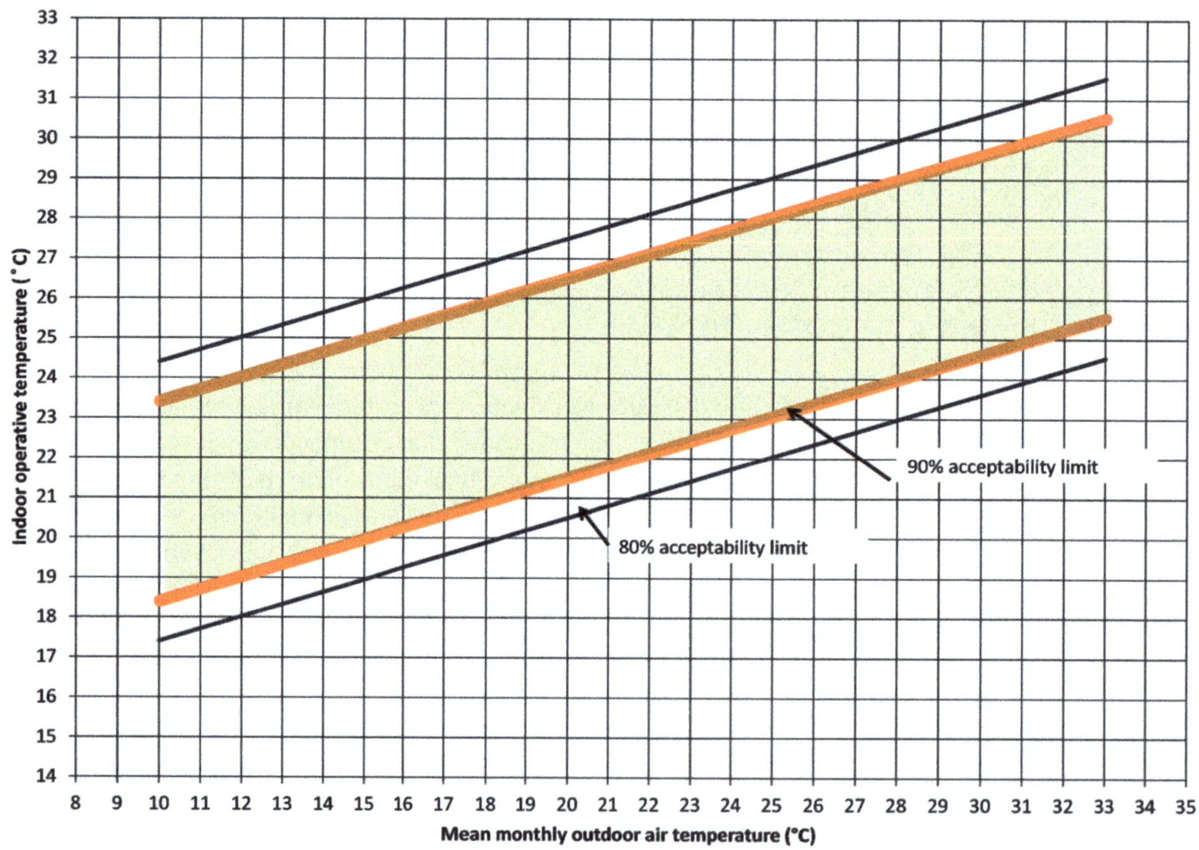

CASE STUDIES

Case Study SOLANOVA Project

SOLANOVA – "Solar-supported, integrated eco-efficient renovation of large residential buildings and heat-supply-systems" started in January 2003 (SOLANOVA). SOLANOVA was the first "Eco-buildings" project of the European Commission in Eastern Europe dealing with a "major renovation" of a large existing building (SOLANOVA 2003-2006). In order to achieve sustainable improvements, in SOLANOVA project three strategies were proposed:

- Design for human needs,
- Optimised resource efficiency of the building,
- Optimised solar supply.

Figure 3 shows analysed building before and after renovation. The originally designed heating power for the whole building was 373 kW and the space heating consumption before retrofit was around 220 kWh/(m^2year). Energy simulations have been used to predict energy demand after modernization. Through detailed analysis, the projected energy demand was reduced to around 20 kWh/(m^2 year). However, actual energy consumption after modernization was about 40 kWh/(m^2 year). Before retrofit, the problem was not only energy consumption but also, for example, too low indoor air humidity or air temperature out of range.

Improving the energy performance of the building was possible through the following measures:

- Decentralization of ventilation units with high real heat recovery efficiency (82%), one ventilation unit in each flat,
- 75 m^2solar thermal area as canopy, providing not only heat but also shade for the shops in the ground floor,

Figure 3. Residential building in Dunaújváros before (left) and after (right) modernization
Source: Solanova, 2003-2006.

- Replacement of heating system with new one equipped with radiators and thermostatic valves, which are easy to use,
- Heat insulation on basement ceiling – 10 cm of polystyrene,
- Roof insulation – 30 cm with extensive green roof,
- Replacement of windows in flats: from south and west side - 2+1 glazing with integrated venetian blinds for shading U_w=1.1 W/(m^{2}K) g-value=0.55 or 0.1 with fully closed blinds respectively; from north side – double-glazing U_w=1.4 W/(m^{2}K),
- Replacement of windows in retail spaces U_w=1.4 W/(m^{2}K) (Hermelink A. 2006a).

The project engineers drew attention to the users and their psychological aspect of building modernization. The building without users does not consume any energy at all. However, the purpose of modernization was to reduce the energy demand with users' acceptance.

The behavior and well-being of the dwellers is a function of personal variables and variables of their environment. Users sometimes spend about 90% of their life in buildings. The dwellers should come to a point, where they consider the "updated" technical surrounding as a comfortable, integral part of their life and not as something exterior, strange and unknown. In SOLANOVA project on the day of the technical handover, more than four hours have been spent to inform and train the dwellers about the new devices. The schooling took place in the flat and it was focused on answering all dwellers questions, instead of holding a lecture about the principles of ultra-low-energy houses (Hermelink A. 2006b).

The SOLANOVA project is an example of a project that has significantly reduced energy consumption of the building and improved user thermal comfort.

Case Study Fredrik Selmers Vei 4, Oslo

The building was originally designed by architect F.S. Platou as a government office building and was finished in 1982 (Figure 4).

Improvement of the energy performance of the building was possible through the following measures:

Figure 4. Office building in Oslobefore (left) and after (right) modernization
Source: www.futurebuilt.no.

- Renovation of the existing façade, and installation a new outer wall of wooden elements, with 350 mm glass wool insulation and clad with two layers of 95% recycled aluminium, was installed on the outside of the existing loadbearing structure,
- Modernisation of windows – new windows mean U-value 0.8 W/(m^{2}K),
- Modernization of heating system - high insulation levels and low heat losses,
- Modernisation of ventilation system – high efficiency of heat recovery,
- Modernisation of lighting system – energy effective and demand controlled lighting,
- Installation of water-saving toilets and fixtures,
- Reduction of internal of heat gains from PCs.

All products in the building have undergone a thorough quality control to ensure that environmental toxins and products that negatively affect the internal air quality are eliminated before selection. Calculated delivered energy after modernisation is 68 kWh/(m^{2}year).

The façades and window areas were analysed and the window quality were a result of a thorough analysis of insulation, daylight versus solar energy penetration and lighting levels in the office spaces. A local energy centre provides the building with heat, cooling and warm water.

Modernization was not limited to the building envelope elements and technical systems. After analysing the existing stage, it was found that the surface of the each storey can be increased. The new areas are connected to the existing loadbearing structure between the five blocks using low-carbon concrete and structural steel with 50 percent recycled content. In this project the total area of the building is now about 4 000 m^2 bigger ten before retrofit.

Important strategies used in the rehabilitation include giving the building a more compact form and introducing new façades. This has contributed to better daylight conditions, reduced energy consumption and reduced greenhouse gas emissions. The façades have been developed in close cooperation with energy consultants. The building´s scale was given by the existing structure, but the light perforated cladding gives the façade a slightly 'dissolved' character.

The project has focused on environmental performance from the outset. An environmental coordinator and energy advisor were contracted, and environmental goals were defined within a variety of themes. A dialogue group was also established with future occupants, meeting regularly throughout the process. The project has been a pilot project for BREEAM-NOR. The project´s environmental plan has been a part of the contracts for all consultants and contractors (FutureBuilt 2010-2020).

Case Study Powerhouse Kjørbo, Bærum

Powerhouse Kjørbo is located outside of Oslo in Norway. It was opened in April 2014 and is the first Powerhouse project to be completed by the collaboration. Office building was finished in 1980s. Now it is renovated to into energy positive buildings. After the renovation, the buildings' energy needs have been reduced by 90% (Figure 5).

Improving the energy performance of the building were possible through the following measures:

- Modernisation of walls, ceilings, and windows which are well insulated,
- Minimalisation of the cold bridges,
- Detail design in order to achieve an extremely airtight climate shell,

Figure 5. Office building Powerhouse Kjørbo, Bærum after modernization
Source: www.powerhouse.no/prosjekter/kjorbo/.

- Installation of solar panels on the roof,
- Installation of low pressure losses ventilation system,
- Installation of exterior automatic sun shading screens,
- Installation ground wells in the park outside the buildings which provide heating for radiators, water and ventilation air, and cooling in the summer,
- Zoning of the floor layout and sensors that control lighting, ventilation, heating and cooling.

A comfortable and attractive indoor environment has been an important factor for the success of the project. The whole building is upgraded into a top modern office space. The energy systems (heating, ventilation, cooling and lighting) are designed to be used only when needed, and the number of sensors and control units is limited to a minimum. The users can control the internal climate themselves by opening windows, and they can also control the solar screening on the windows. The primary energy calculation over the 60 year life of the building results in a surplus of about 200 kWh pr. m^2 heated area. The calculation uses realistic rather than standardised operating data. Delivered energy, excluding technical equipment, is calculated to about 20 kWh/(m^2year). A comprehensive job was done to document the embodied energy of the project (POWERHOUSE).

Case Study KODnZEB: Faculty of Building Services, Hydro and Environmental Engineering

One example of building energy analysis including thermal comfort assessment is a project with the acronym “KODnZEB” (KODnZEB 2015-2017) performed by the Faculty of Building Services, Hydro and Environmental Engineering of Warsaw University of Technology (Figure 6). Case studies were carried out for two buildings belonging to the University. One of them is a building of the Faculty of Building Services, Hydro and Environmental Engineering located on the University Central Campus. Every day, the building is used by more than 2000 people, using 335 rooms of different character (educational, laboratory, technical and office). The building was built in the 1970’s. Over 45 years of use, some modernizations have been carried out, but they have not been comprehensive. Despite those actions, the primary energy demand factor EP is about 150 kWh/(m^2 year). The building requires thus a complex modernization of envelope as well as for the interiors.

The measurements in the class rooms indicated a strong overheating of the rooms located on the south and west side of the building, insufficient ventilation, uncontrolled airflow between rooms, increased concentrations of respirable dust and problems with oxygen content in selected rooms (Sowa J., Noga-Zygmunt J., Ugorowska J. (2017).

An additional problem is the high volume of noise associated with traffic (cars and trams) on the adjacent street. In order to assess the energy demand and thermal comfort, the building model was created. Energy simulations were performed with an hourly step programme in the Design Builder software. The model has been verified using measurements taken during the project. Energy consumption was measured along with the temperature in selected rooms. In a result of an integrated design process, many variants of modernisations were analysed. From many options, the retrofit solutions that were finally selected, not only significantly reduce the energy demand bat also positively influence the improvement of the environmental quality in the analysed spaces.

The final concept obtained in the KODnZEB project includes the following modernization measures: additional thermal insulation of external walls 21 cm mineral wool with a thermal conductivity coefficient λ=0.031 W / (mK), additional thermal insulation of roof 23 cm mineral wool with a thermal conductivity coefficient λ=0.035 W/(mK), double sin façade from south and west side of the building (double skin facades improve the thermal and acoustic properties as well as they are used as additional shading elements), additional atrium from the campus side of the University, replacement of windows for new with a heat transfer coefficient U=0.9 W/(m^{2}K), solar energy transmittance g=0.21 and light transmission coefficient L_t=0.55, furnishing the building with mechanical ventilation with heat recovery (the assumed temperature efficiency of heat recovery is 85%), replacement of lighting system for LEDs, increased use of renewable energy sources (in the final version of the modernization the area of photovoltaic panels installed on the façade and roof of the building was increased to about 600 m^2, rated power is 489 kWp, a ground heat pump was also proposed).

Due to different use schedules during the day and the year, different assumptions have been made for the ventilation system in education rooms, offices, laboratories, corridors and toilets.

Figure 6. View of building from south west side before (left) and after (right) modernization
Source: KODnZEB internal data.

After the described modernization in the rooms, the number of hours does not meet the comfort requirements for category II according to the standard EN ISO 15251. But after modernization the number of hours out of range has been significantly reduced (EN ISO 15251). The results of analysis are shown in Table 4.

Still, in the Summer period the temperature values in the offices exceeded the acceptable value 26°C. For this reason in the final concept of the modernisation, the offices were equipped with the cooling system. In this way, the risk of overheating was eliminated. Figure 7 shows the variations of the temperature in the summer before and after modernization (offices equipped with a mechanical cooling system) (Mijakowski, Rucińska, Sowa, Narowski, 2017). In Table 5 the specific annual primary energy consumption kWh/(m^2year) is presented.

This example shows that significant reduction of energy demand is possible, with improved thermal comfort in different kind of spaces. The total primary energy demand factor is consistent with the definition of nZEB for project KODnZEB and it is less than 20.0 kWh/(m^2year). Overheating of rooms located on Southern and Western elevations was radically limited. However, in the offices, it was necessary to install additional mechanical cooling system.

Case Study KODnZEB: "Muszelka"

Second analyzed building was students' dormitory "Muszelka". Usable area of the building for energy calculations is 3 366 m^2. The building has 4 storeys above ground level and a basement in which technical rooms are located. Students' rooms, common kitchens and toilets are located on each floor and administrative rooms on the ground floor. The students' rooms are designed in groups, in each group there are 2-3 rooms and bathroom with a shower. The building was built in the 1950's. Over the years, some modernization has been carried out, but it has not been comprehensive. In 1989 the insulation of a roof was modernized, and then in 1996 all windows were replaced. External walls are made of bricks without any insulation, estimated heat transfer coefficient is 0.90 to 2.02 W/(m^2K). The building is equipped with a basic natural ventilation system, heating system and domestic hot water system from district heating. The model of the building performed in Design Builder software is on Figure 8.

In the building the thermal comfort conditions have been evaluated according to adaptive comfort model described in EN ISO 15251. In some residential spaces even after modernization air temperature is above the range (Table 6). This indicates periodic overheating of some rooms. In Table 7 there are results of specific annual primary energy consumption for building before and after modernization.

Table 4. Number of hours per year outside the criteria for the category II for thermal comfort according to PN-EN 15251

Room	Before Modernisation	After Modernisation
Office 1	975	491
Office 2	1740	440
Educational room 1	18	70
Educational room 2	65	47

Source: KODnZEB internal data.

Figure 7. Variations of air temperature in office 1 (before and after modernization) and external air temperature
Source: KODnZEB internal data.

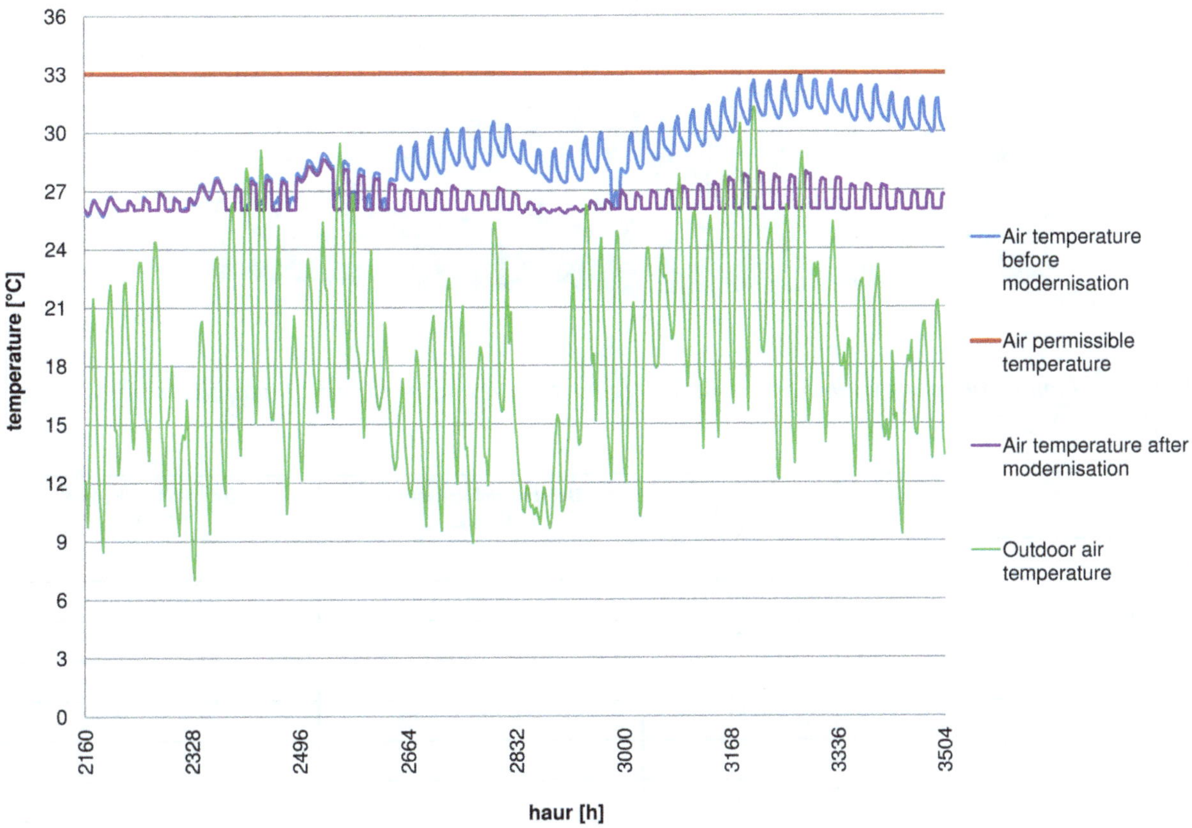

Table 5. Specific annual primary energy consumption kWh/(m²year)

Energy Use	Before Modernisation	After Modernisation
Heating and ventilation	72.2	3.50
Auxiliary appliances in heating and ventilation system	12.1	11.7
Domestic hot water	8.8	2.6
Auxiliary appliances in domestic hot water	0.7	0.5
Cooling	0.9	4.9
Lighting	60.6	10.5
PV	-6.5	-27.5
Sum	148.9	6.2

The total primary energy demand factor is consistent with the definition of nZEB for project KODn-ZEB and it is less than 20.0 kWh/(m^{2}year). It can be assumed that such a decrease in the primary energy demand factor and its value dropping to the level of 17.9 kWh/(m^{2}year) for a building under conservator protection is the result of proving, that even in such building it is possible to achieve an nZEB standard building.

Figure 8. "Muszelka" building model
Source: KODnZEB internal data.

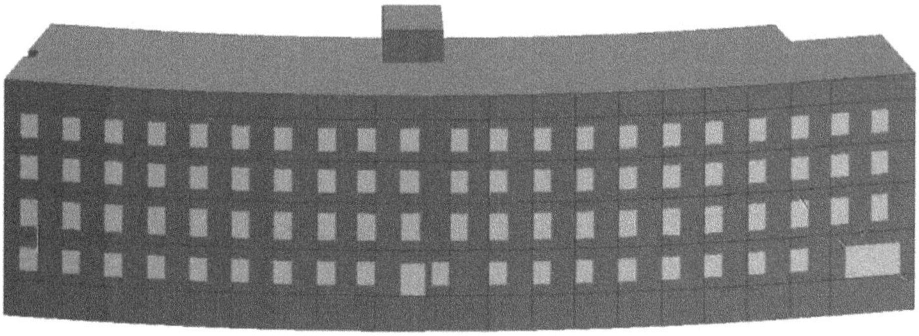

Table 6. Number of hours with a temperature higher than the upper limit of adaptive comfort in accordance EN 12521

Floor	Room	Before Modernisation Klas II	After Modernisation Klas II
Groudfloor	1	0	0
	4	0	0
	7	0	0
	9	0	0
1st	1	0	0
	4	0	0
	7	10	0
	9	7	2
3rd	1	16	0
	4	19	0
	7	52	4
	9	45	17

Table 7. Specific annual primary energy consumption kWh/(m^{2}year)

Energy Use	Before Modernisation	After Modernisation
Heating and ventilation	140.8	19.7
Domestic hot water	71.0	26.0
Lighting	85.3	26.5
Auxiliary appliances	1.4	4.2
PV	0.0	-58.6
Sum	298.5	17.9

CONCLUSION

Presented in this chapter thermal comfort models and problems in modernisation of buildings, shows how important it is to choose the right solution, not only focused on energy demand of the building but also considering the thermal comfort, satisfaction and understanding the influence of the modernisation solutions on users. It is important to define the problem and set a clear goal at the beginning of the process. However, during the initial analysis, the solutions chosen at the beginning might be appropriate. In the case of complex modernization, it is important to choose the right person for the team. They must be open for new nonstandard solutions and ready to change their concept in case of problems with the implementation of the earlier assumptions.

ACKNOWLEDGMENT

We would like to acknowledge support of all designers from Warsaw University of Technology and NTNU Trondheim involved in the KODnZEB grant. This research was supported by the Nordic Financed Mechanism Grant "Design retro-fit nZEB concept for two buildings" [grant number 315/2015/Wn07/OA-XN-04/D no. SAP 511/00023].

REFERENCES

Almeida, M., Ferreira, M., Rodrigues, A., & Bragança, L. (2014). Cost Optimality and nZEB target in building renovation of Portuguese residential buildings. *Proceeding of iiSBE Net Zero Built Environment 2014, 17th Rinker International Conference.*

ASHRAE. (2010). *ANSI/ASHRAE Standard 55-2010.* Atlanta, GA: American Society of Heating, Refrigerating and Air-Conditioning Engineers, Inc.

BPIE. (2015). *Nearly Zero Energy Buildings Definitions Across Europe.* BPIE.

de Dear, R., & Brager, G. (1998). Developing an adaptive model of thermal comfort and preference. *ASHRAE Transactions, 104*(1), 145–67.

EN ISO 15251 Indoor environmental input parameters for design and assessment of energy performance of buildings addressing indoor air quality, thermal environment, lighting and acoustics.

EN ISO 15831:2006 Clothing - Physiological Effects - Measurement Of Thermal Insulation By Means Of A Thermal Manikin.

EN ISO 7726. Ergonomics of the thermal environment - Instrument for measuring physical quantities. (1998). Geneva, Switzerland: International Organization for Standardization.

EN ISO 9920:2009 Ergonomics of the thermal environment — Estimation of thermal insulation and water vapour resistance of a clothing ensemble.

European Parliament's Committee on Industry, Research and Energy (ITRE). (2015). *Boosting Building Renovation: What potential and value for Europe? EU 2050 Roadmap for a low-carbon economy.* Author.

Fanger, P. O. (1970). *Analysis and Applications in Environmental Engineering*. New York: McGraw-Hill Book Company.

FutureBuilt. (2010-2020). Retrieved from www.futurebuilt.no

Giurcaa, I., Aúchileanb, I., Naghiua, G. S., & Badea, G. (2016). *Selecting the Technical Solutions for Thermal and Energy Rehabilitation and Modernization of Buildings. Procedia Technology*, *22*, 789–796.

Hermelink, A. (2006a). A Retrofit for Sustainability: Meeting Occupants' Needs Within Environmental Limits. *Proceedings of the 14th ACEEE Summer Study on Energy Efficiency in Buildings*.

Hermelink, A. (2006b). *Reality Check: The Example. SOLANOVA*.

ISO/FDIS 7730:2005, International Standard, Ergonomics of the thermal environment — Analytical determination and interpretation of thermal comfort using calculation of the PMV and PPD indices and local thermal comfort criteria.

KODnZEB. (2015-2017). *KodnZEB project, supported by Finance Mechanism EOG 2009-2014, within the framework of cooperation Fund Operation Program PL04 Efficient Energy and promotion of alternative energy sources*. Project realized with the cooperation of NTNU Trondheim.

Korsgaard, V. (1949). Necessity of using a directional mean radiant temperature. *Heating, Piping and Air*, 117-120.

Mijakowski, M., Rucińska, J., Sowa, J., & Narowski, P. (2017). *The concept of improving the internal environment in a utility building modernized to the nZEB standard* [Koncepcja poprawy środowiska wewnętrznego w przykładowym budynku użyteczności publicznej modernizowanym do standardu nZEB]. *Building Physics in Theory and Practice*. Retrieved from www.powerhouse.no/prosjekter/kjorbo/

SOLANOVA. (2003-2006). *Project Solar-supported, integrated eco-efficient renovation of large residential buildings and heat-supply-systems, "Eco-buildings" project of the European Commission in Eastern Europe dealing with a "major renovation" of a large existing building*. Retrieved from www.solanova.eu

Sowa, J. (Ed.). (2017). *Nearly zero energy buildings* [Budynki o niemal zerowym zużyciu energii]. Oficyna Wydawnicza PW.

Sowa, J., Noga-Zygmunt, J., & Ugorowska, J. (2017). *Assessment of indoor air parameters in a university building before the expected modernization to the nearly zero energy standard* [Ocena parametrów powietrza wewnętrznego w budynku wyższej uczelni przed przewidywaną modernizacją do standardu niemal zero energetycznego]. *Building Physics in Theory and Practice*.

KEY TERMS AND DEFINITIONS

Comfort Classes: Comfort classes according to ISO 7730:2006.
HVAC: Heating, cooling, and air conditioning.
PMV: Predicted mean vote.
PPD: Predicted percentage of dissatisfied.

Chapter 9
Chosen Case Studies of nZEB Retrofit Buildings

Joanna Klimowicz
Warsaw University of Technology, Poland

ABSTRACT

The reasons why the buildings are named intelligent and the cities are called smart arise from the requirement to achieve effective use of natural resources while maintaining at least current standard of living when faced with global climatic changes and growing scarcity of resources. Now, energy efficient and environmentally friendly urban solutions tend to concentrate on the possibilities of upgrade of already existing buildings that form the majority of the contemporary urbanized landscape. Due to the increasing human population, our world is undergoing rapid urban development. This state overlaps with climate changes and growing scarcity of resources, which has a high impact on the ongoing transformation of our built environment. Many of those issues are mirrored by European legislation, especially in Energy Performance of Buildings Directive, which makes nearly zero-energy buildings a standard by 2020. Many of the technologies are already available. nZEB renovation process will be a challenge for parties involved in the design and construction process.

INTRODUCTION

Sustainable policy is directed towards initiatives which aim for energy efficiency and sustain existence of public sector solutions which aid construction of all new buildings as nZEB by 2019. It also supports measures towards implementation process of low and zero energy buildings characterized with optimum cost coefficients.

Construction sector is responsible for at least one third of world's total energy use. Low technical conditions of the majority of existing buildings and often inefficient energy solutions found in the newly constructed ones have direct impact on the high energy use. This state varies from one country to another, but for example energy performance of a single family house heated with a low temperature gas boiler in Poland is 30% higher than in Sweden (KodnZEB, 2015-2017). This chapter will be dedicated to the

DOI: 10.4018/978-1-5225-4105-9.ch009

pilot projects both on urban as well as individual building scales which allow for the provision of zero emission urban environments. Presented cases are located in Norway, as having one of the best nZEB standards already in place, even if currently accepted as a voluntary option.

BACKGROUND

Recast of EBPD Directive has been welcomed by the members as it defines various issues which for some time were interpreted rather freely depending on the local policies and culture. This includes the definition of an nZEB building and optimum costs. According to some experts (Sartori, et al. 2014), it is not possible to construct a nearly zero energy building without high rise in construction costs. On the other hand, German market prices prove that such buildings may be provided at only 2% of higher investment input (Sartori, et al. 2014). Still even the recast Directive lacks certain precision. Neither calculation methods for the energy characteristics have been included. Applicable in this case Article 7 of the EPBD states that

Member States shall take the necessary measures to ensure that when buildings undergo major renovation, the energy performance of the building or the renovated part thereof is upgraded in order to meet minimum energy performance requirements set in accordance with Article 4 so far as this is technically, functionally and economically feasible". Major renovations, as defined in Article 2 of the EPBD recast, include the renovation of a building where: "(a) the total cost of the renovation relating to the building envelope or the technical building systems is higher than 25% of the value of the building, excluding the value of the land upon which the building is situated; or (b) more than 25% of the surface of the building envelope undergoes renovation.

Member States can follow either or both of those options. Since the definition of major renovations – associated with retrofit procedures differ from country to country, it is currently very difficult to compare the outcome of various renovation choices. Hence, in 2016, the outcomes on observatory of market studies and various data tools used for nZEBs were published (Zebra2020). Due to lack of an official European definition, authors' of Zebra 2020 developed an indicator described as *major renovation equivalent* and defined three renovation levels: *low*, *medium*, and *deep*. It should nevertheless be noted, that any case studies on nZEB buildings new or retrofit, located in different countries and therefore subjected to different legal requirements, do not correspond to the same level of energy savings. Therefore any comparisons should be made to analogous processes rather than uniform outcomes.

Zebra2020 team assumed that with major renovations, building's final heating energy demand can be reduced by 50-80% depending on a country, undertaken measures and current energy efficiency of local building stock (ZEBRA 2020). In general monitoring process of the nZEB market which took place in 17 European countries showed that the annual share of building stock representing major renovations is very low; circa 0.5%in Spain or Poland and only slightly above 1.5% in Germany, France or Austria. It was also found that the heating demand in nZEBs is generally lower for new built than for renovated buildings, but no consistent patterns were found within particular climate zones. It was confirmed that the most common insulation material for the residential building's envelope is expanded polystyrene and stone wool for the non residential ones. The thickness depending on the climatic zones and local building codes, as well as building height and prevailing fire laws. The use of triple glass windows in nZeb retrofitted buildings is the most frequent window type, followed by low-emission double glass solution, the latter being more common in warm climates. The average U-value in windows is clearly lower in cold winter climates (ca. 0.85 W/m^2 K), while in warm summer climates it's ca. 1.15 W/m^2 K.

The most commonly used technology (> 80% of the nZEBs) is mechanical ventilation with heat recovery system. The major difference between countries can be found in the choice of heating systems which depend both on the type of climatic zone as well as the national policy characteristic to various countries (e.g. central Europe has a well-developed district heating system). With respect to the use of renewable energies, the photovoltaic systems, and especially solar thermal systems, are more common in warm summer climates, where radiation is higher. Nonetheless, as with heating media, the use of renewable energies is also highly influenced by national strategies and subsidy schemes. In countries such as Belgium, France, Germany, Italy. Luxembourg, Netherlands, Poland, Slovakia and UK comprising of almost 50% of those examined by Zebra 2020 Team, fossil fuel-based heating systems (coal, oil, natural gas) make up a significant share of the total energy demand for building volume heating. Natural gas is the most common energy fuel covering 50% of the energy demand for heating. The scenario shows a decrease of natural gas demand in almost all European countries. In the current policy scenario, the share of natural gas demand is 35% lower from the total energy demand for heating in 2050.(Zebra 2020). Renewable energy makes up a high share of the total energy demand for heating in countries such as Denmark (85%), Lithuania (80%), Romania (70%) and Sweden (85%). Building related carbon emissions are already low in these countries in comparison to other ones. Coal, a medium which is mainly used in the European Central and Eastern countries, will slowly run out in the long-term scenario in almost all known European mining excavation areas - ca.40-60 years (Zebra 2020). In some countries, like Poland, current policies are very much supporting the coal industry, keeping it as an important fuel for the future development which to certain extent is short sighted. Electricity demand for space cooling is growing in many South European countries. For example in Spain, the share of the electric demand used for space cooling is 11% of the total energy demand for heating and cooling. It is foreseen that the cooling market is going to grow and reach the share of 26% in 2050 in Spain and 16.5% in Italy´s building-related energy demand for heating and cooling (Zebra 2020). In view of this, it should be stated that any retrofitting procedures introduced into country policies are the key drivers for energy savings which will in turn reduce the dependency on fossil fuels.

Due to a low number of constructed and renovated nZEBs, it is very difficult to calculate expected saving. There are different reasons for discrepancies between measured and calculated performance, the most obvious being user impact (users are very often unfamiliar with building's technologies) and lack environmental awareness as well as frequent discontinuity between Client's brief, designer's input and contractor's choice of buildings materials and technologies. These issues are also associated with strategies indicating type of controlling procedures and communications between involved parties. Hence, possible differences depend on a number of different factors, mainly related to the use of building - residential vs non-residential, rather than to the scope of intervention - new vs. retrofit (Zebra 2020).

Zebra2020 team also explored policy instruments which were the most common in European countries. These may be perceived as follows:

- Building codes for new buildings and building renovation;
- Financial and fiscal support policies/programs;
- Increase of renovation rate in public buildings;
- Obligation to install renewable heating systems;

- Compliance with regulatory policies;
- Other instruments like CO2 taxes, mandatory thermal retrofitting in case of façade maintenance or/and during real estate transaction, prohibition of oil and fossil fuel boilers.

The current energy performance of buildings has a strong impact on the level of effective energy use, but cost measures vary depending on countries. In those with high contemporary efficiency of the building stock, any further improvement is so expensive that it has to be supported through strong political and fiscal incentives in order to ensure an effective nZEB policy process. The nZEB transition requires involvement from all investment's participants from the very start of the process. Implementation strategy and choices need to be understood and accepted by all involved parties. In order to start with such upgrade, a national long term policy is required; otherwise all non-standard measures are difficult to implement both from the technical as well as legal point of view.

Some EU members have already put in place regulations prior to the enforcement of the general European requirement, in some cases they implement them more strictly than required (e.g. Denmark on building performance regulation). On the contrary, other countries include only the basic minimum within national legislation, often lacking compliance indicators and with a very limited assessment of implementation of alternative measures. Eight legislative and regulatory recommendations which might be helpful towards establishing a firm benchmark for a new energy effective building type have been found during ZEBRA2020 legal analysis.

At the EU level there is an array of legislation looking to improve the energy performance of buildings. Energy Performance of Buildings Directive (EPBD) and the Energy Efficiency Directive (EED10) appear to be of most import. The EPBD specifically requests Member States to define what is an nZEB building. Also, public new buildings and major renovations are requested to fulfill the nZEB status from 2019, respectively all buildings by 2021. It also contains requirements that the Member States must take necessary measures to ensure that minimum energy performance requirements for buildings are legally set with an aim to achieve cost-efficiency levels.

The EED sets further level goals, requiring Member States to establish long-term strategies to mobilize investment processes within the renovation of the national residential and commercial buildings sectors. The first versions of these strategies were submitted by Member States to the European Commission in 2013. However, assessments (Zebra2020) indicate that more action is needed to ensure that these strategies deliver the required transformation of the building stock. Other procedures can also be undertaken to achieve better policies and financing procedures and promote existing implementation and on-going projects – level of general awareness, promotion of nZEb solution by industries or dependency of property value on energy performance.

According to Zebra2020 there are some issues which may quicken the implementation process of nZEB buildings within EU. These may be defined as follows:

- Clarification of the nZEB definition, it must be ensured that uniform national definition for nZEB is in place in all EU Member States.
- Collection of more and better data – this issue being of utmost import as only 50% of the Member Countries have an opinion on the efficient energy performance of new buildings. A unified European methodology for monitoring improvements on the energy performance of buildings should be introduced. This will be helpful to aid a better understanding and detailed monitoring of

collected data. Guiding rules should be established on the type and format of the data for monitoring improvement of energy efficiency in all buildings and give support for the collection of this data in a central repository. Besides ZEBRA2020, the EU Building Stock Observatory shows correct approach towards providing open source data on Europe's buildings. Various initiatives can be found in individual countries. There is an interesting case found in Ireland, known as the Irish Energy Action, in partnership with the EU-project Episcope (www.episcope.eu), who have developed an EPC mapping tool. The interactive map of Dublin illustrates different building characteristics (including energy poverty indicators). The data is aggregated to defined boundaries, small areas (50-200 dwellings) and electoral divisions (clusters of small areas).

- Improvement of quality and compliance – low knowledge and no skills form barriers preventing the transition to nZEB. This also includes poor workmanship (e.g. poor installation of insulation etc.) which can severely affect the performance of any building including nZEBs. Hence, there is a need to establish an integrated quality framework which addresses knowledge upgrade. Also, a transparent European system that will set a minimum of standards for certifiers' training and their professional experience to allow workers to transfer skills across borders is also a must. Competence standards should also undergo checks; also *declarations of conformity* are crucial and should be offered by independent certification parties. Any demonstration projects that bring together the public and building professionals to illustrate the feasibility of nZEB can also act as examples which form benchmarks for future buildings.
- Implementation a long-term vision, at least for 2050, for improving of the building energy performance with criteria fulfilling COP21 to limit global warming. This should include the potential of already existing volumes, as at least two thirds of the currently existing buildings will still be in use in 2050. Since so far, only some of the nZEB requirements included in EPBD refer directly to existing buildings, therefore there is a need to make modifications within the general directions. It is important that national regulations will distinguish between new building construction process and renovation as to the energy efficiency and other targets to be achieved within each of them. For building renovation nZEB target it might not be reasonable in all cases, yet cost-optimal performance levels should indicate what is achievable within indicated budget.
- Strengthened national renovation strategies should be based on national level definitions and focus on deep renovation of existing buildings with the main aim of effective energy use. These strategies should spur investments and increase the renovation pace.
- Improvement of the energy performance in the public sector can be an important trigger for the overall market for energy performance improvements. Article 5 of the EED requires 3% of public buildings owned and occupied by central governments to be renovated each year to meet the minimum energy performance standards. This 3% condition could be strengthened if applied to all public buildings.
- As energy poverty is linked to the energy performance of buildings, it is vital that this issue is included within actions to be undertaken as one of the tasks throughout all European countries. The budget for subsidies to energy poor consumers is often much higher in some countries than the budget dedicated to building renovation programs. However, shifting these budgets to the renovation of homes aiming to become nZEBs would provide benefits in terms of energy bills, as well as a higher standard of the user health and wellbeing parameters and the many other benefits brought upon by improved energy performance.

- Contemporary buildings are in a transition phase, from being unresponsive and highly-energy-demanding to highly-efficient micro energy-centers consuming, producing, storing and supplying energy. This transition and the new technical solutions which might appear in near future must be taken into account. Many of the technologies required for this transition already exist, but their potential abilities still have to be fully recognized. The development of smart grid requires the integration of buildings with elements of the energy systems including interdisciplinary urban and regional planning solutions.

In general, it must be accepted that a zero-energy building (ZEB) produces enough renewable energy to meet its own annual energy consumption requirements. Nevertheless, it must be remembered that since net energy balance is usually defined on an annual basis at certain times the building will rely on the import of energy from the grid. This allows for reduction of use of non-renewable energy in the building sector. All cost-effective measures to provide more efficient energy use include renewable energy systems that produce enough energy to meet remaining energy needs. There is number of long-term advantages including lower environmental impacts, lower operating and maintenance costs, and better resiliency to natural disasters.

Regardless of the quoted definitions, the fact that there is no definition for a retrofit nZEB building still remains. Those who wish to follow that option should most of all examine the effectiveness of innovative building structural elements and systems, regarding their structural, environmental and energy performance in retrofitting of existing buildings. This should facilitate the choice making process of products and tools allowing for the development of high performance retrofitted buildings in respect of sustainable development, in terms of design process, consumption, building methods, materials, energy efficiency, costs and user comfort.

Legislative and regulatory instruments are at the core of authority measures which should be helpful. Effective use of these instruments, such as setting long-term goals, strengthening building codes or reforming the public procurement processes to focus more on energy efficiency requirements can move nZEBs from a demonstration stage to becoming the new normal.

NORWEGIAN CASE STUDIES

Possibly Norway is one of the countries that is nearer reaching the goal than many other. Through the agreement on climate policy goals for climate policy and measurers were adopted. This first agreement was adopted in 2008, and was based on the J. Stoltenberg government's Report to the Storting on Climate Efforts (Ministry of Climate and Environment, 2014). In April 2012, J. Stoltenberg's second government presented a Report on "Norwegian climate policy". This opened a process to approve 14 proposals for climate measurers which together with the Climate Efforts Report, form the basis for Norwegian climate policy. The main objectives state that Norway should exceed Kyoto commitment by 10% in the first phase. Other initiatives were also recognized. These include policy to reduce the level of greenhouse emissions by 30% before 2020 in comparison to 1990. Carbon neutral level is established as 2050 (Ministry of Climate and Environment, 2014). In 2012 a series of climatic measures was also accepted. These include (Ministry of Climate and Environment, 2014) – implementing of the climate technology investments, funded through the yield from a new fund for climate, renewable energy and energy restructuring; phasing out of the fossil heating oil; stricter energy requirements for the building sector; increasing research on climate issues; maintaining or increasing carbon storage in forested areas; contributing to biogas development; developing passenger city transport as public, bicycle and walking choices, as well as strengthening the role of the railway in transport system. General assumption is that energy requirements should be sourced from alternative energies without any biodiversity losses.

CASE STUDY ZVB

Zero Village Bergen (ZVB) is a pilot project planned by The Research Centre on Zero Emission Buildings (www.zeb.no) whose main aims is development of market competing products and solutions destined for existing and new buildings in order to provide a break through on the market of nZEB buildings in view of their construction, use and dismantling process. The ZEB Centre has divided the research into five working areas – technologies and advanced materials, energy efficient adaptive building's envelope, energy and service sources, energy effective use, concept and strategies for ZEB buildings.

Chief researcher at SINTEF Building Research, Inger Andresen said that, *if we are to reach the climate targets, building a passive house or a plus energy building is simply not sufficient. The whole neighborhood must be carbon neutral.* This issue is currently being solved in described in the planned pilot village. One of the outcomes of the investment planning process of Bergen Village is that many issues and procedures are not included in traditional Norwegian regulations. Therefore, when planning new residential areas with highly ambitious energy and environmental goals, it is crucial that such areas should be planned within already existing plan which focuses on potential energy supply solutions.

Zero Village Bergen (ZVB) is located at Adland, circa 15 km south-east of Bergen and it will consist of buildings with very low energy requirements and minimal CO_2 emissions. The buildings use renewable energy from local energy sources. The energy surplus will be exported to repay the CO_2 emitted during construction, operation and demolition. The urban plan fulfills all important solutions which should allow reaching the goal of zero emissions. Following key issues have been included:

- ZVB is planned as a sustainable living urban quarter – the area was chosen as the one which allows for optimum conditions for solar production, as well as other conditions which are the base to reach zero emission aims. Buildings are situated on the site as to obtain a minimum carbon footprint and a maximum use of daylight. This choice allows for a rational use of infrastructure. Green areas and local urban farm area are located in the central part of the village. This allows for local sustainability and lowers the need to commute on the daily basis.
- Planning – buildings should be designed based on the conditions of energy efficiency. Passive solutions are the choice, strict conditions as to the envelope insulation and airtightness, possibility to storage surplus of heat and energy efficient equipment is also included. All buildings have to be located to allow for a case best solar exposition.
- Additional requirement was to lower the carbon emissions when preparing the site, construction process and choice of building materials (low primary energy). Energy monitoring and calculations were provided during construction process. Timber is considered as the main load bearing building material, as it emits less carbon during manufacture of elements, than concrete or steel
- In order to return carbon dioxide which was emitted during building's life cycle, best alternative source energy systems were considered. These consist of solar energy, heat pumps and a co-generation engine. Surplus energy will be exported to the city grid to be used by external consumers

for electric cars, buses and public buildings. This planning procedure also uncovered that there is a need to develop new business models both for the equipment used as well as for the energy providers and external consumers.

- Energy systems are planned according with individual energy requirements and electric energy needs. CO^2 emissions from manufacturing of the building materials are also accounted for. Within the work package – concept and strategies for zero emission buildings – analysis of two energy concepts for the whole site were provided through cooperation between ZEB researchers (NTU and SINTEF), Skanska (contractor) and Multiconsult (energy consultant). The initial energy equipment base included photovoltaic panels, solar collectors, bio-fuel cogeneration, geothermal heat pumps and wind.
- The site will be developed as a "learning area", giving conditions to construct other buildings. Science and acquired knowledge will be used in building construction business. Bergen Village will be developed in phases, in order to achieve most efficient solutions.

The ZVB will consist of 92,000m^2 TFA with more than 700 dwellings divided between terraced houses (68% TFA) and apartment blocks (25% TFA) consisting 2-4 floors. Remaining area will be allocated to non-residential purposes including retail and offices (7% TFA). For the residential buildings the thermal load is calculated through dynamic building energy performance simulations. Both of the chosen types of envelopes show that the buildings qualify as passive house buildings according to Norwegian standard. Heating system is not yet decided. However the two most probable options are either all electric solutions with heat pumps in the buildings or at a local district heating station or a thermal carrier solution with a local district system (it is still undecided whether it will be connected to the city grid). The overall ambition of the development is that the greenhouse emissions related to the operation of buildings should be zero on annual basis (Sartori I. et al. 28-2016). Due to the long time scale of the development different ambitions levels were specified for different stages in the duration of the project according to the ZEB definition. These include such aims that the area as a whole should reach the ZEB-0 level (emissions related to all operational energy shall be compensated for with renewable energy generation) and the lowest performance level for a single building should be ZEB-0+EQ (emissions related to all energy use in operation "0" except energy use for all equipment and appliances EQ which shall be compensated with renewable energy generation).

CASE STUDY POWERHOUSE KJØRBO

Another pilot project – this time a finalized development - is Powerhouse Kjørbo, modernization of two existing office buildings which were originally constructed in the '80-ties of the former Century, and converted into a plus energy standard. It is a demonstration project currently monitored, confirming that this type of buildings have both commercial and environmental sense even in cold climates. Additionally it fulfils basic developer requirements – financial gains. Powerhouse consortium consists of several entities including Skanska, developer's firm Entra Eiendom, architectonic firm Snøhetta, environmental organization ZERO, consultation firms Asplan Viak and Sapa. This team participates in many new construction projects such as office building Brattørkaia in Trondheim, as well as various modernization projects such as Kjørbo and treats acquired knowledge as a benchmark. Consortium aims at future construction of plus energy buildings, also outside Norway (Powerhouse Kjørbe, 2016)

Figure 1. Kjørbo Building: View from the main path
Powerhouse Kjørbo, 2016.

Figure 2. View of the spiral staircase used as extraction air duct, but also as possible means of communication between floors
Powerhouse Kjørbo, 2016.

Figure 3. Kjørbo building: View from internal court between office blocks
Powerhouse Kjørbo, 2016.

It should be noted that Consortium decisions evolve around two conditions. The first one is well known and describes a plus energy building which produces more energy annually than was used within the building's life cycle. On the other hand, Powerhouse definition complies with Nordic technical and environmental norms where produced energy compensates energy used during building's energy life cycle, including primary energy contained within building materials and used during construction process, production of building materials, demolition works and building use. Calculations do not include energy used by the tenants (Sowa, 2017).

Case buildings are located in Sandvika Business Park area and date back to the '80-ties. It is a complex of nine buildings, out of which four storey Building no 4 and three storey Building no 5 were chosen for the retrofitting with a total heating floor area of 5.180m^2. The site is owned by Entra Eiendom. Prior to modernization average energy used was estimated at 240kWh/m^2/a including energy for equipment. Delivered energy was divided on electricity (125kWh/m^2), district heating (75kWh/m^2) and cooling (40kWh/m^2). Heat losses for windows, ventilation, infiltration and thermal bridges were high. The facades were mainly covered with glass and black aluminum profiles. This façade in combination with no solar shading devices caused low indoor climate conditions (Sørensen, et al, 2017). Modernization works were conducted in 2013-2014. Success was achieved through a high level interdisciplinary cooperation between partners, as well as scientific input from Research Center on Zero Emission Buildings, who prepared analysis of the most effective solutions. The main goal related to the building form and indoor climate was not to be sacrificed with energy aims, even if every single choice made was measured against energy impact. (Sørensen, et al, 2017). In fact, design emphasized a relationally sensitive holistic approach, where known solutions were used and optimized.

Building Information Modeling (BIM) method was used for 3D inventory of existing buildings, as well as technical installation systems and surrounding site – density, dimension and location of green areas was used as data in the calculations supporting sun operation levels and screening of the facades, in order to achieve the best case layout of the roof sun panels. Laser scan was also found useful during preparation of a 3D model of existing load bearing elements which were fully reused in the design and during later construction process.

Newly fitted Kjørbo building was awarded BREEAM-NOR[1] Outstanding certification at Design level with best categories energy, use area and ecology, transportation and waste. It also fulfills all criteria expected in Norwegian Passive House standards for commercial buildings. It is a pilot project constructed in Norwegian Zero Emission Building standard, constructed as one of the achievements within a grant conducted at Research Center on Zero Emission Buildings, with the main aim to promote solutions allowing for low carbon emissions during the building's life cycle (KodnZEB, 2015).

Used building and installation systems solutions allowed for construction of a building which uses ca. 80% of usable energy associated with a standard (Norwegian) building with energy class C. Additionally, when including electricity generated by the photovoltaic panels, reduction is over a 100%, which allows a position of a plus energy building. Carbon analysis proved that proposed modernization

Figure 4. Kjørbo building: view of open office area
Powerhouse Kjørbo, 2016.

will allow achieving a zero carbon level during buildings usable life. New systems allow for a 10% less water use than in typical Norwegian contemporary buildings. Beside dual sanitary tanks, designers included placement of additional water meters as well as an audible water sensor system reacting in case of extreme water usage and appearance of potential water leaks.

Good standard working environment is promoted. During design process high emphasis was used to choose environmentally responsible building materials, including alternative ones with low emission of toxic chemical substances. Additional value is brought through location of the complex within a picturesque landscape and allowing for direct view out to a park and a fjord. Investment's values are associated with sustainable urban development – it dwells not only on the modernization of existing buildings, but also includes alternative means of transport. In the basement level there is a bike parking area with adequate sanitary support. Priority is given to electric cars.

This retrofit procedure allowed the building to lower annual energy use to 25.1kWh/m^2, and when including sun energy input from photovoltaic panels, energy produced surpasses actual needs. The team worked together from the very start of the project creating an interdisciplinary design process, where energy efficiency was achieved through the integration of mechanical systems and architectonic solutions. The main aim was to achieve effective energy measures through high level air-tightness of the building's envelope as well as good insulation solutions. The "U" value for walls is 0.13W/m^{2}K, roof 0.08W/m^{2}K, and windows 0.80W/m^{2}K. Achieved air tightness parameter is 0.23 air exchanges per hour with pressure of 50Pa (Powerhouse Kjørbo, 2016).

The radiators are used only during the coldest months. Ventilation air is delivered through a system of displacement uptake ventilators and circulating inside the building under the condition that door is kept open in office rooms not used. System is fed with two heat pumps taking energy from ten 200m deep boreholes. Recuperation of the excess of heat from the server room has also been included. In case of installation systems and heat recuperation it is indispensable that the heating and cooling power will be transmitted in total to the treated air flow. This can be achieved through best case scenario of the equipment, as well as correct insulation of the air condition unit walls and the in-coming and outgoing pipe system. Ventilation channels passing through unused and unheated areas, as well as all pipes connected to the heating and cooling systems also must be insulated and waterproofed. This particular system uses central spiral staircase (also used as normal communication route) as the main exhaust route (Sowa J., 2017).

Cooling requirements are lowered through the use of external screens, low electric lighting gains and use of untreated surfaces of reinforced concrete building elements acting as local temperature modifiers due to high thermal inertia. Users can manually open the windows during the summer months. Newly installed window systems allow for a high transmission of daylight wavelengths and limit the need to use artificial light. All working areas have been located parallel to the external walls. Additionally a system of energy efficient light fixtures individually controlled within every 15m^2 of the office area has been provided.

Building is also connected to a photovoltaic system located on the roof and also on the roof of a nearby parking lot belonging to the Powerhouse Kjørbo complex, and providing 1556m^2 of PV modules in total. Accepted parameters shall maintain a balance for energy requirements during building's life cycle, including primary energy contained in the building materials and used during construction process.

According to design calculations system produces over 221 MWh/year during first year and 232 MWh during second (45,2kWh/m^2 per heated areas). If not including appliances and server room the need for delivered energy was consequently circa 25.1kWh/m^2 over two years. This means a surplus, in comparison with the requirements calculated as 21.6kWh/m^2 (not including individual tenant needs). Described system provides electricity directly to the building's installation systems and the surplus is delivered do the city grid to be used by other consumers (Powerhouse Kjørbo, 2015).

One of the important ideas within this project was establishing the smallest possible carbon footprint. Design team used tools available at www.klimagassregnskap.no which was very helpful when calculating the level of primary energy contained in reinforced concrete. Additionally designers chose building materials with low primary energy characteristics – such as timber façade cladding and re-use of original glass façade elements as internal partitions. This choice allowed achieving carbon emission lower by 70% from the reference building. Existing black glass façade was exchanged for a nearly black surface charred aspen paneling (it was the investor's wish to keep the color as close to the original as possible), and aluminum framed windows. New solutions include alternative energy sources, heating, ventilation and electric light systems. Environmental friendly durable materials were also preferred as the first choice. Amongst them, the timber façade must be mentioned once more - a natural material with environmentally friendly upgrade treatment foreseen within a period of time far longer than for other facades build from different building materials. Strict conditions found for BREEAM NOR Outstanding, initiated choice of other alternative building and finishing materials, including Bisphenol A2 free paints used as coating layer for basement walls and floors, as well as insulation with low emission of toxic substances. An interesting choice was made for the interiors, where part of the ceiling and walls is clad with vertical baffles produced from recycled plastic bottles (KodnZEB, 2015). This solution was implemented as an acoustic measure and proved to be a highly aesthetic feature, giving all areas a light and airy outlook. During construction process waste management was one of the most important issues as over 97% of waste mass divided into twelve waste streams was redirected from landfill. Simultaneously the same time both investor and contractor lead a media campaign concerned with environmentally friendly buildings and explaining innovative approach to this building investment (Powerhouse Kjørbo, 2016).

User acceptance is another side of the story. Eight months after moving into new premises, the team responsible for the running-in period of the investment received a number of complaints from the tenants. The first issue arose in regard to the lighting system, the energy saving ambition and produced quality of lighting which shut off when users were not detected in the premises. This was due basically to insufficient number of sensors and their placement. After consultations with lighting professional, it was determined that 45 minutes of inactivity before shut off would be sufficient. This was a relief to the tenants as the auto-off setting was initially set to seven minutes of inactivity. Acoustic satisfaction was another of area of discussion. Baffles used on exposed concrete provided results within limits in most places, but it was more difficult in landscape areas. Acousticians continued to work on soundproofing improvements in various areas, except for exposed concrete which had to stay due to energy reasons. Other solutions such as silent areas were also provided.

Refurbished buildings will reduce energy costs by circa 80% compared with a new building with the energy label C. When the generated PV electricity is included, the reduction will be over 100%, promoting profitability to the landlord and tenants (Sørensen, et al, 2017).

CASE STUDY FREDERIK SELMERS VEI 4

The second very interesting building belongs to the Norwegian Tax Authorities. It was also initially constructed in the 80-ties of the former Century. The building was originally designed by architect F.S.

Platou as a government office building, later on changing owners but still belonging to the government class. Fredrik Selmers vei 4 is a fine example, which like Powerhouse Kjørbe proves that rehabilitation process of existing building can be cost effective also in cold climates. In this case the approach was different, as the building was retrofitted and extended - total area increased by ca 4000m^2. Currently the building has 38000m^2 and houses approximately 1500 office spaces. Initial energy use was 165kWh/m^2/year, with the present data falling around 80 kWh/m^2/year. This building also received BREEAM-NOR Very Good Certification and is also considered to be a passive building within A energy class (Nordic Innovation, 2015).

Initial design concept consisted of three buildings, connected together by galleries. The main rehabilitation strategy was to change the volume of the building to achieve a more compact form. New designers made "in-fills", extending the gallery area into usable areas used for different purposes. This idea also was followed by a new design and outlook for facades which were consulted with energy specialists. The final solution consisting of finely perforated white aluminum sheets gives the existing structure certain lightness and a fine aesthetic outlook. Modified solutions also contributed to better access to daylight, lower energy consumption and reduced greenhouse emissions.

The new continuous floor on level two is organized as a meeting place for Fredrik Selmers vei 4 with a coffee bar, canteen area, auditorium, and large meeting center used by the occupants from all of the office blocks.

The project idea was focused on environmental performance from the very outset. An environmental coordinator and energy advisor were contracted, and environmental goals were established. Contracts signed with contractors also included the requirement to pursue environmental issues. Additional data was received during workshops with future occupants, meeting regularly throughout the modernization process. The project has been a pilot project for BREEAM-NOR. Major parts of the analysis and design process were dedicated towards the use of low carbon strategies. The greenhouse gas emissions for the

Figure 5. View of former façade, Fredrik Selmers vei 4
KodnZEB, 2015.

Figure 6. View of retrofitted façade, Fredrik Selmers vei 4
KodnZEB, 2015.

retrofitted building were reduced by 48% compared to the reference building that was calculated in line with TEK 102. Since, this was a rehabilitation project and the foundations and primary structure were already existing therefore their impact was set to zero in the greenhouse gas calculations. Only the new added building materials that produce CO_2 emissions were included in the calculation analysis.

Fredrik Selmers vei 4 is an area with a good commuting standard, located only a two-minute walk from Helsfyr station which is a large public transit hub. The nearby Fyrstikktorget has a good selection of services and acts as a local center for Helsfyr area. The new design includes a facility for the parking of electric cars and a large number of bicycle places. A mobility plan was developed together with the building's occupants in accordance with BRREAM-NOR requirements and user expectations.

Following the rehabilitation process, the energy use in the building was lowered by 50%. The heating requirements are minimized due to high insulation levels and low heat loss. The façades and window areas and the window quality are a result of a thorough analysis of best case insulation, daylight versus solar energy penetration and high lighting levels in the office spaces. Energy savings are also achieved with energy effective and demand controlled lighting (16 kWh/m^2) and reduced internal loads from PCs. A local energy center provides the building with heat, cooling and warm water. Heating demands are covered by district heating (ca. 30%), as well as the reuse of surplus heat from the server area using an air-to-air heat pump (ca. 70%) (Nordic Innovation, 2015).

Figure 7. Detail of retrofitted façade perforated cladding, Fredrik Selmers vei 4
KodnZEB, 2015.

Figure 8. View of the canteen located in extended part of the building, Fredrik Selmers vei 4
KodnZEB, 2015.

The new office areas are joined with the existing loadbearing structural elements and situated between the formerly existing five blocks. These new areas were constructed using low-carbon concrete and structural steel with 50% recycled aggregate content. Additional insulation layers were added in the basement level and on the roof. The existing façade was removed, and a new insulated outer wall was constructed. The main structural choice was wooden elements, with 350mm glass wool insulation. External cladding with two layers of 95% recycled aluminum, was installed on the outside of the existing load bearing structure (Nordic Innovation, 2015). The outermost layer of cladding is of finely perforated steel and lacquered white. The internal walls consist of drywall, veneer sheeting, timber and insulation. In addition, glazed and opaque system walling has been used. Most of the façade elements were prefabricated. The interior doors are of thin, lacquered steel plates over an insulated core, veneer, laminate and glass. Recycled sheet drywall and environmentally friendly paint and sealants were chosen. All products in the building have undergone a quality control to ensure that toxins and products that negatively affect the internal air quality are eliminated. The main materials are documented with EPDs (Environmental Product Declaration) and timber products are certified to be from sustainable forest plantations.

Water-saving toilets and fixtures were installed in the building and the calculated water use is 4.7m^3/person/year. Refurbishment allowed achieving 39,859 m^2 (ca. additional 2,000m^2) of office areas destined for 1400 full time staff. Heated area is 34832 m^2, all other areas being heated indirectly. Glass to floor area ratio is 13% (Nordic Innovation, 2015).

Green house calculations as well as after construction monitoring proved that design was much better than the reference building data. Net energy consumption in Norway is set at 72kWh/m^2 per year for a passive house standard. For Fredrik Selmers vei 4, calculated delivered energy was 68kWh/m^2 per year, surpassing passive house standard and awarding the building with energy label class A. Energy is sourced from district and local heating system, a heat pump (water-water) is included within installation including recuperation of the waste heat from server room (Nordic Innovation, 2015).

Achieved U values are as follows: U-value roof – 0.12 average, U-value floor – 0.07 average, U-value wall – 0.16 average, U-value windows/doors – 0.8 average, also - specific fan power 1.5 kW/(m^3/s) and heat recovery efficiency at 85%. The choice of materials and technologies used gave the cost 500 M NOK – this being 40 M NOK extra due to conversion of the building from standard TEK'10 to a passive house standard. This can be converted to a price per square meter ca. 13,100NOK, with extra expenses energy 1047 kr/m^2. The new technical system applied allowed for the heat recovery ventilation at 0.86 efficiency level, with SFP-factor ventilation at 1.5 and SPP factor pumps at 0.3. Lighting system is fitted with sensors for daylight and movement with light fixtures at 5W/m2. Warm heat recovery unit is installed in the cooling system in the server room with coefficient – 95% used for heating purposes. Carbon dioxide calculations were also included in the design – these include 90% of recycled aluminum and 50% of recycled steel and gypsum (Nordic Innovation, 2015).

CONCLUSION

It should be noticed that current policy scenarios are driven by existing policies which include energy performance requirements, available financial instruments and the level of environmental awareness amongst policy makers. In some countries such as Norway or Denmark, more intensive policies are being developed, leading to a higher number of renovation and more efficient new building construction. These policies also include incentives towards a higher share of renewable energy sources and lower emission of carbons. Still, even this approach does not mean that certain climate or energy targets are actually met.

Based on existing case studies, it should be perceived that constant assessment and review of the building investment process during all stages (from concept, design, through construction and maintenance procedures), as well as verification of data is an indispensable part of all procedures. Reduction of building energy consumption in new building construction or renovation process can be accomplished through various means, including integrated design, energy efficiency retrofits, reduced plug loads and energy conservation programs. Reduced energy consumption makes it simpler and less expensive to meet the building's energy needs with renewable sources of energy.

REFERENCES

KodnZeB. (2015-2017). *Unpublished grant outcomes and analysis from KodnZEB project, supported by Finance Mechanism EOG 2009-2014, within the framework of cooperation Fund Operation Program PL04 Efficient Energy and promotion of alternative energy sources. Project realized with the cooperation of NTNU Trondheim Ministry of Climate and Environment*. Retrieved from https://www.regjeringen.no/en/topics/climate-and-environment

Nordic Innovation. (2015). *30 Sustainable Nordic Buildings. Best practice examples based on the Charter principles*. Nordic Innovation.

Peterson, K., Torcellini, P., & Grant, R. (2015). *A Common Definition for Zero Energy Buildings*. Prepared for the U.S. Department of Energy by The National Institute of Building Sciences, US Department of Energy.

Sartori, I., Merlet, S., Thorud, B., Haug, T., & Andresen, I. (2016). *Zero Villager Bergen. Aggregated loads and PV generation profiles*. ZEB Project report 28-2016. SINTEF Academic Press and Norwegian University of Science and Technology.

Sartori, I., Santamouris, M., Cipriano, X., & Lopez, J. S. (2014). *How to promote "nearly" Zero-Energy Buildings for new and retrofitting buildings in the municipal context*. World SB14 Barcelona, Spain.

Sørensen, A. L., Andresen, I., Walnum, H. T., Justo-Alonso, M., Fufa, S. M., Jenssen, B., & Fjeldheim, H. et al. (2017). *Pilot Building Powerhouse Kjørbo. As Built Report*. The Research Centre on Zero Emission Buildings. NTNU.

Sowa, J. (Ed.). (2017). Budynki o niemal zerowym zużyciu energii. Oficyna Wydawnicza PW.

Throndsen, W., Berker, T., & Knoll, E. B. (2015). *Powerhouse Kjørbo. Evaluation of construction process and early use phase*. SINTEF Academic Press and Norwegian University of Science and Technology.

ZEBRA 2020. (2016). *Nearly zero-energy building strategy 2020 - Strategies for a nearly Zero-Energy Building market transition in the European Union*. Co-funded by the Intelligent Energy Europe Programme of the European Union. Retrieved from www.zebra2020.eu

KEY TERMS AND DEFINITIONS

BREEAM-NOR: A type of a classic BREEAM certification modified to fulfill the expectations of Norwegian technical standards; it has five levels, with Outstanding being the highest grade.

EBPD Directive: Directive 2002/91/EC of the European Parliament and of the Council of 16 December 2002 on the energy performance of buildings, with further amendments.

Energy Poverty: State when large numbers of people in developing countries and some people in developed countries are negatively affected by very low consumption of energy and start using dirty or polluting fuels.

EPD: Environmental product declaration.

Plus Energy Building: A building which produces more energy from renewable energy sources, over the course of a year, than it imports from external sources. It uses a combination of microgeneration technology and low-energy building techniques.

Powerhouse: An alliance of companies dedicated to building PlusEnergy buildings.

Storting: Norwegian Parliament, supreme legislature established end of nineteenth century by Constitution of Norway.

ZEBRA 2020: A project which gathered eight partners from academia, research and private consultancy; 17 European Member States were covered by ZEBRA2020 (Austria, Belgium, the Czech Republic, Denmark, France, Germany, Italy, the Netherlands, Norway, Poland, Lithuania, Luxemburg, Romania, Slovakia, Spain, Sweden and the United Kingdom). It monitored the market uptake of nZEBs across Europe.

ENDNOTES

1 BREEAM-NOR is a type of a classic BREEAM certification modified to fulfill the expectations of Norwegian technical standards, it has five levels, with Outstanding being the highest grade.

2 TEK 10 is a Norwegian Standard, Building Technology Regulations, being a Guidance on technical requirements for construction work.

Chapter 10
Existing Buildings:
How to Meet an nZeb Standard – The Architect's Perspective

Urszula Kozminska
Warsaw University of Technology, Poland

Elzbieta D. Rynska
Warsaw University of Technology, Poland

ABSTRACT

This chapter is dedicated to the modernisation process of existing buildings aiming to achieve the nearly zero-energy standard. The process is described from the designer's perspective. Related issues, requirements, constraints, design options, and local determinants are analysed, and optimal architectural solutions for selected case studies are also presented. The analysis is based on the KodnZeb project, which included the modernisation of two existing buildings, located in Warsaw (the Faculty of Building Services, Hydro- and Environmental Engineering and Student Housing Muszelka), which differ in architectural features function, location, and needs. Thus, two diverse approaches to the modernisation process are examined. The chapter concludes with general guidelines and recommendations for similar architectural projects.

INTRODUCTION

In the 20th Century it became obvious that most of the existing building stock in Poland is in need of major redevelopment (Rynska, 2008). The re-cast of EPBD (EPBD, 2010) requires from 2019 that all new buildings occupied and owned by public authorities are nearly zero-energy buildings (nZEB), and by the end of 2020 all new buildings should also have the same status. Therefore, a new questions arise: is it possible to modernise existing buildings to meet a nearly zero-energy standard? How it can be done and what should be considered during the design process? Answers to these questions can be found in the analysis of the project "Design retro-fit nZEB concept for two buildings – KODnZEB", which was conducted in co-operation between Warsaw University of Technology and Norwegian University of Sci-

DOI: 10.4018/978-1-5225-4105-9.ch010

ence and Technology in Trondheim in years 2016 – 2017. The basic aim of the project was to develop an interdisciplinary step-by-step management in case of providing nZEB modernisation procedures of existing buildings (Sowa, 2017). Norwegian know-how and the technology were successfully transferred. However, this transfer had to include a "made to fit" aspect based on the joint knowledge of researchers from various disciplines and countries.

The KODnZEB project aims to modernise two existing buildings, owned by Warsaw University of Technology, to achieve an nZEB standard. Contemporary urban and architectonic features of Warsaw University of Technology complex are the outcome of after-war rebuilding process and construction of various new buildings. Interventions within historic buildings were triggered by inadequate space to house growing didactic requirements. Main emphasis was placed on efficiency and intensification requirements. However, often inadequate technical skills and low quality materials were used. The last years of the former decade, especially after receiving EU member status and possibility to draw on international funds, were filled with revalorisation works on the WUT campus. WUT management requested an urban and architectural analysis, which pointed out buildings to be extended and plots to be invested in. Prior to the start of KODnZEB project the research team and WUT management representatives analysed possible choices. WUT authorities proposed three university buildings, which required modernisation. The ones chosen as case studies differ in type, function, time of construction and used technologies. The first chosen site is dormitory "Muszelka" (Shell), one of the student housing buildings located in Narutowicza complex in Warsaw. The plan of this student social housing dates back to 1922. However, it was built in 1950 (arch. Z.Dytkowski). The second site is the seat of the Faculty of Building Services, Hydro and Environmental Engineering (FBSHEE) located in the main WUT campus area and constructed in the 70-ties based on design by S. Jaczewski and J. Reda. These two different case studies required diverse approach to the design and modernisation process. Various functions, structures, architectures, needs and requirements for selected case studies enabled the designers to investigate how existing buildings can be retrofitted not only to achieve nZEB standard but also to create modern, functional, aesthetic and environment-friendly architecture. The analysis of design processes for both selected KODnZEB case studies, presented in this chapter, aims to present main problems and challenges as well as possibilities and solutions to achieve mentioned goals. The detailed description of both case studies, design processes, selected architectural solutions can be treated as an inspiration for similar projects developed in the future. Thus, the chapter concludes with some general recommendations and advises concerning design solutions for other modernisation projects.

BACKGROUND

The first task of the design process was to provide a common definition of nZEB for KODnZEB project based on primary energy factor (PEF) value. According to EPBD Recast (EPBD, 2010): "nearly zero-energy building means a building that has a very high energy performance" and "energy required should be covered to a very significant extent by energy from renewable sources". The nZEB standard is achieved when PEF value is higher than in a zero-energy building and lower than in a building, which meets minimum requirements. In this project an nZEB building was defined as the one, which meets 90% of requirements for a zero-energy building (Mijakowski et al., 2015). This definition was chosen, as both of the buildings were undergoing retrofit development and it was decided to lower the benchmark. Thus, PEF values for nZEB are:

- For collective dwelling building (without cooling): 9,5+5,0kWh/m^2/year
- For collective dwelling building (with cooling): 9,5+5,0+2,5x$A_{f,c}$/A_f kWh/m^2/year
- For public use building (without cooling): 6,5+10,0kWh/m^2/year
- For public use building (with cooling): 6,5+10,0+2,5x$A_{f,c}$/A_f kWh/m^2/year

where A_f – heated usable area, $A_{f,c}$ –cooled usable area

Energy balance is calculated yearly for the whole building (building area is calculated to the outer façade with installations) and includes primary energy from heating, cooling, ventilation, warm water heating, lightening and auxiliary energy. Calculations are made in accordance with PN-EN 137900 (Mijakowski et al., 2015) and with the use of Design Builder Software.

THE DESIGN PROCESS

After defining the nZEB standard usage profiles were created and available technologies analysed. The architects were provided with guidelines concerning necessary insulation and new installation systems including: ventilation ducts, photovoltaic panels, heating and cooling systems. Their placement and dimensions were set out.

Then buildings' location as well as architecture, structure, functionality, technical and aesthetic state was analysed. Detailed urban and architectural analysis, literature research (Addis, 2006; Bregier, Kronenberg, 2010; Idem, 2010; Marchwinski, Zielonko-Jung, 2012; Roberts, 2005; Yudelson, 2009), documentation review and field trips were followed by a multiple-choice survey, which was conducted among the FBSHEE users. Gathered data provided the designers with information about real needs and problems of the building's users. The results of this survey were taken under consideration during modernisation process. The assessment was followed by a profound history analysis. The project was repeatedly consulted with experts in this field as design works proceeded. Moreover, the selected sites were analysed by the landscape architect, who prepared detailed tree and vegetation surveys.

In the next step, several European case studies were examined to find an inspiration for Warsaw University of Technology buildings' renovation. The aim of the majority of investigated modernisation projects was to meet current energy, safety and fire resistance standards. Moreover, their program, layout and aesthetics were improved. Various solutions were used to achieve a better energy standard. The most common practice in analysed case studies was to place a layer of insulation on the outer side of the building's envelope and cover it with plaster or other cladding (e.g. Panelaky Housing in Sobota, Slovakia, GutGut, 2014). Sometimes, mostly due to historic conservation requirements, insulation was added on the inside. More complex examples included the addition of curtain walls (Student Housing in Munich, Germany; Knerer und Lang Architekten, 2010-2013), loggias, balconies (Europarei Housing, Uithoorn in the Netherlands, Atelier Kempe Thill, 2004-2005; Aparicio, Alonso, 2012), shutters and other elements (C10 college building in Darmstadt, Germany; Staab Architekten, 2007-2011). Extensions were introduced to improve living conditions and to increase floor area (Tour Bois-le-Pretre in Paris, France, Druot, Lacaton & Vassel, 2011). Spatial rearrangements created within an existing structure included demolition of partition walls or the use of movable systems. New elements often significantly improved building's appearance. External additions highlighted the details of original elevations or were designed as a second skin, screening dilapidated façades. Extensions were joined to existing structures or were designed as independent, self-supporting elements. At times, new volumes were provided to highlight the entrance or to introduce a new function. Moreover, thermo-modernisation almost always required old windows, doors and installations to be exchanged to ensure adequate energy standard. Legibility and aesthetics of interior space was often improved by the replacement of finishing materials and lighting equipment or by the use of specific colours. The analysis of selected projects provided valuable information and inspiration for modernisation of both Warsaw University of Technology buildings – student housing "Muszelka" and the FBSHEE (Rynska et al., 2017).

During the course of the project designers from WUT were trained by Norwegian experts. Workshops were organised in Warsaw in June 2016. Two days trainings aimed to provide detailed information about the nZEB buildings' design. Experts presented the data from the Norwegian research concerning architecture, structure, installations, lighting and environmental impact of materials. Moreover, Scandinavian know-how was pointed out in selected case studies built during last years in Norway, including: an office building Frederik Selmens VEI4 in Oslo, single family ZEB Pilot House in Larvik and Zero Emission Neighbourhood Development at Adland in Bergen. But the most inspiring and relevant for KODnZEB project was Powerhouse Kjobro - a 5200m^2 office, building from the 1980s, which was redesigned in 2014 by Snohetta. This pilot project, led by Norwegian Research Centre on Zero Emission Buildings, produces more energy than it consumes (Salvesen, 2015). The energy balance was calculated for the whole lifespan of the building (i.e. for the next 60 years). The total energy demand was reduced by 80%. Highly insulated walls, windows with good insulation performance, limited thermal bridges resulted in a good thermal performance of the building's envelope. Photovoltaic modules placed on the rooftop produced more energy for electricity (200 000kWh/year) than the building required. In comparison ventilation, lighting, heating and cooling requires 145 000 kWh/year. Ventilation pre-heating, cooling and water heating in the building were run on geothermal energy. Lighting, fans and materials were optimised to reduce energy consumption. Moreover, energy losses were taken under consideration in calculations of PEF. Thus, a holistic project of Powerhouse Kjobro exceeds the nearly zero-energy standard but – to some extent - was instructive for KODnZEB project in Warsaw. However, direct implementation of Norwegian design solutions was not possible due to local requirements and diverse determinants. Norwegian case studies were treated as inspirations for Polish projects and recommended solutions were adjusted to local legal requirements. Moreover, all calculations (concerning energy, ventilation, lighting, heating etc.) were made in accordance with Polish standards and regional weather conditions.

The analytical part of the project ended with general recommendations and guidelines for the modernisation projects of both buildings. Preliminary projects of the student housing "Muszelka" and the FBSHEE were consulted with Polish and Norwegian experts during two days workshop, which took place in October 2016 in Warsaw. All recommendations and suggestions from these meetings were taken under consideration in further design process. Next six months were used to develop coherent and adequate modernisation concepts for both buildings. During that time multiple internal consultations, changes and adjustments were introduced – the design process was conducted in close collaboration with different consultants. The interdisciplinary design team consisted of architects, environmental engineers, electricians, historic and landscape professionals.

Due to local conditions and historic conservator's requirements, modernisation of the FBSHEE was a different and more complex process than the renovation of student housing "Muszelka". However, both pilot projects required extended analysis and longer preliminary design phase. The transfer of Norwegian knowledge had to include a "made to fit" aspect. Project strategies, standards and materials were chosen according to a specific local context and existing possibilities. Polish economic, environmental and social conditions vary from Norwegian situation. Thus, a direct implementation of foreign solutions was not possible and selected solutions had to be transformed, consulted, calculated and tested during design process to meet Polish standards and answer local needs.

CASE STUDY: STUDENT HOUSING MUSZELKA

Current State

Main data:

- **Gross Covered Area:** 837,25m^2
- **Height:** 15,82m
- **Usable Floor Area:** 3013,54 m^2
- **Cubature:** 13303,42 m^3

The dormitory "Muszelka" (Figure 1), one of the student housing buildings located in Narutowicza complex in Warsaw, dates back to 1922, when site between Grojecka, Mochnackiego, Uniwersytecka and Narutowicza Square was given the status of Academic Colony. The architectural project was done by architect K. Tolloczko. It was then planned that the Colony would consist of housing pavilions, a kitchen and canteen facilities, swimming pool, gym area, student's club, library, health buildings and student organization offices. Three dormitories were built before 1939 (Bratniak, Pineska and Akademik). These formed an "U" shaped complex with internal courtyard which was closed only in 1950 (arch. Z.Dytkowski) according to the original design assumptions from 1922. "Muszelka" is four stories high, with one underground level. It houses 150 students. Main load bearing elements are reinforced concrete, full ceramic brick walls with no insulation. It was modernized in the late 80-ties. The roof slab was insulated and window framework was exchanged in the late 90-ties. Architectural features characteristic to this complex are monumental but modest with predominant simplified classical features. Detail is scarce, especially on the buildings constructed after 1945. "Muszelka" fits into the initial urban context, and basically follows the site ownership lines. Hence, the South and North façades are curved forming a street line. The urban layout is much more interesting than the architectonic features used. Possibly that is the reason why this area is listed as valuable urban historic surroundings. The site where Academic Colony is situated is subject to a valid Master Plan, which will undergo changes within next few years due to a modernisation of the adjacent public square area and emphasis to create a new quality of life within busy public area. "Muszelka" is a part of the historic complex. Its simple architecture remains in strong relation with surrounding buildings. Thus, the respect for the existing form is necessary and modifications should be integrated with changes in other academic buildings.

Project

The aim of the project was to transform the student housing "Muszelka" into a functional, aesthetic, user-friendly and nearly zero-energy standard building. Local urban conditions and historic requirements limited the scope of the modernisation works. The building is a part of a student housing complex located between Narutowicza square and Mochnackiego street. The urban layout is valuable and under historic protection. All modifications of the building's outer shell and its façades should be coordinated with

Figure 1. The façade of student housing "Muszelka" from Małachowskiego street
Source: KODnZEB internal data.

adequate changes in neighbouring dormitories. However, the renovation of other student housing is not yet planned. Hence the transformation of the exterior of "Muszelka" needs to be limited to the cleaning and insulation works. New details cannot change the building's appearance and its relation with the surroundings. Thus, to achieve the nZEB standard, elevations are covered with a minimal layer of insulation. The majority of renovation works, which aim to improve the comfort of usage of the building, takes place in its inside. The structure remains untouched. Only small spatial rearrangements are introduced to improve functionality. All installation systems are exchanged. New elements such as photovoltaic panels are placed on the roof in a location that makes them invisible from the street and backyard level. The most noticeable change is the new interior design project. All finishing materials, lighting and furniture are exchanged. The colour, material and aesthetic solutions are unified. Those modifications aim to transform the degraded student housing into a modern, nearly zero-energy, functional, aesthetic and user-friendly collective residential building.

The modification of the outer shell of the building is limited to the minimum. The façades were cleaned, insulated with the thin layer of aerogel (4cm) and plastered. Delicate lines of monumental elevations with modest details remain untouched. All windows are replaced with the new ones, which must keep the character, colour and internal divisions of the existing ones. The gates leading to the backyard are cleaned and renovated. Existing roof is insulated with 20cm of mineral wool placed on the technical level, in the attic. Photovoltaic panels are positioned flat 70cm off from the roof's edge. They are invisible from Mochnackiego street and from the backyard. They do not interfere with the building's form. Cleaned and modernised façades of student housing "Muszelka" harmonise with the neighbouring historic complex of dormitories. Courtyard, green areas and urban infrastructure, which surrounds the building, is also redesigned to create modern, multipurpose and lively public space (Figure 2). The modernization includes the arrangement of an existing internal courtyard and the refurbishment of space along Mochnackiego street.

Figure 2. The façade and the courtyard of the student housing "Muszelka": View from the backyard
Source: KODnZEB internal data.

The student housing "Muszelka" requires minor spatial rearrangements. Its structure and layouts are functional and remain intact. Necessary modifications aim to improve the standard of the building, its accessibility and comfort of usage. The layout of the dormitory (Figure 3) is modified on the ground floor in the entrance area. Glass wall systems are used and the function of rooms is changed to create spacious entrance hall with reception, a common room and indispensable facilities (e.g. laundry room). New lift is located next to the existing staircase; sanitary areas for handicapped people are added on every floor to improve accessibility of the building. Also common space, kitchens, laundry and living room are redesigned. Moreover, student units are modernised. The number of people living in every area is defined as 150 persons who inhabit single, double and quadruple units. There is a family room and an apartment for 5 people in the ground floor. Every student unit consists of a small entrance hall with wardrobe, kitchen annex (incl.: a fridge, a sink, an electric kettle, space for food preparation – no cooking facilities). Moreover, the new layout proposes functional sanitary units with toilets, showers and doors. Each room has a designated sleeping and working space. All necessary modifications of student units are introduced within an existing layout – only several partition walls have to be modified to meet the current standard.

The degraded, chaotic and dark interiors of student housing "Muszelka" are completely redesigned (Figure 4). The unified system for the modernisation was created to organise and brighten existing space and to improve living conditions. All finishing materials, claddings and furniture are exchanged. Minor spatial rearrangements are introduced. On the ground floor, common room is created and it includes space for work and recreation and laundry. Modernised kitchens (Figure 5) on every floor provide the place for users interactions. All installation systems and light fitouts are replaced. Finishing materials and colour

Figure 3. Muszelka: Ground floor plan
Source: KODnZEB internal data.

Figure 4. Muszelka: Student units (proposed solutions)
Source: KODnZEB internal data.

Figure 5. Muszelka: Entrance hall and kitchen (proposed solutions)

scheme are unified. Interiors are mostly finished with white and warm yellow (NCS: S05580-Y10R). The vibrant colour, which appears in floors, doors and furniture, improves the legibility of the existing layout. White and yellow surfaces brighten up and revive dark rooms. Moreover, the introduction of bright colours reduces the energy consumption, because the power of luminaires can be lower and both daylight and electric light are well dispersed. Thus, the choice of bright colours for interiors contributed to achieve nearly-zero energy standard. All walls in the building are plastered and painted in white. In sanitary areas orange, ceramic tiles are introduced. The walls above countertops in kitchens and annexes are finished in float glass panels (e.g. AGC). The system of suspended ceilings is introduced in the corridors (e.g. mineral ceiling Armstrong Perla OP 0,95 in white). The floors in common areas are made in yellow epoxy raisin (NCS: S05580-Y10R). The terrazzo on the staircase is renovated. In the sanitary rooms the floors are covered with orange, ceramic tiles and in the student units - with light-gray fitted carpet (e.g. Shaw Contract, colour: 20111). The doors, window sills and furniture are made from MDF and acrylic boards in white or yellow. The light fitouts are selected to provide adequate lighting and to save energy. Mostly linear and square luminaires are used. All frames are painted in white. There are: inbuilt luminaires in the corridors and surface ones on the staircase (e.g. Zumtobel Z_SLOT); square ones in the kitchens, common rooms and in the lodge (e.g. Zumtobel Z_OMAXI) and similar, smaller ones in the student units (e.g. Zumtobel Z_OMINI). Moreover, there are linear, wall-mounted luminaires in the kitchens and annexes and technological ones in technical spaces.

The finishing materials and systems selected for this project create user-friendly, bright and aesthetic interior space. To achieve this goal pro-environmental, low-emission, healthy, non-toxic, reused and recyclable materials were chosen (Ashby, 2013). Such materials with the Cradle-to-Cradle certificate (Braungart et al., 2002) as fitted carpet Shaw Contract, suspended ceilings Armstrong Perla OP0,95 or float glass panels AGC are used. Furthermore, the whole design process was conducted in line with the rules of 3R (Reduce-Reuse-Recycle) (Petzet, 2012) concept: modifications are limited to necessary renovation works, the original structure of the building remains untouched and new elements are created rationally as durable and energy saving resources.

Challenges and Opportunities

The main problem concerning the modernisation of student housing “Muszelka” is the fact that the building is under historic protection. It limits the scope of works on building’s external shell and the modernisation needs to happen mostly inside “Muszelka”. The conservator requires that building’s façades remain untouched and it was also suggested that necessary insulation should be added on the inside of the external wall. Such placement of the insulation layer would not protect sufficiently from the energy loses and it would not permit to meet a nearly zero-energy standard. Moreover, it would reduce already small living area. Thus, other solutions were discussed with the historic consultant and other experts. It was decided that a thin layer of external insulation, which does not change the building’s appearance, would be added. The student housing “Muszelka” is covered with a 4cm layer of aerogel. The façades after the modernisation keep existing details and harmonise with neighbouring complex of historic dormitories. The downside of the use of aerogel is its higher price in comparison with a regular insulating material. Thus, this solution is not always economically feasible.

The requirements of conservation protection limit the use of photovoltaic panels. Because it is forbidden to interfere with the building’s façade outlook, all necessary photovoltaic panels are placed on the roof. Moreover, they have to be arranged in an adequate distance from the building’s edge to be invisible from the street and backyard level. According to the conducted urban analysis, photovoltaic panels need to be placed 70cm from the roof's edge. Thus, the space for photovoltaic panel is limited and maximum potential energy gains are reduced. In some cases, where the roof area is small, the requirements of historic protection can be a significant constraint, which blocks the possibility to meet an nZEB standard.

The other challenge of the design process is "Muszelka"'s low technical and aesthetic state. However, this disadvantage can be treated more as an opportunity than a real problem. Contemporary technologies and a well-balanced project of modernisation can improve the building's technical and aesthetic state significantly. New insulation, windows, finishing materials, colours and a coherent system of visual information transform outdated interiors and create pro-environmental, user-friendly, healthy, aesthetic and functional living space.

CASE STUDY: THE FACULTY OF BUILDING SERVICES, HYDRO AND ENVIRONMENTAL ENGINEERING

Current State

Main data:

- **Gross Covered Area:** 2202,31m^2
- **Height:** 30,4 – 41,36m
- **Usable Floor Area:** 17476,95m^2
- **Cubature:** 76751,94 m^3

The Faculty of Building Services, Hydro and Environmental Engineering (Figure 6) is located in the Main Campus area, which is also listed as a historic preservation zone (Werner, 2015). Nevertheless the building itself is not under historic care as its location was initially pointed out as a potential place for development in 1955. The building was constructed in the 70-ties, based on design by S. Jaczewski and J. Reda. It has an "L" shape layout and forms an internal open atrium flanked by existing historic building. The wing located parallel to the street is eight stories high, the other one – has eleven levels. It is used on everyday basis by different groups of students (app. 2000 people). Main load bearing elements are in reinforced concrete, monolithic in the basement level and prefabricated framework skeleton elements on the upper floors. Slabs are also prefabricated. Its architecture is in a strong contrast with surrounding historic eclectic buildings dating back to the turn of 20th Century. In fact, this modernistic building has no architectural linkage with existing historic details, as it was purposely designed to form a contrast both in scale as well as in choice of tectonics. The detail of decorative, ceramic mosaic modernised in various shades of blue still exists on West, East and North façade. The South elevation was insulated in 2007 and mosaics were mostly hidden behind insulation and solar photovoltaic batteries. The contrast between the FBSHEE and surrounding buildings is enhanced by poor technical condition and questionable aesthetics of façades, typical for this type of architecture (Basista, 2011). Last renovation works included thermal insulation of the roof of the lower building, replacement of old windows, photovoltaic panels installation on the southern façade and some changes in the interior. Unfortunately, new elements do not improve architectural features. Energy audit showed that new insulation needed to be placed on

Figure 6. The FBSHEE: Current state
Source: KODnZEB internal data.

building's roof and elevations. But necessary modernisation should not be limited to thermal insulation. The visual perception of old façades and building's technical condition should be improved as well as the linkage of the faculty with picturesque surrounding could be provided.

Project

The modernisation of the Faculty of Building Services, Hydro and Environmental Engineering (FBSHEE) aims to achieve nearly zero-energy standard, improve the building's external and internal appearance, comfort of usage and functionality. To meet required energy standard it is necessary to introduce new insulation and to exchange all windows and installation solutions. To improve the comfort of usage the adequate temperature and lighting conditions should be created. Moreover, shading and acoustic protection elements from Nowowiejska street side are needed. It is also indispensable to improve learning and working conditions in seminar rooms, auditoriums and offices as well as to introduce comfortable working and social areas in corridors. Due to bad technical and aesthetic state the FBSHEE's facades are renovated and all finishing materials are exchanged and unified. These mandatory modifications of the building's form and spatial layout aim to create a linkage between the degraded FBSHEE, a green courtyard and historic buildings of main WUT campus. Multiple goals and the scale of the building cause the modernisation of the FBSHEE to be a complex process, which requires coherent transformation procedures. It has to integrate such indispensable (for nearly zero-energy standard) technologies and elements as: ventilation ducts, photovoltaic cells, shutters and acoustic protection. Moreover, those elements have to be designed in a way which improves the building's appearance and links it with the picturesque WUT campus. Due to diverse factors and requirements for each façade, it was decided that the project should be developed in line with the "flexible second skin" concept (Figure 7). Such skin should integrate all necessary elements and harmonise with neighbouring buildings and greenery. Thus, South and West façades, located close to busy Nowowiejska street and the main entrance, are developed

Figure 7. The FBSHEE: Design schemes
Source: KODnZEB internal data.

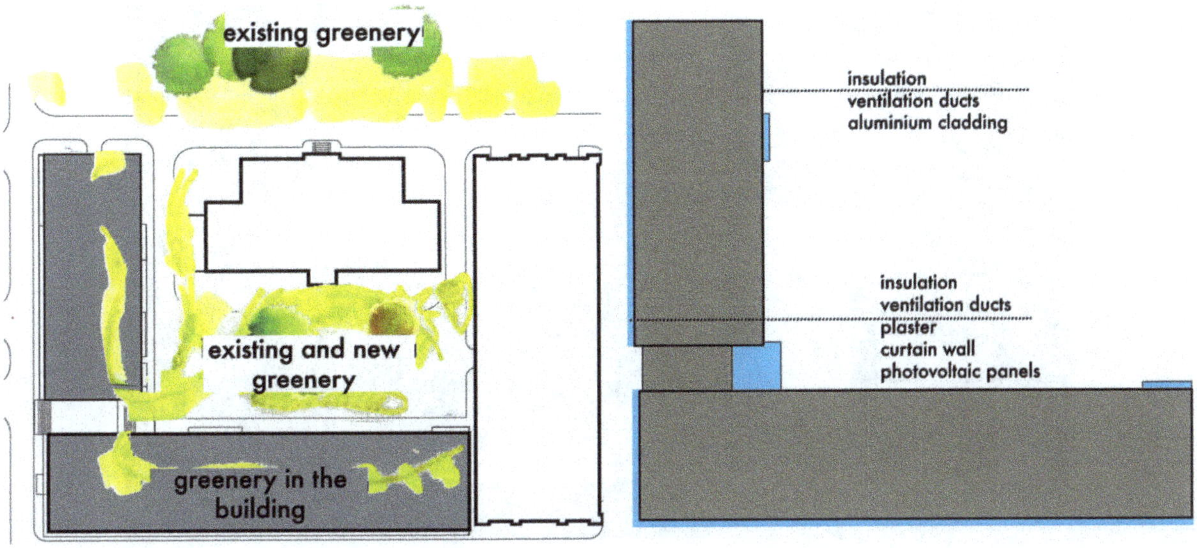

as curtain walls with functional details and technological features. East and North elevations, which face the historic WUT campus, are covered with greenery and create soft, natural background for an internal courtyard - a public recreation space. Surrounding greenery breaks the FBSHEE's monumental and rectangular volume. It creates a linkage with historic WUT campus and encroaches the faculty building through a newly designed glass cubic atrium – from there it spreads around internal recreation and working space. The green pockets appear on every floor to break the outscaled space and to improve the comfort of usage – they create a user-friendly internal space.

To achieve a nearly zero-energy standard it was necessary to introduce new layers of insulation and multiple technological elements. This requirement can be seen as a possibility to improve the degraded, existing façade. New layers are introduced in form of a second skin, which integrates such necessary insulating and technological elements as: a 21cm layer of mineral wool, ventilation ducts (width: 20-40cm), photovoltaic cells, shading and acoustic protection elements. The location of those elements is related to the sun, street and campus – differs depending on the building side. Diverse conditions required that every façade is designed independently. The new second skin is flexible. The layer of insulation and ventilation ducts are present on every elevation. Photovoltaic cells, shading and acoustic protection elements are located on the West and South façade (Figure 8) – where the building neighbours a busy street and tramway tracks. Thus, southern and western façades are covered with new curtain walls. This glass shell with independent steel structure is placed 1m away from the existing reinforced concrete wall. The existing wall is insulated, plastered and covered with ventilation ducts. The curtain wall hides installations, protects from the noise and integrates elements, which shade and produce energy. Photovoltaic cells, which are embedded in the glass curtain wall, create fine detail of the façade. Their placement is related to the features of the existing wall – photovoltaic cells are more densely spaced in horizontal area between windows, where they cover ventilation ducts. There are fewer cells in the windows area, upper part of the façade and on the first floor level to create blurred edges of the effective (in terms of energy production) elevation zone. Similar solution is used on the southern façade, close to the Old

Figure 8. The FBSHEE: View from Nowowiejska street
Source: KODnZEB internal data.

Boiler House. Photovoltaic cells are placed less densely in the direction of the historic building – this underlines the contrast between the old and new substance. This detail is designed to meet requirements of the historic conservation protection, which includes condition to keep the detail of blue-green mosaics existing between window openings. Thus, these elements are covered with transparent double-glazing to keep façades insulated.

Elevations from North and East (Figure 9) are surrounded with the greenery of the historic WUT campus. Thus, there is no need to use shading and acoustic protection elements or photovoltaic panels. For this reason curtain wall is not used on these façades – only some elements are introduced in staircases and new atrium. Elevations are insulated with a layer of 21cm of mineral wool. Vertical ventilation ducts (dim.100x100cm) are located beside staircases – they are hidden in the slight interpositions of the façade (the outer wall of the staircase is moved forward by 100cm). Horizontal ventilation ducts, placed between windows, are flattened. Their width is reduced to 20cm when an adequate cross-section area is kept (required relation between width and height of the duct is 1:5). Due to the reduction of ducts it is possible to hide them and the layer of insulation under the customized aluminium cladding (e.g. Rheinzink panels). The surface of aluminium panels is lightly curved under the ventilation duct and the space above is used as a plant pot. In the future, the greenery from the pots will overgrow the façades. The renovation of space around the FBSHEE requires complex approach, as it is a part of main WUT campus. The existing courtyard is now mostly used as a parking place. In future it will be transformed to a lively public square with wooden furniture, eco flooring and other organic green arrangements. The external planted areas are linked with internal building arrangement through a new glass atrium. This linkage opens the Faculty to a new plaza and historic surroundings. Moreover, the atrium connects the external greenery with the new internal green solutions: a multi-storey green wall in the atrium, plants in concrete pots on the balconies and other small forms of greenery. These enhance the building's common space natural environment and create comfortable and healthy areas for work and rest.

Figure 9. The FBSHEE: View from the WUT campus
Source: KODnZEB internal data.

The spatial layout of the FBSHEE is simple and legible (Figure 10). Seminar rooms, laboratories, offices, auditoriums and toilets are located along "L" shaped corridors. According to the conducted user survey the size and location of these rooms is correct. But there is a requirement for recreation and working space for students. Thus, they are introduced on every floor – in variety of the new atrium and in the newly designed bay windows in the northern part of the building. The vertical atrium hides an entrance to the green courtyard. The new glass cubic form transforms unused balconies into student work and recreation areas, underlines the back entrance and links the building with the historic surroundings, public space and greenery. The new linkage is also created between the FBSHEE and the Old Boiler House. An unused, narrow space between buildings was filled with random and low-quality materials. In the project, they are replaced with light glass - steel structure. The transparency of the joining element creates an invisible linkage between buildings and introduces the insight to the WUT campus from Nowowiejska street. Moreover, due to fire protection regulations, it was necessary to add a new staircase in the upper part of the building. It starts from the top floor of the lower building and does not influence the building's appearance.

Figure 10. The FBSHEE: Fragment of the floor plan
KODnZEB internal data.

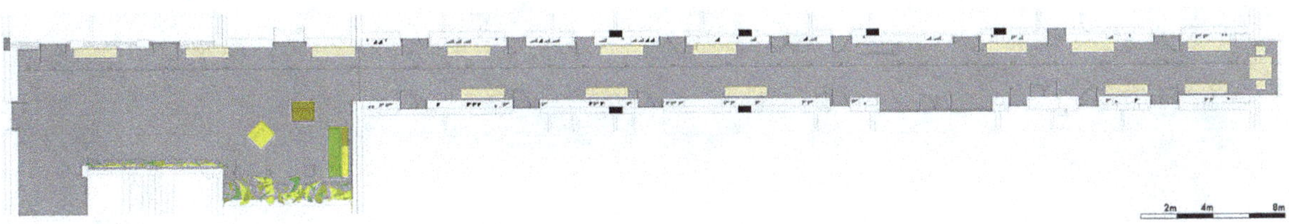

Currently, interiors of the FBSHEE are chaotic, unaesthetic and finished with low-quality materials. Thus, the aim of the project is to unify multi-material interior fit-out. The use of simple forms and consistent design of lighting enables necessary energy savings and improves the aesthetics of the buildings' interiors Figure 11). The white and gray colours are used in the interiors of the FBSHEE to improve legibility of existing layouts and to enhance building's aesthetics. Low-key colours chosen for the interiors help to calm and unify numerous existing aesthetic choices. Walls are painted in white, floors (quartz flooring, tiles in sanitary rooms), laths and doors are grey (NCS: S4000N). The same colours are visible in doors, plinths, shelves and toilets. Monochromatic interiors are brightened with wooden and green elements. Furniture is made from light, pinewood and the only colour in interiors is that of natural greenery (vertical walls, plant pots) and furniture finishing and upholstering (NCS: S1070-G30Y, S3060-G60Y, S1070-G60Y). All benches, railings, information boards, window sills and acoustic claddings in auditorium are made from pinewood (Figure 12). Suspended ceilings are introduced in corridors (e.g. mineral ceilings Armstrong Perla OP 0,95 in white).

Lighting system is changed and optimised to provide energy savings necessary to achieve nearly zero-energy standard. New lighting system is chosen to reduce energy consumption and to improve the existing learning and working conditions. Surface-mounted, in-built, spot and linear luminaires as well as technical lights with extra protection are designed (e.g. Zumtobel Z_SLOT_1200, Z_SLOT1200_NB, Z_SLOT_2000, DARK, Z_OMAXI). The choice of light fittings enhances the aesthetics of interiors. All frames of the luminaires are painted in white. Moreover, there are decorative luminaires in the atrium. Transparent, spherical lights (L45, diameter: 40cm) are hanged irregularly filling the whole atrium space.

The finishing materials and systems selected for this project create user-friendly, bright and aesthetic interior space (Ashby, 2013). To achieve this goal pro-environmental, low-emission, healthy, non-toxic, reused and recyclable materials were chosen. Such materials with the Cradle-to-Cradle certificate as suspended ceilings Armstrong Perla OP0,9 were introduced. Only certificated, locally sourced wood was used. Furthermore, the whole design process was conducted in line with the rules of 3R (Reduce-Reuse-Recycle) (Petzet, 2012) concept: modifications are limited to necessary renovation works, the original structure of the building remains untouched and new elements are created rationally to last and to save energy and resources.

Figure 11. The FBSHEE: Interiors - corridors and atrium
KODnZEB internal data.

Figure 12. The FBSHEE: Interiors - auditorium and office room
KODnZEB internal data.

Challenges and Opportunities

The FBSHEE building is not covered by conservation protection. Thus, the scope of modernisation works does not have to be limited to inside areas. Degraded façades can be redesigned to create a linkage between the faculty - an out-dated, large-scale building - and historic buildings of the main WUT campus. Currently, the architecture of the FBSHEE remains in a strong contrast with the neighbouring complex. Its inhuman scale and poor technical and aesthetic state underline this contradiction. Moreover, local conditions and character of the surrounding vary on every side of the building. From South and West the FBSHEE neighbours a busy Nowowiejska street. From North and East it borders the WUT historic campus and existing greenery. Different are also technological requirements for each façade. On South and West elevation it is indispensible to introduce photovoltaic panels, shutters and elements that provide acoustic protection. Those elements are redundant on other sides of the building. Therefore, the goal – to meet a nearly zero-energy standard and to modernise the FBSHEE so that it will harmonise with the surroundings – can be seen as a challenge. Diverse conditions, requirements, large-scale of the building, its poor technical state, aesthetical chaos of outer and inner shell show the complexity of this modernisation project. During design process not only a problem of achieving the nZEB standard should be solved but also the architecture, interiors and functionality of the building needs to be improved.

However, the complexity in this case study is not only a challenge - it is also an opportunity to meet a required - nearly zero-energy - standard. In case of an out-dated building, which is not covered by conservation protection, it is possible to undertake complex modernisation actions. Façades can be redesigned, installations and materials – exchanged and technological elements – implemented on the outer shell of the faculty. All these necessary modifications would not be possible if the building was under strict conservation protection or could be unjustified and more difficult, if it was already modernised. If that was the case, for such large-scale building as the FBSHEE, it could not be possible to achieve an nZEB standard. Thus, seemingly negative features of the faculty in reality work in favour of the renovation project. The building's poor technical and aesthetic state leaves space for unquestionable modernisation works. Its strong contrast with neighbouring surroundings and lack of conservation protection does not limit the renovation to a preservative updating but creates a functional example of modern architecture. Finally, over-scaled dimensions of the FBSHEE result in a large space on elevations and roofs for elements, which - in the end - produce more energy than the building uses. Thus, through a complex, collaborative and interdisciplinary design process a nearly zero-energy, user-friendly and aesthetic education building is created.

FROM CASE STUDIES: RECOMMENDATIONS FOR SIMILAR PROJECTS

Two presented case studies: the student housing "Muszelka" and the Faculty of Building Services, Hydro- and Environmental Engineering differ in location, urban context, architecture, scale, materials and function. Various local determinants, conservatory requirements and needs result in diverse approach to the design process and miscellaneous architectural solutions. The modernisation of the FBSHEE is a different and more complex project than the renovation of student housing "Muszelka". Both buildings require refurbishment which is not only limited to adequate insulation. The design process of selected case studies was conducted in close collaboration with Norwegian designers and their experience in nearly zero-energy architecture design. However, Polish economic, environmental and social aspects vary from Norwegian situation. Thus, a direct implementation of foreign solutions was not possible – the transfer of knowledge had to be adjusted to local conditions. Project strategy, standards and materials were chosen according to a specific local context and existing possibilities. The same issues should be taken under consideration, when the outcome of the KODnZEB project is used as a reference for other projects. When this is acknowledged, selected case studies may be inspirational and a valuable source of knowledge for similar projects. Moreover, some general recommendations for a nearly zero-energy existing buildings can be formed:

- Every building is different so the approach to the design process should be each time unique. The direct transfer of foreign know-how is not possible. It is necessary to acknowledge a "made to fit" aspect and to take under consideration such local determinants as: urban context, local climate conditions, architectural features, building's function, structure, materials and its technical and aesthetic state. Therefore, a profound analysis of urban, architectural, structural, environmental, economic and social conditions should be conducted. The assessment needs to include literature and documentation review, field trips and tests as well as other expert opinions. Moreover, it is recommended to conduct a survey among the building's users to evaluate real needs and problems, which can be solved during the modernisation process.
- Existing buildings are often covered by conservation protection. Even not listed building can be a part of or in relation with an urban complex, which is under conservation protection. In that case, some limitations can also apply. Thus, it is necessary to check and know the conservator's requirements for the modernisation on an early stage of the design process to avoid possible complications. It is advised that consultations with the historic expert take place in the preliminary phase of the design process.
- The nearly zero-energy projects need to be developed in close collaboration with various experts. The interdisciplinary team should include: architects, environmental engineers, electricians, historic and landscape consultants. Indispensible consultations, tests and meetings should be scheduled in advance. The team should closely collaborate since the beginning of the design process

and during the whole process. The later stakeholders are included, the more complications appear. Design solutions should be discussed and approved by other participants of the process.

- Different technologies can be used to achieve a nearly zero-energy standard. The main elements which should be taken under consideration in the project are: new insulation, the exchange of windows, claddings and installations to improve building's energy performance and the use of energy producing elements such as photovoltaic panels. It is important to choose technologies and installation system in the beginning of the project. All necessary equipment, ducts and other technological elements should be defined (with their dimensions) on an early stage of the design process. It is advised to check ventilation ducts' sections and assess the possibilities of their modification with experts – often the sections may be significantly reduced. Moreover, it is recommended to check different locations of photovoltaic panels or their placement on the building's façade. Options should be consulted with the conservator to avoid possible complications (e.g. it is good to check if the drawing created by photovoltaic cells harmonise with the existing façade or neighbouring buildings).
- Broader approach to modernisation project is recommended: the renovation should not be limited to insulation and implementation of energy producing technologies. It should also include common refurbishment practices. The functionality and comfort of usage of an existing building can be improved by small spatial rearrangements. The legibility of layout can be updated by the introduction of colour or coherent information system. The technical and aesthetic state of the outer shell and interiors can be modernised, when finishing materials, claddings, installations, lighting and furniture are exchanged. Furthermore, the linkage with an existing urban context and surrounding greenery can be also updated through an adequate landscape design.

CONCLUSION

Presented selected case studies, the student housing "Muszelka" and the Faculty of Building Services, Hydro- and Environmental Engineering, show that a nearly zero-energy standard can be achieved also for existing building stock. Thus, requirements of the Re-case of EPBD may be fulfilled. However, it can only happen if the project is conducted by the interdisciplinary team and when all stakeholders closely collaborate during the whole design process. No direct implementation of foreign know-how and solutions is possible. It is necessary to take under consideration all local urban, architectural, structural, environmental, economic and social conditions in the renovation project. The modernisation process should not only aim to meet a nearly zero-energy standard but also to improve overall qualities of the buildings: its functionality, comfort of usage, external appearance, internal environment and relation with surroundings.

REFERENCES

Addis, B. (2006). *Building with reclaimed Components and Materials*. London: Earthscan.

Aparicio S., Alonso J. (2012). Reestructuracion de la vivienda social en Holanda: Caso Europarei, Uithoorn. *Revista de Architectura, 27.*

Ashby, M. F. (2013). *Materials and the Environment. Eco-Informed Material Choice*. Oxford, UK: Elsevier Inc.

Basista, A. (2001). *Betonowe dziedzictwo. Architektura w Polsce w czasach komunizmu*. Warszawa, Poland: Wydawnictwo Naukowe PWN.

Braungart, M., & McDonough, W. (2002). *Cradle to cradle: Remaking the way we make things*. London: Vintage.

Bregier, T., & Kronenberg, J. (Eds.). (2010). *Wyzwania zrownowazonego rozwoju w Polsce*. Kraków: Fundacja Sendzimira.

EPBD Recast, Directive 2010/31/EU of the European Parliament and of the Council o 19 May 2010 on the energy performace buildings (recast). (2010). *Official Journal of the European Union*.

Idem, R. (2010). *Uspolecznianie procesu zrownowazonego projektowania architektonicznego*, Gdańsk Wydawnictwo Politechniki Gdańskiej Knerer und Lang Architekten. Retrieved February 3,2016 from http://www.knererlang.de/

Marchwiński, J., & Zielonko-Jung, K. (2012). *Wspolczesna architektura proekologiczna*. Warszawa: Wydawnictwo Naukowe PWN.

Mijakowski, M., Narowski, P., & Sowa, J. (2015). Definicja budynku nZEB na potrzeby projektu KODn-ZEB. Warszawa: Academic Press.

Petzet, M. (2012). *Reduce, Reuse, Recycle: Rethink Architecture. German Pavilion*. Ostfildern: Hatje Cantz.

Roberts, J. (2005). *Redux. Designs that reuse, recycle and reveal*. Layton, Gibbs Smith.

Rynska, E. D. (2008). Rehabilitation and adaptive reuse of historic buildings in Poland. In *Harmonisation between Architecture and Nature. Eco-Architecture II*. WIT Press.

Rynska, E.D., Kozminska, U., & Oniszk-Popławska, A., Szubert, Klinowska, D, & Tofiluk, A. (2017). Sustainable interdisciplinary transformation of Warsaw University of Technology Building. KodnZeb case study. *International Journal of Sustainable Development and Planning*, *12*(4).

Salvesen, F. (2015). *Powerhouse Kjobro-Renovation to plus energy standard*. Asplen Viak AS, Company presentation for KODnZEB project.

Sowa, J. (Ed.). (2017). *Budynki o niemal zerowym zuzyciu energii*. Warszawa: Oficyna Wydawnicza PW.

Staab Architekten. (2007-2011). *Darmstadt University of Applied Science. Overhaul and Extension of the C10 Highrise*. Retrieved February 3, 2016 from http://www.staab-architekten.com/

Wagner, A. (2015). *Architektura Politechniki Warszawskiej*. Warszawa: Oficyna Wydawnicza Politechniki Warszawskiej.

Yudelson, J. (2009). *Green Building through integrated design. A GreenSource Book*. New York: The McGraw-Hill Companies, Inc.

KEY TERMS AND DEFINITIONS

3R: A concept based on European waste management hierarchy that promotes reuse and recycling; it states that architectural interventions should be limited to indispensable actions and favours smart strategies over elaborated forms.

Building Retrofit: An act of modernisation, renovation, or refurbishment of existing buildings to improve their state and to meet current requirements.

Cradle-to-Cradle: A theory and an approach to architecture and industry that sees products and materials as parts of closed, continuous cycles (biological and industrial ones); it eliminates the problem of waste and promotes renewable resources, the reuse of materials and substances; aims to create fully resilient urban settlements and buildings.

Energy-Efficient Retrofit: An act of modernisation of the building that reduces the amount of energy required for the building's operation.

NZEB Standard, Nearly-Zero Energy Standard: A standard of very high energy performance that uses significant extent energy from renewable sources; an nZEB standard is achieved when PEF value is higher than in a zero-energy building and lower than in a building that meets minimum requirements.

Sustainable Modernisation: An act of renovation or refurbishment that enables a building to maintain resources at certain rate or level and that uses a resource so that it is not depleted and permanently damaged; a pro-environmental, healthy, energy-efficient, resource-efficient modernisation process.

Urban Regeneration: A program of redevelopment of a degraded urban area or its part to reverse its decline, including the redevelopment of physical structure, environment, economy, and social relations.

Chapter 11
Single-Family Residential Building Energy Retrofit:
A Case Study

Michał Pierzchalski
Warsaw University of Technology, Poland

ABSTRACT

This chapter is a case study for the energy retrofit of an existing single-family residential building. The main assumption of the project was creating a model example for an energy retrofit with the aim of achieving the nZEB standard in existing residential building. The discussed building was built between the 1960s and the 1970s. The building was built using mixed technologies. The flooring on the ground floor was replaced; the foundation, external walls, and roof were thermally insulated. The windows and doors were replaced with higher parameter ones. Moreover, a modern biomass boiler was installed in the building along with the installation of a mechanical bidirectional ventilation unit with a heat recovery. Before the renovation, the building used about 133.4 GJ final energy for heating annually. After the renovation, the building uses about 8.89 GJ annually. The author describes all the stages of the renovation, the technical solutions, the calculations of economic and environmental benefits of the conducted renovations.

INTRODUCTION

Single-family detached houses are by far the most commonly found buildings in Poland. Single-family and multi-family residences constitute 88% of Poland's residences, and as much as 94% of all residential buildings when taking into account net usable area (Building Performance Institute Europe, 2012).

According to statistical data (Ministry of Infrastructure and Development, 2014a) and analyses by Żurawski (2012) these are primarily buildings with poor thermal insulation and low quality windows, which is connected to the high amounts of energy required to heat these buildings. Table 1 presents a comparative list of existing buildings in Poland. It is important to note that according to Żurawski (2012) the real demand could be higher than the one calculated and presented below by as much as 60%.

DOI: 10.4018/978-1-5225-4105-9.ch011

Table 1. Comparative list of residential buildings in Poland (up to 2014) with their respective per unit of energy demand indicators

Year of Construction (Year)	Number of Buildings		Per Unit Energy Demand for Final Energy FE (kWh/m²/year)	Per Unit Demand for Primary Energy PE (kWh/m²/year)
	Thousands	%		
Before 1918	404.7	7.3	>300	>350
1918-1944	803.9	14.5	260-300	300-350
1945-1970	1363.9	24.6	220-260	250-300
1971-1978	659.8	11.9	190-220	210-250
1979-1988	754.0	13.6	140-190	160-210
1989-2002	670.9	12.1	125-160	140-180
2003-2007	321.6	5.8	90-120	100-150
2008-2011	205.1	3.7	-	-
Under construction	27.7	0.5	-	-
Undefined	332.7	6.0	-	-
Total	5544.3	100.0	-	-

Adapted from Ministry of Infrastructure and Development (2014a).

Therefore, apart from ensuring high-energy efficiency in new buildings, the key and primary priority of actions should be the improvement of energy efficiency in Poland's existing buildings.

At the beginning of 2015 the Polish National Energy Conservation Agency (KAPE SA) began an energy retrofit so-called "deep energy retrofit" project with the aim of creating a model example of renovating single-family residential buildings. A building built between the 60's and 70's of the last century located in the Dąbrówka-Wyłazy village of Poland's Mazovian Voivodeship was chosen. The main assumption of the project was adapting the building to requirements, which will apply to all buildings in Poland designed after 2020. The requirements for buildings designed after 2020 are the same as the standard and definition of near Zero Energy Building (nZEB).

During the realisation of the project the "Ładny Dom" magazine cyclically published articles describing each stage of the renovation. The articles appeared in issues between October 2015 and June 2016. The chapter describes the entirety of the decision, design, and implementation process of the building's renovation. Moreover, the author has compared the building's parameters before and after the energy retrofit. The cost of the entire endeavour is also presented, along with a description of all the formal problems encountered during the projects realisation. General recommendations for similar renovation projects are presented.

In Poland renovations consisting of improving a building's energy standard have been going on for years. Their main aim, as well as the reason for why this type of work is required, is lowering the cost of operation by decreasing energy and fuel consumption. The second motivation for this type of renovation work is increasing the comfort of a building's use. It is important to note that there might be many more reasons for making such a decision, e.g. the desire to improve the owner's social status, or more recently environmental reasons. The latter one is most often connected to observing poor atmospheric air quality during heating season and smog alerts in recent years.

Most energy retrofits rely on additional insulation of the walls and roof, and replacing windows. The possible scope of energy retrofit actions is conditioned by the investor's financial ability. A large role affecting the scope of renovation may also be the build owner's lack of knowledge on possible energy retrofit work or on the effect on each of the building's elements on the use of heat energy. This is why comprehensive energy retrofits of all elements of a building, which can affect the buildings energy consumption, are not as popular.

An energy retrofit of building concerns two areas of action:

- Limiting the demand for heat energy required to ensure appropriate air temperature inside the building. This is achieved by additional insulation of partitions, replacing windows and outside doors, renovation of the ventilation installation, recovering waste heat or capturing solar thermal energy.
- Improving effectiveness of thermal energy generation and distribution in the building. This is done by modernisation or replacement of the heat source or of particular elements of the heating installation.

Most energy retrofits performed rely only on additional insulation of the outside walls and the roof, or possibly replacing the windows, which leads to a decreased demand for heat. In these cases, modernisations and adjustments of the heating system installations, adapting their operation to the decreased demand for heat are rarely performed, which does not give the possibility of using all the potential effects of additionally insulating a buildings partitions.

Sometimes another situation takes place: performing the energy retrofit in the reverse order, beginning with the heating system installation. In these cases, heating costs also decrease after the modernisation or replacement of the heating system. Problems appear if and when afterwards the building's owners decide to add additional insulation to outside partitions. When this happens it appears that the installation is oversized, a boiler with too much power was used, which is now working at too short increments with increasing standing losses, oftentimes connected with using radiators, which are too large (or too many). In these situations apart from a poorly operating heating system the investor also has to bear much higher renovation costs.

In this case it is easy to reach the conclusion that when it comes to energy retrofits It is better to be comprehensive. The benefits and effects of a complete approach could be large enough that it might be economically feasible to bear the increased costs or even take out a loan along with paying the additional interest.

DEEP ENERGY RETROFIT

The term "deep energy retrofit" often appears in many different elaborations published in Poland, however there isn't a single cohesive and biding definition of the process concerning a comprehensive, total energy retrofit.

In an elaboration titled "Financing building thermo modernisations from funds available for cohesion policies" (Ministry of Infrastructure and Development, 2014b) (based upon Technical guidance: Financing the energy renovation of buildings with Cohesion Policy funding, Final report: A study prepared for the European Commission DG Energy (European Commission, 2014) defined the term "ground-up renovation" as follows,

According to the Energy Efficiency Directive (see motive 16) economically feasible bottom-up renovations lead to modernisation, thanks to which both the consumption of delivered energy and the consumption of final energy of buildings are reduced by a significant percentage in comparison to levels before the renovation, which has the effect of giving the building a very good energy characteristic. These ground-up renovations can also be conducted by stages. The Commission's Offices (see SWD (2013) 143 final), that significant improvements of energy efficiency coming from ground-up renovations allow achieving more than 60% of energy savings.

The term "thermo modernisation action" is currently in use and defined by the Law on supporting thermo modernisations and renovations (Marshal of the Sejm of the Republic of Poland, 2008),

... thermo modernisation actions – endeavours with the object of:

a) *improvement, resulting in decreasing the demand for energy delivered for heating hot domestic water and heating to residential buildings, communal residence buildings, and buildings belonging to territorial government entities meant for public use,*
b) *improvement, resulting in decreasing primary energy losses in local heating networks and heat sources supplying local heating networks, if the buildings mentioned above, using energy from these sources meet energy savings requirements as defined under buildings codes, or if actions meant to decrease the consumption of energy delivered to these buildings were undertaken,*
c) *connecting to a centralised heating sources, when removing a local heating sources, which results in decreasing the cost of delivering heat to buildings mentioned above,*
d) *partial or complete replacement of the heat source to renewable energy sources or high efficiency CHP*

Regional Operational Programmes, sometimes define deep energy retrofit as a thermo modernisation which results in the buildings meeting the applicable requirements pertaining to the energy standard of new objects, i.e. Detailed Description of the Priority Axes of the Regional Operational Program of the Pomeranian Province for years 2014-2020 (Marshal's Office of the Pomeranian Voivodship, 2015):

Comprehensive, "deep" energy modernisations of buildings are actions affecting the improvements of a building's energy efficiency, aimed at decreasing the value of a building's annual usable energy demand, the annual demand for final energy, or the annual demand for non-renewable primary energy. In cases of improvements concerning energy savings, thermal insulation as defined in technical/building regulation. These regulations are understood as the Ordinance of the Minister of Infrastructure of April 12, 2002 on the technical conditions which a building and its placement should fulfil.

Cooperation between the Environmental Economy Institute, the Polish National Energy Conservation Agency (KAPE SA), and Buildings Performance Institute Europe (BPIE) attempted to choose the main assumptions and create a definition of "deep energy retrofits". Five definitions were proposed, depending on the situation and the beginning assumptions. The most demanding definition assumes a renovation adequate for a building with a near zero energy use (Zaborowski, 2013).

NEARLY ZERO-ENERGY BUILDINGS

As part of fulfilling the obligations of art. 9 paragraph 1 of Directive 2010/31/EU of the European Parliament and of the Council of 19 May 2010 on the energy performance of buildings (recast) according to which from 31 December 2020 all new buildings should have a near zero energy use, the Ministry of Infrastructure (currently the Ministry of Infrastructure and Construction) began changing polish building codes. The Ministry announced Act 91 of the Council of Ministers of 22 June 2015 on accepting the "National plan aimed at increasing the number of buildings with low energy use" (2015).

In national circumstances a "near zero energy building" is understood and defined as a "building with low energy use" (Ministry of Infrastructure and Development, n.d.; Ministry of Infrastructure, 2015), which was defined in paragraph 3 of the above mentioned Act:

Definition; a "building with low energy use" should be understood as a building meeting the requirements of energy savings and thermal insulation under the technical-building regulation mentioned in art. 7 paragraph 1 of the Law of 7 July 1994 – The Construction Law (Journal of Laws of 2013, item 1409, consolidated text, in Polish), that is specifically chapter X and attachment 2 to the Ordinance of the Minister of Infrastructure of 12 April 2020 on technical conditions which a building and its placement should fulfil (Journal of Laws of 2002, No. 75, item 690, consolidated text, in Polish) in effect from 1 January 2021 for buildings occupied, and belonging to the government from 1 January 2019.

Therefore the implementing the Directive on the national level is primarily changing the regulation concerning the technical requirements for the new buildings constructed in Poland. The most important regulation concerning the requirements pertaining to energy efficiency is found in the above mentioned Ordinance on technical conditions, commonly called "technical conditions" (Ministry of Infrastructure, 2002). The above-mentioned Ordinance's Attachment 2, titled "Thermal insulation requirements and other requirements concerning energy savings", contains the minimal thermal insulation parameters for partitions, e.g. effective heat transfer coefficients for walls, roofs, ground floors, partitions between heated and unheated parts of buildings, as well as windows and outside doors. Further sections of the Attachment also contain other requirements concerning energy savings in buildings – among others, the parameters of windows and glass surfaces, surface condensation of water vapour, and the buildings air-tightness.

Apart from this chapter X – "Energy savings and thermal insulation" contains requirements concerning maximal values of PE indicators (non-renewable primary energy) for different types of buildings. The exact method of calculating the energy demand is defined by a different Ordinance of the Minister of Infrastructure and Development of 27 February 2015 on the methodology of determining a building's or part of a building's energy characteristic and energy performance certificates (Ministry of Infrastructure and Development, 2015).

The regulations assume a gradual introduction of the nZEB standard – the requirements are gradually increased and changed for buildings designed from 1 January 2014, next for those designed from 1 January 2017, and finally those designed after 1 January 2021. In the case of public use buildings of public property the last change to the regulation should take place on 1 January 2019.

Another way of implementing Directive 2010/31 is a change to the Ordinance on the detailed scope and form of building design (Ministry of Transport, Construction and Maritime Economy, 2012), which expanded the obligation of conducting and analysis of the possibilities for feasible use of high efficiency alternative systems, and changed the scope of this analysis.

In 2012 Buildings Performance Institute Europe developed an analysis (BPIE, 2012) of implementing nZEB buildings in Poland. In was based on simulations of different variations with respect to compliance to nZEB buildings defined earlier in an elaboration titles "Principles for near zero-energy buildings (BPIE, 2011). It took into account economic, legal, technical, and ecological aspects, as well as its effect on the job market, the development of industry and technological development.

Based on the conducted simulations of different variants of a single-family residence three most cost-effective proposals were chosen. These were the basis for proposing a definition of near zero-energy buildings in Poland. Table 2 shows recommended minimal requirements for single-family residential buildings.

Based on Table 2 it is possible to say that a deep energy retrofit of an existing building up to the nZEB standard should consist of comprehensive actions connected to, among others, improving the thermal insulation of outside partitions, improving the air-tightness of a buildings, and modernising the heating installation to meet the requirements for new buildings designed after 1 January 2021. In these buildings part of the energy should be generated from renewable energy sources (RES). One of the aims of these actions should be decreasing CO_2 atmospheric emissions.

FINANCING

Between 2015 and 2023 The National Fund for Environmental Protection and Water Management (NFEP&WM) planed launching a priority programme called RYŚ – energy retrofits of single-family residences. The program was to be implemented alongside an ecological educational programme. Its effect was supposed to be decreasing final energy use by 300k GJ/year, decreasing CO_2 emissions by 25k Mg/year, and reducing PM10 dust emissions by 50 Mg/year and PM2.5 dust emissions by 45 Mg/

Table 2. Minimal requirements for single-family residential buildings in the nZEB standard

Minimal Requirements	Year	
	2015/2016	2020
Primary energy [kWh/m^2/year]	70	30-50
Share of renewable energy [%]	>20	>40
CO_2 emissions [kg CO_2/m^2/year]	<10	<3-6

According to the BPIE (2011).

year. In the end, after a change in the Fund's management in 2016 the RYŚ programme was scrapped just before it began.

NFEP&WM announced the launch of a different priority programme called Region, which is meant to support the use of renewable energy sources and energy retrofits. The programme will be implemented through the Voivodeship Funds for Environmental Protection and it will be them, who will precisely set the rules and the final offer for natural persons. Right now, these are only announcements with no concrete details.

Only one countrywide programme aimed at single-family residence owners is in operation. It is a programme financed from the public budget and consists of energy retrofit premiums currently amounting to 20% of a loan used for the realisation of an energy retrofit, however it cannot amount to more than 16% of the renovations costs and double the annual energy savings as determined by and energy audit. A requirement for receiving financing is performing and energy audit (along with a technical and economic analysis) and beginning work from bank loaned funds.

Apart of the above local public institutions (including the Voivodeship Funds for Environmental Protection and Water Management (VFEP&WM)) launch their own programmes supporting energy retrofits, though these usually focus on their local area.

ENERGY DEMAND CALCULATIONS

Developing a methodology for the need of statements of energy performance of buildings is a consequence of realizing by the Republic of Poland the points of the 2002/91/EU Directive of the European Parliament and the European Council, enacted on December 16, 2002. This document requires the EU member states to impose an obligation within their law to carry out energy performance analysis of buildings and flats in case of new construction commissioning, building renovation or properties being subject of trade. A first version of the methodology was published as the Ordinance in 2008. In this moment a slightly modified version is force, published in 2014, which considers the EPBD 2010/31/EU Directive.

The developed energy performance analysis methodology considers annual energy demand for heating and cooling, necessary to achieve required thermal comfort in an analyzed building.

The building energy performance is determined according to the ISO standard 13790:2008, calculations consider the following indicators of annual energy consumption: primary energy, final energy and usable energy.

The calculation methods are based on monthly energy account in case of flats and residential buildings, and a simplified hourly energy simulation in case of public buildings and residential buildings with cooling.

The calculations use weather data (typical meteorological year) published in the Internet by the Ministry of Infrastructure and Construction, based on statistical and climatic measurements carried out for the territory of Poland. The primary energy is computed on basis of the w_i coefficients, indicating the non-renewable primary energy input necessary for producing and providing the energy carrier or energy for technical systems. The w_i coefficients, which are published in the Ordinance mentioned above, are listed in Table 3.

Table 3. Non-renewable primary energy input coefficient (w_i) necessary for producing and providing the energy carrier

Building Energy Supply System	Type of Energy Carrier or Energy	W_i
Energy production within the building	Heating oil	1.10
	Natural gas	
	Liquefied petroleum gas	
	Black coal	
	Lignite	
Local renewable energy sources	Solar energy	0.00
	Wind energy	
	Geothermal energy	
	Biomass	0.20
	Biogas	0.50
District heating from cogeneration	Black coal or gas (In case of lack of data concerning the energy produced in cogeneration, it is acknowledged that w_i =1.2)	0.80
	Biomass, biogas	0.15
District heating from a local heating plant	Black coal	1.30
	Natural gas or heating oil	1.20
Electricity grid	Electrical energy	3.00

Ministry of Infrastructure and Development, 2015.

MAIN FOCUS OF THE CHAPTER

Description of the Building

As mentioned in the introduction this energy retrofit project concerned a single-family residential building built between the 60's and 70's of the previous century, located in an eastern Poland village. It is an example of typical village architecture from that period. The building was constructed during Poland's socialist period (Polish People's Republic 1944-1989). At that time construction materials were in short supply and investors had difficulty acquiring adequate supplies leading to the, often, accidental use of construction materials.

In the past, the building lost heat quickly - both due to the lack of adequate thermal insulation and also due to the air leakage of the external partitions and the floor.

It was heated using a tile stove located in each room. Wood and coal was used as fuel.

During over 40 years of use the building was renovated in the following ways:

- In the 70's water from a deep well was connected to the building,
- In 2012 the building was connected to the village's municipal water supply,
- Bathroom renovation with a partial floor replacement using 10 cm styrofoam insulation,
- Replacing fibre asbestos-cement roof tiles with metal roof tiles,
- Recently the foundations were insulated along with a vertical hydro-insulation of foundation walls.

Figure 1. Photograph of the house before the energy retrofit
KAPE SA, 2016.

All of the buildings elements remained unchanged until 2015.
Construction inspection established that:

- Building set on poured concrete,
- Wooden floor on wooden joist supported on fieldstones, originally the space between the stones and the joists was ventilated, however recently the vents in the foundation walls were covered up,
- Brick walls built from grey (fly ash additive) cell concrete block, 25 cm thick, with a ventilation space 7 cm thick and walls 12 cm thick walls out of perforated brick,
- Steel joist ceiling above the ground floor (steel beam ceiling with ceramic cellular brick infill),
- Roof construction – gable framing supported on wooden ceiling beams.

Figure 2. Taking apart the floor: The stones visible were used as support for floor joists, wall vents are also visible.
KAPE SA, 2016.

Figure 3. One of the tile stoves used as a heat source in rooms; Fired with wood and coal

Figure 4. Visible roof construction and steel joist ceiling above the ground floor

Along with energy retrofit of the building the investor decided to adapt the unused attic for living purposes. This increased the useable area from 92.7 m^2 to 197.8 m^2.

Moreover a technical evaluation also showed that the existing ceiling wouldn't be able to bear the designed loads. It was necessary to reinforce the ceiling by welding additional IPE 80 beams along the entirety of the existing beams lengths. 3x40 flat bars were installed between the beams 80 cm apart.

The rafter framing was in a very bad technical condition and also required reinforcement. With this in mind some elements were replaced or additionally reinforced by adding 38 mm x 160 mm boards to each rafter. Additionally 100x250 purlin was constructed with 120x120 mm brackets. Posts were propped directly on load bearing walls. The buildings sitting, foundation walls and outside walls were deemed in a good technical condition.

In terms of energy efficiency none of the buildings elements fulfilled existing requirements. Table 4 presents existing requirements and the calculated parameters of the building based on the inspection.

Together with the energy retrofit the following reconstruction of the building was planned:

- Widening of the indoor door openings,
- Adapting the attic for residential purposes, which required building stairs, an opening in the existing ceiling, constructing walls on the first floor and roof windows,
- Widening the day room windows on the south side of the building.

The following energy retrofit measures were undertaken:

- Additional insulation of foundation walls,
- Removing the existing floor and building a new one thermally insulated with extruded polystyrene sheets (0.037 [W/mK]) 18 cm wide,
- Replacing windows with ones of higher thermal parameters (heat transfer coefficient U_w=1.1 [W/m^2K]). "Warm" window installation performed on steel consoles in the thermal insulation layer of the walls. Maximal air-tightness between the window and the wall was ensured,
- On the south side, installing outside windows with solar roller blinds equipped with photovoltaic cells requiring no outside power. The blinds were installed to prevent overheating. It is important as it is an often overlooked, yet important element – energy efficient buildings shorten the heating period which can oftentimes be a discomfort during the transition period due to overheating of the building,
- Replacing all doors, including outside doors with ones of a coefficient of U_d=1.3 [W/m^2K],
- Additional insulation of outside walls with λ=0.031 [W/mK] styrofoam sheet 28 cm thick using External Thermal Insulation Composite System (ETICS),

Table 4. Comparison of outside building partitions along with regulation dictated heat transfer coefficients and the coefficients of existing elements. Requirements in each year in accordance with polish building codes. In 2015 when the project took place the coefficients of 1 January 2014 were the regulation.

No.	Type of Partition	Heat Transfer Coefficient U_c(max) [W/(m^2K)]			
		In Force From 1.01.2014	In Force From 1.01.2017	In Force From 1.01.2021	Calculated Existing
1	Outside walls	**0.25**	0.23	0.20	**1.01**
2	Floor on ground floor	**0.30**	0.30	0.30	**0.57**
3	Roof over unheated attic	**-**	-	-	**7.13**
4	Ceiling under unheated attic	**0.20**	0.18	0.15	**0.36**
5	Windows	**1.3**	1.1	0.9	**ok. 3.6**
6	Outside doors	**1.7**	1.5	1.3	**ok. 2.6**

Ministry of Infrastructure, 2002.

- Insulating the roof without replacing existing metal tiles. After reinforcement mineral wool sheets with a coefficient of 0.030 [W/mK] and a total width of 28 cm, along with a waterproofing membrane,
- Removing existing tile stoves and replacing them with a new central heating installation using a modern wood gasification boiler. Space heating using steel sheet radiators. Heating installation with a water tank,
- Installing mechanical ventilators with an 86% efficient recuperator.

Figure 5. Attic after reinforcement of rafter framework and ceiling; Roof windows already installed
KAPE SA, 2016.

Figure 6. Window installation in the wall's thermal insulation layer on steel consoles
KAPE SA, 2016.

Figure 7. New heating and domestic hot water installation; The new wood gasification boiler is visible.
KAPE SA, 2016.

Figure 8. Building after the energy retrofit
KAPE SA, 2016.

Purchase of the materials was covered by the project's sponsors – producers cooperating with KAPE SA on the pilot project. Key energy retrofit work costs are presented in Table 5.

Based on the information available on the used materials and the project designed the projected coefficient Uc of the designed partitions was calculated. Table 6 presents the current requirements and the calculated designed coefficients.

Table 5. Summary of the building's energy retrofit costs. It includes the cost of installing roof windows (performed by the window suppliers), cost of adapting the roof (partition walls) and finishing works – e.g. plasterwork, flooring, and indoor doors. The prices are based on information delivered by the suppliers and the owner.

No.	Scope of Work	Cost in EUR (€)
1	Preparatory work	2,594.00
2	Roof windows	3,840.00
3	Insulation of the roof and the ceiling	5,053.00
4	Equipment and installation of central heating and domestic hot water	5,832.00
5	Equipment and installation of ventilation	5,984.00
6	Insulation of the walls and floor	5,573.00
7	Vertical windows	5,349.00

Source (Ładny Dom).

Figure 9. Comparison of the costs of each element under energy retrofit: 1 – Preparatory work, 2 – roof windows, 3 – insulation of the roof and the ceiling, 4 – equipment and installation of central heating and domestic hot water, 5 – equipment and installation of ventilation, 6 – insulation of the walls and floor, 7 – vertical windows.

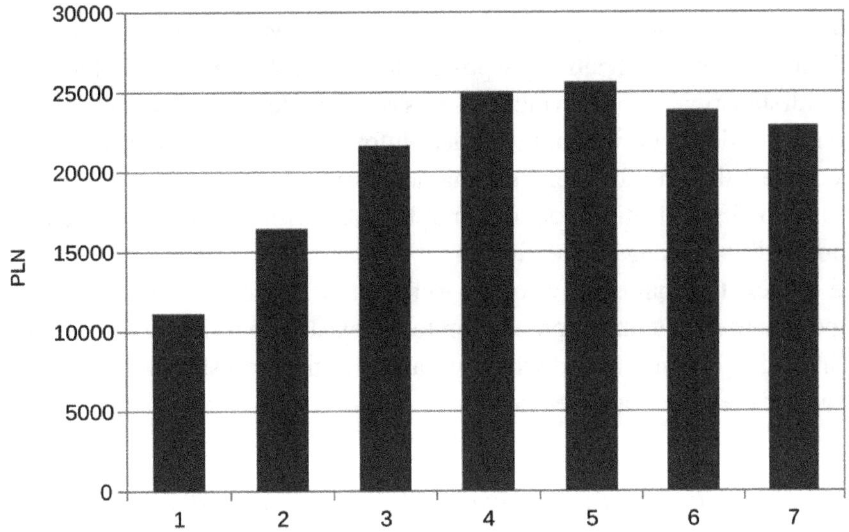

Energy Demand Calculations

Based on the conducted expertise and the construction inventory an energy audit of the existing building was performed. The project's construction design was used for calculating the projected energy characteristic. The calculation was done using the Audytor OZC 6.5 Pro software in accordance with existing Polish regulation Ministry of Infrastructure and Development (2015)[1]. The calculation used the assumption that the tile stoves were fired using wood and coal in a ratio of 50/50.

Table 6. Comparison of the designed outside partitions of the building with the regulation dictated U_c coefficients. Requirements in each year in accordance with polish building codes. In 2015 when the project took place the coefficients of 1 January 2014 were the regulation.

No.	Type of partition	Heat Transfer Coefficient U_c (max) [W/(m²K)]			
		In Force From 1.01.2014	In Force From 1.01.2017	In Force From 1.01.2021	Calculated Designed
1	Outside walls	0.25	0.23	0.20	**0.10**
2	Floor on ground floor	0.30	0.30	0.30	**0.14**
3	Roof over unheated attic	-	-	-	**0.11**
4	Ceiling under unheated attic	0.20	0.18	0.15	**not applicable**
5	Windows	1.3	1.1	0.9	**1.1**
6	Outside doors	1.7	1.5	1.3	**1.3**

Ministry of Infrastructure, 2002.

The biggest thermal energy losses were noticed through the outside wall, which had a share of 40.9% of the total loses. The existing building's energy losses (before the energy retrofit) with respect to each of the building's elements are shown in Table 7 and Figure 3.

Using a numerical model the solar gain (via windows) was calculated together with the estimated inside gains. The gains are presented in Table 8.

The per unit demand for usable energy (heating domestic hot water) was 267.5 kWh/m²/year. Per unit demand for usable energy for space heating purposes was 243.3 kWh/m²/year. Calculations of energy demand and in line with statistical date from table 1 – It is an example of a very energy demanding building.

Next, based on the building design and technical information provided by the suppliers of the material used in the project the buildings energy characteristic was calculated. The calculation used the assumption that the wood gasification boiler cannot be fired with anything other than wood, therefore eliminating the possibility of using coal entirely.

This time the highest thermal energy losses were noted in the ventilation and outside windows. However it is important to note, that in comparison to the existing building there was an additional 10.2 m² of windows (including 6.4 m² of rood windows, and 3.8 m² of outside windows. Table 9 shows the energy loses of each of the building's elements.

Table 7. Heat losses by partitions in [kWh/year] and the percentage share in the building before the energy retrofit

Building Element	Heat Loss [kWh/year]	Share [%]
Outside doors	557	1.8
Ceiling over heated section	3399	11.2
Ground floor	3814	12.6
Ventilation	4664	15.4
Outside windows	5495	18.1
Outside walls	12416	40.9

Figure 10. Energy losses in partition [kWh/year] and their share; Left to right: Outside doors, ceiling under attic, ground floor, heat in ventilation, outside windows, outside walls

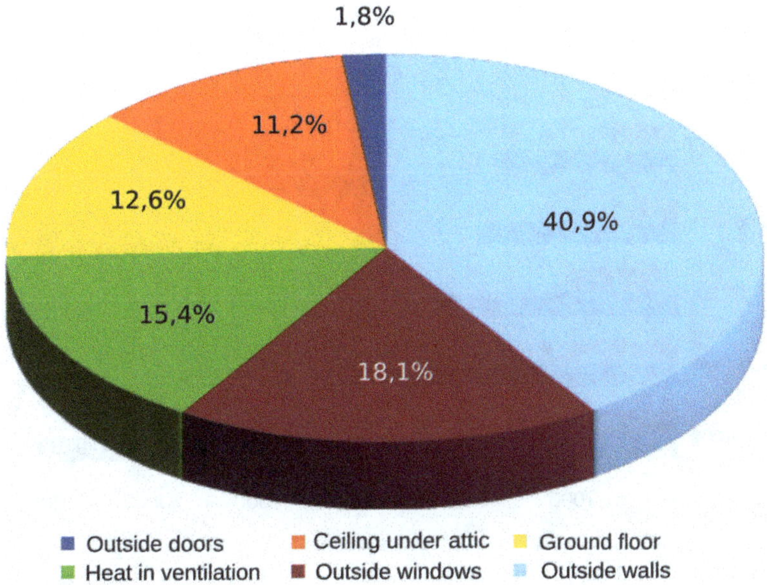

Table 8. Energy gains during heating seasons in the building before the energy retrofit [kWh/year]

Energy Gains	kWh/Year
Gains from the sun	4916
Inside gains	4123

Table 9. Heat loses by partition in [kWh/year] and their share in the building after the energy retrofit

Building Element	Heat Loss [kWh/year]	Share [%]
Outside doors	278	3.2
Roof	1243	14.4
Ground floor	870	10.1
Ventilation	2053	23.7
Roof windows	729	8.4
Outside windows	1885	21.8
Outside walls	1593	18.4

Figure 11 presents the comparison between energy losses before and after the energy retrofit. The biggest difference is the energy loss due to outside walls. It shows a significant reduction of energy losses.

Calculating the effect brought on by the energy retrofit and the final investment cost it was established that insulating the outside wall is the most cost-effective solution.

Figure 11. Comparison of energy losses by building elements before and after the energy retrofit [kWh/year]. Y axis, top to bottom: External doors, Roof, Ground floor, Heat for ventilation, External windows, External walls; X axis, left to right: kWh/year, Original state, after energy retrofit.

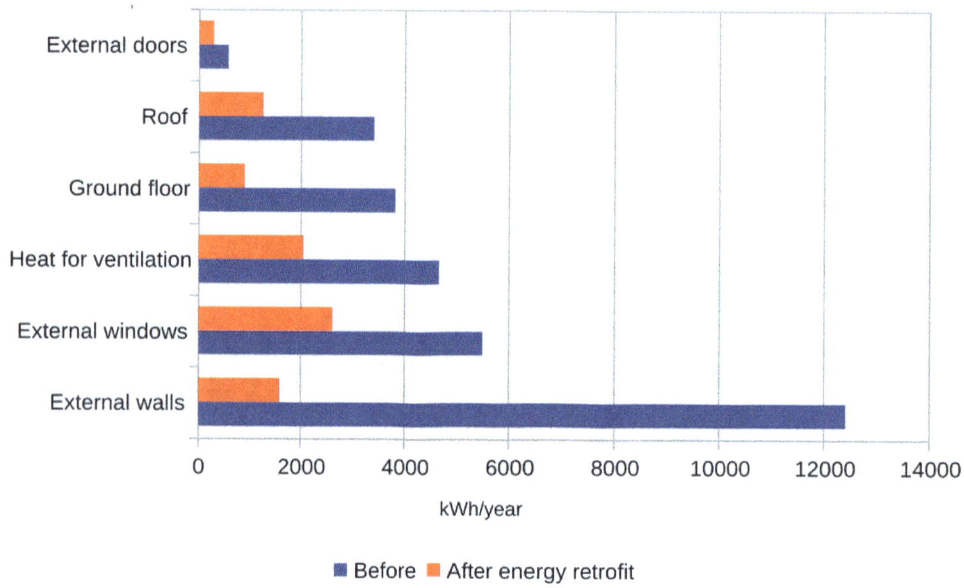

Summarizing the energy demand balance a clear reduction for usable energy has been show. The results of the comparison are shown in Table 10 and Table 11. It is an indicator of the building's energy efficiency level.

Ecological Effect

Based on the simulations ran the cumulated energy and carbon footprint during the buildings 50-year lifecycle before the energy retrofit was calculated. The emissions indicator for the chosen energy carriers was taken from research of the National Centre for Emissions Management (KOBiZE, 2016). The CO_2 emission indicator for final consumers of electrical energy is 798 kg/MWh (KOBiZE, 2017). The CO_2 emission indicator for wood based fuels is 112 kg/GJ, and for coal based fuels it is 94.72 kg/GJ.

Table 10. Comparison of the building's annual demand for final energy for heating purposes $Q_{K,H}$ (without supplementary equipment), the annual demand for final energy for heating domestic hot water $Q_{K,W}$ (without supplementary equipment), the annual demand for final energy together with supplementary equipment and the annual demand for non-renewable primary energy Q_P

Annual Demand for Final Energy for Heating	Existing Building	After Energy Retrofit
$Q_{K,H}$ [kWh/year]	37056.5	2470.5
$Q_{K,W}$ [kWh/year]	3875.7	7609.2
Q_K (total) [kWh/year]	40932.2	11288.8
Q_P (total) [kWh/year]	37130.6	5040.8

Table 11. Comparison of the per unit demand for usable energy (UE), final energy (FE) and primary energy (PE) [kWh/m²/year] of the building after the energy retrofit and presenting the maximal values of the EPH+W indicator for heating, ventilation and heating domestic hot water [kWh/m²/year].

Per Unit Energy Demand	Existing State [kWh/m²/Year]	State After Energy Efficiency Renovation [kWh/m²/year]	Requirements for Single-Family Residences [kWh/m²/year]		
			In Force From 1.01.2014	In Force From 1.01.2017	In Force From 1.01.2021
Demand for usable energy - UE	267.4	35.8	-	-	-
Demand for final energy - FE	442.4	57.1	-	-	-
Demand for primary energy - PE	401.3	25.5	max. 120	max. 95	max. 70

The annual emission connected with building use was 5029 kgCO_2 annually (according to Table 12), that is 25.43 kg CO_2/m²/year.

In the past 50 years of use the amount of CO_2 emissions avoided is 1068.73 MgCO_2.

Calculations of Built in Energy and Emissions

For the purpose of calculating the built-in energy and emissions the most important construction materials used in the energy retrofit were taken into account. The calculations did not include finishing work materials and construction process. The construction materials taken into account for the calculation are presented in Table 13.

To calculate the built-in energy and emissions for construction products averaged coefficients from the ÖKOBAUDAT, Inventory of Carbon and Energy (ICE), and SBS Building Sustainability (SBS) databases was used together with type III Environmental Product Declarations.

For construction materials listed in Table 12 the amount of built in energy due to their production is equal to 314615.8 MJ (87393.28 kWh), while their Global Warming Potential is 13792.79 kgCO_2.

Economic and Environmental Savings

Based on energy simulations it was calculated that the building originally used 37056.5 kWh (133.40 GJ) of energy as wood and coal for heating, and 3875.7 kWh (13.95 GJ) of electrical energy for heating domestic hot water, giving a total of 40932.2 kWh (147.36 GJ) final energy annually.

Table 12. CO_2 emissions comparison of the building before and after the energy retrofit

Energy Carrier	Existing Building [kgCO_2]	After Energy Retrofit [kgCO_2]
Electrical energy	3092.81	964.86
Biomass (wood)	16331.95	4064.14
Coal	6978.74	0
Total annual CO_2 emission kgCO_2	**26403.50**	**5029.00**

Table 13. Construction materials used in the energy retrofit

Construction Materials	Quantity [m³]	Quantity [m]	Quantity [pieces]	Quantity [m²]
Extruded polystyrene (foundation)	4.22			
Extruded polystyrene (floor)	16.56			
EPS 032 Styrofoam (outside walls)	42.62			
Glass mineral wool (roof)	42.76			
Glass mineral wool (ceiling)	7.99			
Outside plaster	0.85			
Wood boiler			1	
Steel sheet radiators	7			
Central ventilation*			1	
Ventilation ducts		35.6		
Windows (outside walls)			7	17.68
Windows (roof)			5	6.5

*Information on small central ventilation units with heat recuperators could not be found. Calculations used data for a large central unit weighting 369 kg. The building uses a central unit weighting 78 kg – a proportionally smaller built in energy and emission coefficient was used in the calculations.

After the energy retrofit the building uses 2470.5 kWh (8.89 GJ) for heating and 7609.2 kWh (27.39 GJ) for heating domestic hot water, totalling at 10079.7 kWh (36.29 GJ) energy per year. As this is done using a single wood fired boiler it was assumed that all of the energy is produced from biomass.

CO_2 emissions were avoided due to the energy retrofit. Over a 50-year operation period the amount of CO_2 emissions avoided was equal to 1054.94 $MgCO_2$ (after taking into account built in emissions from building materials used for the energy retrofit.

Figure 12 shows a comparison between the built-in emissions connected to producing building materials and CO_2 emissions over a 50-year operation period.

Figure 12. Cumulated CO_2 emissions over a 50-year operation period for the existing building and the building after the energy retrofit in $MgCO_2$. Top, bottom, left to right: Existing building, Designed building, Embodied CO_2 emission, Operational CO_2 emission, avoided CO_2 emission.

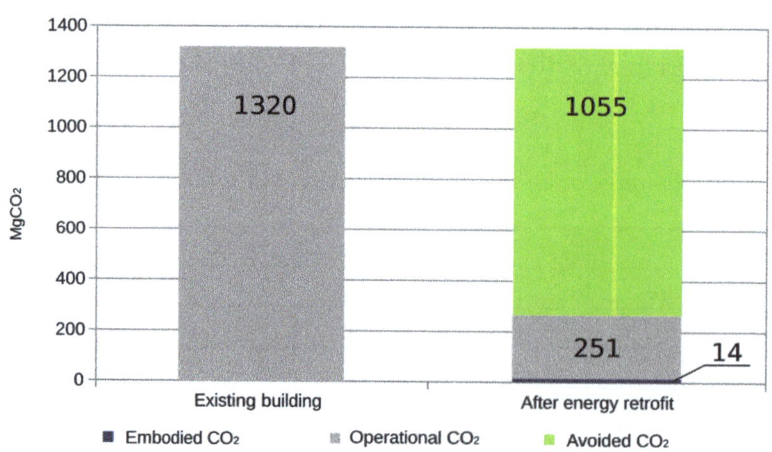

According to the Infor.pl internet service the mean average annual price of 1 tonne of coal in 2016 was 182.42 EUR. According to the ogrzewamy.pl internet service the average price of fire wood (birch) was 37.40 EUR/m^3. The calorific value of birch wood is about 2200 kWh/m^3. By this account annual heating and heating domestic hot water in the existing building cost 959.40 EUR, and after the energy retrofit it will cost 340.94 EUR. This means an annual saving of 618.46 EUR. Table 14 presents a cost comparison.

Based on Simple Pay Back Time calculations of the investment will be achieved would be after over 55 years. It is important to note that heating using coal is the cheapest solution in Poland. The payback period would be shorter if the only energy carrier was electricity, oil, liquid gas, or natural gas.

SOLUTIONS AND RECOMMENDATIONS

During the realisation of the energy retrofit it met with difficulties of a formal nature.

One of the most troublesome, connected to the investor's with to adapt the attic for living proposes as part of the energy retrofit. This type of changing the purpose requires fulfilling administrative formalities. First it was necessary to ascertain in the planned function was compliant with the Local Spatial Management Plan for the area. In this case the area didn't have a Spatial Management Plan. In these situations (where there is no Plan in place) it is necessary to submit a motion for a decision on zoning permits. This procedure can take months, as it was in this case. Additionally, both changing the purpose of an area as well as the construction required for the renovation of the building demanded the development of building project and obtaining a building permit. All these procedures delayed the realisation by multiple months.

An energy retrofit consisting of insulating outside partitions, replacing windows or replacing indoor installations in single-family residences can be conducted by only notifying the relevant public office, without the necessity of developing a project or obtaining a building permit. However the regulation is not unequivocal in treating these cases as each action, which changes the building's technical parametres, or changes spatial management requires a building project and a building permit. From a formal point of view installing windows with better parameters, or increasing the width of walls (which alters the area of the building and the distance between the building and the plot's boundary) also causes these changes. Cases in which officials demanded full procedures of obtaining a building permit have been known.

Table 14. Comparison of heating and heating domestic hot water before and after energy retrofit in EUR/year

<table>
<tr><th>Type of Energy Consumption</th><th>Building Before Renovation [EUR]</th><th>Building After Renovation [EUR]</th><th>Savings [EUR]</th></tr>
<tr><td>Heating</td><td>415.83</td><td rowspan="2">171.36</td><td rowspan="4">618.46</td></tr>
<tr><td>Heating domestic hot water</td><td>543.58</td></tr>
<tr><td>Supplementary energy</td><td>-</td><td>169.58</td></tr>
<tr><td>Total</td><td>959.40</td><td>340.94</td></tr>
</table>

Another issue is connected to the fact that in recent years regulation has changed repeatedly – in this insulating a roof built many years ago it may turn out it does not meet the required bearing capacity (it is now loaded with additional insulation layers). In this case it is necessary to reinforce or rebuild the roof's construction, and that requires a building project and a building permit.

Key Issues Concerning an Energy Retrofit

The most important element, preceding any energy retrofit should be performing an energy audit of the building. Poland has many certified auditors, who will prepare energy simulations and recommendations for optimal renovation work based on existing documentation and an inspection of the building. These should include the following elements:

Measures of renovating outside partitions, including additional thermal insulation. In Poland EPS Styrofoam, rock or glass mineral wool are most often used, with polyurethane PIR/PUR sheets more rarely used. The optimal insulation layer width is based on economic and technical calculations (simply put the cost of the additional layer and its effect on the partitions heat transfer coefficient U). Up to a certain boundary adding thermal insulation layers costs less than the cost of the energy saved of the assumed operation period. Using a slightly thicker insulation layer than in the calculations is a small risk if allowed by the technical conditions.

When insulating partitions it is essential to remember about the partitions between heated (residential) areas and unheated areas of the building. It might be a ceiling under an unheated attic, the floor over an unheated cellar or a wall between the garage and a living room. It is important to attempt to eliminate thermal bridges between connections with unheated areas. The effects of thermal bridges are described below (in section 5 of this list).

Windows and other transparent partitions have a big impact on a building's energy consumption. In the case of these types of elements we can see significant technological advancement in recent years, and a significant improvement in insulation. By far most existing buildings have composite double glazed windows. Still, around 15% of residential buildings have box windows and around 7% have single glazed windows (Żurawski, n.d.). Table 15 shows the types of windows.

Existing windows have a high heat transfer coefficient U_w. Current regulations impose the obligation of using windows, balcony doors and other outside transparent surfaces with a coefficient U_w=1,1 [W/m^{2}K] or less, and in some cases (with a lot of glazing in relation to floor area) below 0,9 [W/m^{2}K]. From 1 January 2021 U_w=0,9 [W/m^{2}K] will be applicable to all new windows. It is not possible to improve the thermal insulation of existing windows, only to improve air-tightness between the window and the wall. Due to the above it is important to include the replacement of windows in the energy retrofit. Using windows meeting current parameter requirements is recommended, or alternatively ones meeting requirements in force after 2020. When replacing windows it is worth noting how they are installed, which should be placed in the wall's insulation layer using steel consoles or special hot installation methods.

Table 15. Share of each type of window in existing residential buildings in Poland

	Single Glazing [%]	Double Glazing [%]	Triple Glazing [%]
Share in residential buildings	7.44	90.50	2.06

Adapted from Żurawski (n.d.)

Modernisation of ventilation systems consists of installing equipment, which allows control over the amount of air (ventilators, exhaust fans) or installing a bidirectional ventilation unit with a heat recuperator. The second solution is much more difficult as it requires finding room for a central unit and ventilation ducts in practically the entire building. This is connected with significant additional construction costs. Often in (older) buildings all rooms have ventilation ducts leading to the roof – then vents in the bathroom and the kitchen can be used as exhaust vents while vents in rooms can be used as air supply vents. In this situation placing the ventilation central unit in the attic is can be connected to the proper existing ducts in the buildings.

Modernisation of the heating installation could consist of modernising an existing installation, replacing some elements, or replacing the entire heating system. The choice of the solution is dependent on the scope of the renovation, the technical conditions of the existing equipment and the local possibilities of using particular energy carriers. In cases of a comprehensive energy retrofit it might turn out that a discarding the existing heating system and replacing with a different one, e.g. one based on electrical energy, would be justified due to lower investment and operation costs. This can be the case if the building is meant to achieve the nZEB standard – which includes a low demand for thermal energy. From an ecological point of view using electrical energy as the main energy carrier can seem counterintuitive. It is important to remember that when analysing these solutions using a life cycle analysis methodology, that is costs incurred during the production of the boiler, the pipes, pumps, etc. could be higher than when using e.g. heating wires, or simple and small electrical heaters. These solutions could be especially beneficial when combined with a local heating source such as a biomass boiler.

In the case of existing installations, the modernisation could also consist of replacing the source of heat (boiler), air-tightening the installation, installing thermal valves in case of overheating, installing a weather or room regulator, or installing fireplace insert resistant to the aggressive effect of exhaust fumes. Properly installed and operated heating installations using thick steel pipes and massive cast iron heaters are resistant to corrosion. After their air tightening and adjustment, they might as well operate for the next 20 years. It is important to remember that by maintaining many operational elements of the existing installation we are positively affecting the ecological effect – unnecessary replacements of heating installations are connected to negative effects on the environment through the processes involved in their production, transport, etc. In cases where the installation wasn't correctly operated – if it was characterised by significant water losses which could lead to often top-ups using pipe water, most probably the inside of the pipes will be filled with sediments difficult to remove. Attempts at chemical removal could also lead to further loss of containment. Evaluating pipes consists of taking a couple of slices of pipe from the installation, if their bad condition is confirmed the entire installation must be replaced along with the installation of new radiators with a lower thermal inertia.

When replacing the heat source the most favourable energy carrier should be chosen. This could for example be network natural gas, electrical energy, biofuels, or the municipal heating network. The last one is beneficial from the ecological point of view. Generating both heat and power has a highly beneficial indicator of primary energy supply. In Poland, central heating networks are very developed and therefore it is worth to use this option if only the technical conditions of a connection (network close to the building) are suitable. It is worth noting that e.g. Warsaw has the largest urban heating network in the EU (Włoch & Brzeziński, 2010). Using renewable energy sources such as biofuels will obviously be the most beneficial due to environmental protection.

Making the decision on modernising the heating system and a choice of an optimal fuel should be preceded by an economic and user comfort analysis. Such an analysis should be part of an energy audit.

It can also include calculations of a Simple Pay Back Time (cost of the modernisation and annual energy savings). This indicator does not take into account fluctuations in monetary value, energy savings value and changes in energy carrier costs. Taking into account all of these facts could be overly complicated and too burdened by a high degree of uncertainty. For the sake of simplicity it can be assumed that inflation affects all elements (costs and savings) to the same degree and proportionally – therefore it doesn't affect the length of the payback period. Despite its simplicity SPBT can indicate valuable information. If it turns out that the assumed project will pay itself back in a couple of months it is a clear signal such a project should be undertaken. It has been assumed that for single-family residential buildings measure with a 10-year payback period are worth undertaking due to the forecasted real increase in energy carrier cost being higher than inflation.

Insulating ceilings, flat roofs and attics is another important element, which could provide a large challenge. Oftentimes ceilings over unheated basements in single-family houses are practically uninsulated or insufficiently insulated. The heat transfer coefficient could be around 1.2 – 1.3 [W/(m^{2}K)]. Insulating a ceiling should be done from the basement side. This can be made difficult due to the basements height, which is often not high enough and each centimetre less could limit its functionality. Due to this it might be valuable to use thinner materials with higher lambda coefficients. These are graphite styrofoam λ=0.031 [W/mK] and polyurethane PIR/PUR sheets λ=0.020-0.026 [W/mK]. Ceilings under unheated attics should be treated the same. Insulating such a partition is easier that insulating a partition over a basement. It is worth to secure the ceiling with a waterproof membrane first (humidity delayer) and use mineral wool as an insulator. From the top it should be secured with a windproof, water permeable membrane. The area of an unheated attic or loft should be ventilated. It is a key component of the buildings and should be implicitly renovated as due to favourable conditions for heat transfer (natural heat transfer upwards) it is a source of significant energy loss – particularly when partitions are not insulated enough. It is also element which can be energy efficiently renovated using little financial contributions.

Removing thermal bridges is one of the most important tasks, one often overlooked in analyses performed for energy retrofit plans. In the case of buildings using significant amounts of energy the share of thermal bridges can be unnoticeable in the energy balance. Their effect is observed in cases where water vapour condenses on partitions inside buildings. Investors also observe a noticeable increase in cold in certain sections of the building e.g. in ceiling corners, by balcony floors. In these places, if humidity remains, it creates an environment for fungi. In nZEB standard buildings thermal bridges can constitute a large share in heat losses and a significant deterioration of the energy balance. Thermal bridges constitute the most difficult, and sometimes the most expensive elements of an energy retrofit.

Below are a couple of examples of the most common thermal bridges:

- Thermal bridges connected to existing balconies. Balconies were most often constructed out of reinforced concrete or steel elements acting as supports. These types of construction are a significant thermal bridge. Without radical measures such as taking a balcony apart (and replacing it with an independent separated construction), the only possibility is insulating the entire balcony sheet from all sides, along the entire length of the balcony. The width of the insulation layer is best chosen by performing a thermal bridge simulation, the best results could be achieved by achieving the PSI coefficient of the linear bridge to C2 class PSI = 0.1- 0.25 [W/(mK)] which according to Kaliszuk-Wietecka (2017) qualifies as a small impact linear bridge.
- Another type of thermal bridges is found in existing buildings with outside concrete stairs. In this case it is difficult to insulate the stairs. The best course is a separation (dilatation) between the

stairs and the building's construction. Thermal bridges also appear at the place where the heated and unheated sections of the building meet. Heat transfer despite partition insulation between these rooms occurs within the walls. To limit the effect of these bridges, apart from the separating partition it is necessary to insulate a section of the wall as far as 1 m away from the partition on the unheated sections side.

A building's air-tightness is an important element, which has a significant effect on the results of an energy retrofit. National regulations recommend that buildings have an air-tightness *in buildings with gravitational or hybrid ventilation -* $n_{50}< 3.0$ *1/h, and in buildings with mechanical or climate control ventilation -* $n_{50}< 1.5$ *1/h.* It is recommended that, if possible, air-tightness below 1.0 should be achieved. To achieve this it is necessary to ensure that all joints of outside partition elements and in cables and pipes have a high air-tightness. The same applies to the installation of windows and outside doors.

CONCLUSION

The energy retrofit design presented in this chapter was carried out by the Polish National Energy Conservation Agency and its design partners. It transformed a building with very large energy consumption (final energy use for heating amounting to 133.4 GJ per year) and CO_2 emission (annual emission equaling 26,403.50 kgCO_2) into a nearly Zero Energy Building (with final energy consumption as low as 8.89 GJ per year).

In Polish conditions performing a deep energy retrofit to achieve the nZEB standard means improving of an existing building's parameters, so that is meets the requirements on energy characteristics as required by Ordinance for buildings designed after 1 January 2021. As the case study proves, it is possible to fulfill at least the requirements for new buildings designed after 2020. All types of retrofitting actions should be preceded by an energy audit and a cost analysis. The first step should be limiting losses of energy delivered for heating due to its dispersion through outside partitions. The method of supplying heat should be adapted to suit individual needs and local conditions. In some situations a good solution will be using biomass boilers, in another using low-condensation natural gas boilers, and in yet another heat pumps or electrical heating.

As it may be observed on the example described in this elaboration, not always does this type of deep energy retrofit bring about simple returns in a reasonable time. In the described energy retrofit project the payback time is as long as after 55 years of operation. This is especially the case when the energy carrier is firewood or coal, which are cheap in Poland. However, these investments created a huge environmental profit in avoiding CO_2 emissions, which over a 50-year life cycle will amount to 1054.94 MgCO_2. Another benefit of these operations is reducing the emission to the atmosphere of other harmful substances, such as sulphur oxides, nitrogen oxides and particulates responsible for winter smog, which is increasingly frequent in Poland.

It is important to notice that the calculations reflect the comparison between the original usable area of 92.7 m^2 and the post-energy retrofit along with an adapted attic with an area of 197.8 m^2. An additional benefit is increasing the living area. Another benefit achieved is the improved comfort of use – mainly due to a stable perceivable temperature and good air quality inside the building. These factors positively influence the conditions associated with health. These types of actions, apart from enhancing the use value and aesthetics of the building, also increase its value on the property market. From a social point of view, the increased social and financial status of the energy efficiency renovated building's owner could also be taken into account. These factors, apart from financial benefits, should be considered in the analysis of costs and profits of energy retrofitting.

Per unit emission of CO_2 before the energy retrofit was 133.49 kg CO_2/m^2/year, and after the renovation it was 25.43 kg CO_2/m^2/year, which is still above the BPIE recommendation (above 10 and 3-6 kg kgCO_2/m^2/year after 2020). Reducing the emission indicator would require the use of a different energy carrier or using other renewable energy sources (e.g. solar collectors or photovoltaic panels).

In Poland, there could be about 80% (Żurawski, 2012) of buildings with a similar energy demand for heating purposes. Therefore, implementing these types of measures is very important. Performing energy retrofits of existing buildings limits energy consumption, greenhouse gas and dust emissions- and as a consequence improves Poland's air quality. This is why public financial support is important, particularly since not in each case will a deep energy retrofit bring about the return of the costs incurred as part of the renovation. These operations should include funding programs and informing actions.

REFERENCES

Buildings Performance Institute Europe. (2011). *Principles for nearly zero-energy buildings - Paving the way for effective implementation of policy requirements*. Author.

Buildings Performance Institute Europe. (2012). *Implementing nearly zero-energy buildings (nZEB) in Poland – Towards a definition and Roadmap*. Author.

European Commission. (2014). *Technical guidance: Financing the energy renovation of buildings with Cohesion Policy funding, Final report: A study prepared for the European Commission DG Energy*. Author.

European Parliament and of the Council. (2010). Directive 2010/31/EU of the European Parliament and of the Council of 19 May 2010 on the energy performance of buildings (recast). Author.

Kaliszuk-Wietecka, A. (2017). *Sustainable Building - Selected issues in building physics*. Wydawnictwo Naukowe PWN SA. (in Polish)

Marshal of the Sejm of the Republic of Poland. (2008). Act on the promotion of energy retrofit and renovations. *Journal of Laws item 130*. (in Polish)

Marshal's Office of the Pomeranian Voivodship. (2015). *Detailed Description of the Priority Axes of the Regional Operational Program of the Pomeranian Voivodship for years 2014-2020*. Author. (in Polish)

Ministry of Infrastructure. (2002). *Ordinance on technical conditions to be met by buildings and their location of 12 April 2002* (Journal of Laws of 2002, No. 75, item 690, consolidated text). Author. (in Polish)

Ministry of Infrastructure. (2015). *Act 91 of the Council of Ministers of 22 June 2015 on accepting the "National plan aimed at increasing the number of buildings with low energy use"*. Author. (in Polish)

Ministry of Infrastructure and Construction. (n.d.). *Energy efficiency of buildings*. Retrieved from http://mib.gov.pl/2-567ab849c4abe.htm (in Polish)

Ministry of Infrastructure and Development. (2014a). *Programming the Financial Perspective 2014-2020 - Partnership Agreement*. Author. (in Polish)

Ministry of Infrastructure and Development. (2014b). *Financing the energy renovation of buildings with Cohesion Policy funding*. Author. (in Polish)

Ministry of Infrastructure and Development. (2015). *Ordinance of the Minister of Infrastructure and Development of 27 February 2015 on the methodology of determining a building's or part of a building's energy characteristic and energy performance certificates* (Journal of Laws of 2015, item 376). Author. (in Polish).

Ministry of Transport, Construction and Maritime Economy. (2012). *Ordinance of 25 April 2012 on the detailed scope and form of construction design* (Journal of Laws of 2012, item 462). Author. (in Polish)

The National Centre for Emissions Management (KOBiZE). (2016). *Fuel Values (WO) and CO_2 Emissions (EC) in 2014 for reporting under the Emission Trading Scheme for 2017*. Author. (in Polish)

The National Centre for Emissions Management (KOBiZE). (2017). *CO_2, SO_2, NO_x, CO and TSP emission factors for electricity based on information in the National Greenhouse Gas and Gaseous Emissions Database for 2015*. Author. (in Polish)

Włoch, M., & Brzeziński, P. (2010). *Warsaw heating network system, and selected European systems*. Academic Press. (in Polish)

Zaborowski, M. (2013). *Definition of deep energy retrofits, Working Draft No. 2*. Institute of Environmental Economics. (in Polish)

Żurawski, J. (2012). *Zero- or almost zero-energy construction in Polish conditions*. Conference "Insulation". (in Polish)

Żurawski, J. (n.d.). Energy efficiency of buildings in Poland present state and development perspectives. [in Polish]. *Lower Silesia Agency for Energy and Environment.*

KEY TERMS AND DEFINITIONS

Built-In Energy: The energy that is consumed by the processes involved in producing all the building products, their transport to the construction site, the building processes, the energy consumed during the operation, the repairs and the demolition of the building. This is the energy consumed throughout the file cycle of the building (cradle-to-gate).

Carbon Footprint (the Built-In Emissions): The sum of all greenhouse gas emissions caused indirectly or directly by a person, organization, process or product. The value is expressed in tonnes or kilograms of carbon dioxide equivalent (tCO_2eq. or $kgCO_2eq$.). The carbon dioxide equivalent allows for the compilation of influence of various greenhouse gases.

Final Energy (FE): The energy supplied to the building. The final energy demand takes into account the losses associated with the efficiency of the installation systems. In case of buildings it is the energy that the users have to pay for.

Heat Transfer Coefficient (U_c): A property of a wall, roof, or other partition, describing its ability to transfer heat. To calculate the heat transfer coefficient of a building envelope, the thickness and thermal conductivity of each layer of the envelope have to be considered. The better the thermal insulation of the building envelope, the lower is the heat transfer coefficient.

Primary Energy (PE): The energy contained in raw materials such as black coal, lignite, natural gas, heating oil, or in renewable sources. In the case of natural energy carriers, it is defined as chemical energy accumulated within a source without human processing. The demand for primary energy depends on the amount of energy lost in the process of generation and transmission, as well as the type of energy carrier. Primary energy demand is calculated as a product of final energy demand and the W_i coefficient.

Thermal Conductivity Coefficient (λ): A feature of a given material, indicating its capability to conduct heat. Good insulation materials are characterized by low thermal conductivity coefficients. The thermal conductivity of styrofoam is 0.031 – 0.038 W/(m·K); in case of mineral wool it is 0.034 – 0.050 W/(m·K).

W_i Coefficient: Indicates the non-renewable primary energy input necessary for producing and providing the energy carrier or energy for technical systems. Wi coefficient is used to calculate the primary energy demand of buildings. The values of the W_i coefficient valid in Poland are published in Ordinance of the Minister of Infrastructure and Development of 27 February 2015 (Table 3).

ENDNOTE

1 The indicated ordinance implements the Directive of the European Parliament and Council of 19 May 2010 on the energy performance of buildings (Official Journal of the European Union, L 153, 18 June 2010).

Chapter 12
Integration of Vegetation With Architecture Forms

Kinga Zinowiec-Cieplik
Warsaw University of Technology, Poland

ABSTRACT

One of the most interesting ways of returning nature to the city is green architecture, which embraces hybrid buildings at the base of the coexistence of the natural and building worlds. By definition, green architecture does not have to be green at all. Nevertheless, it takes on a number of hybrid forms integrated with vegetation. Built and natural form in the composition is inseparably "tangled" – one follows other to form a cohesive whole. Green architecture becomes part of a larger green infrastructure system. The idea of green infrastructure leads to passage from passive protection, to the active-present in every aspect of human life. It is a development tool respecting the laws of nature. Architecture enters the world of nature as never before, clings to it through the cooperation of designers with specialists in environmental sciences from cellular microbiology to macro scale processes in ecosystems. This requires designers to be particularly sensitive to the natural world, understanding and accepting its rules.

INTRODUCTION

According to the definition provide by Encyclopedia Britannica, "*Green architecture*" (Wines, 2016), is *"a philosophy of architecture that advocates sustainable energy sources, conservation of energy, re-use of building materials and the concern of safety, and the placement of a building with consideration for the impact it exerts on the environment"*. By definition, what is known as green architecture hardly has to be green at all – or referring literally to plants. Nevertheless, such type of architecture assumes a number of hybrid forms integrated with vegetation. In every composition built and natural forms are inseparably "tangled" - one follows the other in order to form a cohesive whole.

This applies to newly designed buildings as well as to architectural objects that are undergoing modernization, especially in intensive urban areas where shortage of greenery occurs. In such cases, façades and roofs represent a great potential for green technology in terms of EPBD requirements (EPBD, 2010).

DOI: 10.4018/978-1-5225-4105-9.ch012

In the context of the study entitled "Design retro-fit nZEB concept - KODnZEB" on modernization and adaptation of two university' buildings to nearly zero-energy standard (nZEB) that was conducted in 2016-2017 in cooperation between Warsaw University of Technology and the Norwegian University of Science and Technology in Trondheim, author presents a contemporary history of green architecture and attempts to outline directions for its development. The work yields the benefits and potential of green forms in building a healthy, sustainable, friendly and resilient urban environment. The collected and analyzed data was used to carry out research projects on green compositions for two buildings under modernization, namely the building of The Faculty of Building Services, Hydro and Environmental Engineering (designed by S. Jaczewski and J. Reda 1970-ies) and a dormitory house "Muszelka" (designed by Z. Dytkowski 1950).

BACKGROUND

The basis of research work on the vegetation integrated with architecture was to delineate the sources of modern ideas concerning green architecture. Planting greenery on the building surfaces is dictated by, inter alia, the need to improve the quality of the urban environment, concern for biodiversity, the need to combat climate change, etc. Therefore, the next stage in the research was to carry out analysis of functions such as biotic, climatic, hydrological, soil, phytoremediation, social, health and cultural functions of greenery that form the foundations for the contemporary resilient ideas. Afterward, widespread recognition of the problem concerning services provided by the environment in light of the concept of green infrastructure which incorporates buildings integrated with vegetation has become a major aspect of the research (Costanza, et al. 1997, McMahon 2000).

The next phase of the work was the recognition and analysis of the biodiversity in architectural forms and the potential for their application. Polish realizations and possibilities are presented in the context of green architectural forms in connection with particular climatic conditions, especially those to be found in Warsaw. The studies conducted have become the basis for KODnZEB's research in the field of green composition, which provides an integral part of the case study and the design process.

GREEN ARCHITECTURE IDEAS

Pillars, upon which today's sustainable design is based, should be sought in the works of eminent designers who created in the second half of the 20th century by their critical development to the modernist order. Their works pointed to the threat of dehumanization the technological society was facing. Ian McHarg's projects, especially his cohesive spatial development strategy that was described in *"Design with Nature"* published in 1969, gave man a responsible role of the subject consciously influencing the state of the natural environment. It is worth mentioning another theory that is controversial from the point of view of science, yet has the capacity for influencing the imagination to a large extent, namely the 1979 Gai hypothesis put forward by a British ecologist, James E. Lovelock (1979). Another publication that aroused great distress was *"Silent Spring"* (1962). The publicity and controversy surrounding the appearance of the book effectively focused public attention on environmental issues.

One of the first issues to affect the principles and guidelines concerning green building design were issued by The Landscape Institute, UK's *"Landscape Design with Plants"* (1977), where one of the chapters was devoted to the design of roof gardens. The same year, Roy Manson's article on the relationship between architecture and nature was published in the 3rd issue of the Futurist. Another significant publication, especially in terms of the development of the concept to combine greenery with architectural structures was the Italian Lotus issue no. 31 of 1981, in which the term "vertical garden" was probably used for the first time (Ungers 1981).

But it was not until 1970s when the initial signs of ecological revolution in architectural realizations were first to be seen. Small scale ideas that yielded great power, begun to be realized by radical creative groups such as Ant Farm and Jersey Devil. Earth projects were carried out by Peter Noever and Gustav Peichi, while Arthur Quarnaby built geodesic lands. In 1970, the SITE studio began its operation. James Wines, the founder of the studio and the later publisher of Green Architecture, has become the father and a propagator of green design. The collaboration between SITE and BEST Products Company Inc. that took place in the 70s resulted in the emergence of certain appealing projects. At the BEST Forest Building in Huston, Texas, spontaneous scrub vegetation combined tall trees was used as an intrinsic part of the building. As seen from the given example, James Wines' studio was ahead of its time, as they focused on naturalism and spontaneity at the same time being able to notice charms and ecological functions of wasteland.

Another vital name for green architecture is the one of Emilio Ambasza. After working for several years at the MoMA, Ambasza established his own studio in 1976. Emilio Ambasza's designs and projects embodied his motto "The Green Over the Gray." His ACROS Fukuoka Prefectural International Hall (project1990) is the most recognizable realization of the creative credo of Ambasz (Bauchanan 1992), at the same time providing the materialized definition of what green architecture is in essence. Public park, greenery and the building itself became united in case of this project. In Poland, there exists one of the most interesting examples of close reference to the idea of the Ambasz – namely, the building of the Warsaw University Library created by Marek Budzynski and Zbigniew Badowski, together with the surrounding garden climbing onto the roof (the project of Irena Bajerska (1993).

The contemporary green architecture seen as coexistence of Nature and Architecture in a hybrid design would not have emerged if it had not been for two pivotal figures, namely Friedensreich Hundertwasser considered the father of the contemporary green roof, and Patrick Blanc, who revolutionized the idea of greening the façade.

In September 1980 Hunderwasser introduced his architectural manifesto, presenting a model of a house (now called the Hunderwasser house) supplied with his own commentary. The artist paid attention to the crying need for human coexistence in close contact with nature. He claimed his project to be an attempt at a dialogue with nature, which he saw as an equal partner. Being an advocate of wild unrestrained nature, he maintained that the

... growth of the trees should proceed as naturally as possible, without man interfering in a gardening capacity, with the aim of letting a piece of truly untouched nature come to be. Not a manicured, clipped, evergreen, affected garden. In this regard, the grass must not be cut, the foliage which falls on the grass not be removed" (Hunderwasser 1980).

Looking at Hunderwasser's ideas, it must be noted by promoting the marriages of nature and architecture, he was 10 years ahead in developing ideas of biodiversity, spontaneity and variability as a permanent component of a building project.

Yet another influential person, a botanist Patrick Blanc fascinated by wild rainforest plants, built the first prototype of it starting from the second half of the 1970s (Fronty 1978). The already technologically advanced planted wall was constructed in a hall of the Cite des Sciences et de l'Industrie in Paris in 1986. The spectacular external vertical garden appeared in 1994 in the park surrounding the Chaumont-sur-Loire palace. For over 20 years, the International Garden Festival in Chuamont-sur-Loire assumed the leading role in the mapping of contemporary design ideas. The garden created for the purpose of the festival is understood as a lens focusing on important problems concerning shaping of the contemporary landscape. The first edition of the festival entitled "*Delight*" took place in 1992 and was attended by such designers as Emilio Ambasz with the creation of *"Nymphfeum in the embrace of the earth"* and James Wines and his SITE with *"The Garden Between Geometry and Freedom of Nature"*. In its 23-years history, the festival touched upon themes coherent with the development of ideas, becoming the place of the evolution of these ideas. In 1993, the topic under discussion was *"Imagination in Times of Crisis"*, ten years later in 2003 designers were pondering on the issue of *"Weeds"*, and in 2011 they developed ideas for the gardens of the future and their biodiversity.

The 1980s and 1990s were the period when the ideas that shape the modern environmental awareness could find their realization on a broader scale. Three important projects were conducted in the Parisian post-industrial districts of the 1980s. Bernard Tschumi won the competition for the park de la Villette by reinterpreting the concept of the public space based on the deconstruction idea by the philosopher Jacques Derrida. Citroen Park competition was won by two teams: modern artist - Jean Provost and Gilles Clement, who made no attempt to hide his pro-environmental roots. Their project was based on the idea of Serial Gardens - small interiors revealing the richness of plant forms influencing the human senses, while the *"Garden in Motion"* - a wildlife apology came last in the series.

Jan Le Caisne, on the other hand, won Bercy's competition by means of using and highlighting the remembrance of the place. From the three spectacular designs of the Paris parks of the 1980s presented above, concrete conclusions have emerged, namely the essence of nature and history of the place (Bercy), the importance of environmental issues (Citroen) and openness to experience (de la Villette).

The innovation in case of the *"Garden in Motion"* was to see the beauty of wasteland and spontaneity in the world of plants – the factor previously used in the work of SITE (Building Forest) and the factor which Hunderwasser paid attention to. This time the topic was presented to a wider audience (1991), becoming a turning point in contemporary plant-based design thinking. Opening for change, acceptance nature dynamics and spontaneity began to form the basis for project activities. At the root of such thinking was the motto the designers should follow: do as much as possible, making as little as possible. It is surprising that the slogan created in the opposition to modernist ideas is so close to the famous "less is more" Mis van de Rohe.

Gilles Clement based his theory on three priorities:

- Maintain and increase biodiversity as a resource guaranteeing the future of the natural world;
- Maintain and enhance the biological quality of life: water, soil and air;
- Limitation of raw materials and energy consumption.

"Garden in Motion" owes its name to the spontaneous movement of plants. Easy-to-see and visible changes in this range are shown by herbaceous plants with short life cycles – annual and biennial beautifully blooming meadow plants. According to the author, the art of garden and landscape design should be more focused on observation and understanding of the dynamic world of plants. It is noteworthy that

the motto, as well as the postulates of "Garden in Motion" are in line with the assumptions of Agenda 21, which were elaborated a year later in Rio de Janeiro, and remain a design canon up to this day.

Over the following 20 years, Gilles Clement clarified the development stages of the ideas he presented at the exhibition *"Planetary Garden"* and in the *"Third Landscape"* (Clement 2003). The 1999/2000 Exhibition of *"Planetary Garden"* was the highlight that presented the development of the Garden in Motion into planet scale. In the exhibition, references to Gai Lovelock's theory and Ian McHarg's postulate formulated 30 years earlier can clearly be noticed. The Earth this time is understood as a Garden - a closed, finished and quantifiable space, which man is obliged to care of and cultivate with respect and consideration. At the heart of the concept of the *"Planetary Garden"* was the aim to draw public attention to the diversity of life forms on Earth and to provide responsible management of nature by man. Much attention was paid to the dangers associated with the anthropogenic devastation of the natural environment. The author of *"Planetary Garden"* (Clement, Eveno, 1997) posed open questions: how to exploit environmental resources in such a way as not to destroy it, how the Planetary Garden works to allow nature and man to survive? *"Third Landscape"*, on the other hand, was understood as a part of the *"Planetary Garden"* by humans- places abandoned by humans and re-infected by nature, which provide a biodiversity source that has to be taken special care of.

Gilles Clement has raised important questions about the future of our planet and its natural resources, and during the London exhibition of 2009 called *"Radical Nature, Art and Architecture for a Changing Planet 1969-2009"*, it was attempted to provide answers of architects and artists to the questions posed. Radical Nature showed the pioneering works of the 1970's Ant Farm collective, as well as Richard Buckminster Fuller's visionary projects, and the projects by artists such as Joseph Beuys, Hans Haacke and Robert Smithson, accompanied by younger artists: Heather and Ivan Morison, Philippe Rahm and Tomas Saraceno. Many works/projects/installations were directly used and/or designed to support Nature. Such projects included Simon Starling's Island for Weeds, Wolf Hilbertz's Reef Builders, Henrik Hakansson's Fallen Fores, A Confrontation of Agnes Denes or projects of the French architectural and experimental team R&Sie(n) and the famous biomimetic architect, Michael Pawlyn.

As it was shown, the intensive search for the integration of vegetation with architectural forms we witness today began in the 1970s. For nearly 30 years, technological, architectural, natural, aesthetic, artistic solutions and philosophic support, have been developed and perfected and today they are slowly beginning to form a new designing canon. The need to utilize attainable areas of dense urban space under the greenery, including such areas that came into being as architectural surfaces, is currently being strengthened by results of research into the functions and importance of vegetation for the creation of a healthy living environment.

THE IMPACT AND IMPORTANCE OF VEGETATION ON ARCHITECTURAL FORMS

Research in various fields proves the need to increase green space within the urban landscape.

At the beginning of the 21st century the concept of green infrastructure was introduced (Benedict, McMahon, 2000). The European definition was published in 2013 (COM (2013) 249 final). The idea of green infrastructure leads to passage from passive protection (national parks, nature reserves, etc.), to active protection that is present in every aspect of human life. Such protection makes a development tool that respects the laws of nature. It combines concepts of green systems functions with ecosystem services that allow estimating and presenting the value of such systems in a measurable way. The better the environment, the better its performance and the higher the benefits of green infrastructure understood as a network of natural and semi-natural areas. Such infrastructure is a tool to help improve the quality of the environment. In the long run, it proves to be a cheaper and a more sustainable solution than the conventional methods. Adoption of the EU strategy for 2020 has intensified efforts to support the construction and expansion of green infrastructure, also in Poland.

Research into the functions of greenery, especially biotic, climatic, hydrological, ecological soils, pollution removal ones (Szumacher I. 2005) has been conducted since the 1970s (Wojcicka 1971, Wysocki, Myczkowski 1976, Kawon, Zmuda 1977, Bednarek 1979, Zimny Bytenrocz, 1979). The subject is well recognized.

Biotic functions include:

- Food production increasingly returning to cities in the form of urban farming and gardening (Szczurek, Zych 2012, Philips 2013, Urban Gardening 2013);
- Production of raw materials - e.g. biofuels (Frac et al. 2009, Koziel, Wlodarczyk 2011);
- Shaping the living conditions of organisms, e.g. the biocenotic trees which provide the habitat for other organisms - one old tree can house up to 2041 animals belonging to 257 species (Wohlleben 2016 after Grossner et.al. 2009).

Climatic functions are represented by:

- Regulation of bioclimatic conditions, including the reduction of solar radiation (shading) up to 90% by dense tree crowns, and thus lowering the air temperature by more than 10°C and the generation of delicate air movement (Szczepanowska 2001);
- Regulation of air quality by absorbing CO_2 - one 25m high tree over the vegetation season may absorb the amount of CO_2 equal to the amount of it produces as an effect of the activity of two single family households (Borowski 2013);
- Regulation of air quality by O_2 extraction - 10m high tree produces an average of 118 kg of oxygen per year, and an individual man consumes 176 kg of oxygen over that period of time, so the production of oxygen by two medium-sized trees meet the needs of one person per year (Borowski 2013);

Hydrological functions specify:

- The so-called "little water cycle" where the vegetated surfaces support the water cycle in the environment, slow down the flow of water allowing infiltration of precipitation water to deeper layers; the wetland and marshy communities are not overestimated - they become reservoirs, retaining water creating "hydrological windows" (Wolski 2013);
- Counteracting degradation of water by self-cleaning, which is revealed by the root deposits used in the draining rushes and swimming pond technologies (Gasiorowski 2011).

Ecological functions of soils consist of:

- Counteracting physical degradation, e.g. the protection of slopes by vegetation (root systems);

- Counteracting chemical degradation, among others thanks to detoxification based on the purification of soils from harmful constituents in fitoremediation processes (Gawronski, Gawronska, Rokosza 2000);

Pollution removal functions:

- The absorption of particulate pollutants (PM10, PM2,5) and gaseous (polycyclic aromatic hydrocarbons, dioxins, benzene, etc.) that become picked trapped from the air by plants and collected on the leaf surface in the form of wax by, e.g. such trees as ashlar or birch (Gawronski, Szmit 2011).

Research related to social functions is becoming increasingly important. Green areas not only create recreational and leisure facilities of cities but also exert a very positive impact on the human health. Research conducted in the Netherlands and the United States prove greenery planted near residences lowers the level of stress and aggression (Szczepanowska 2001, Borowski 2013). Parks squares and gardens provide an important meeting place for social interaction, or development local communities (Gawryszewska 2009).

It turns out that lack of contact with nature can lead to nature deficiency syndrome (Louv 2014) manifesting itself in the form of depression, dementia, lack of creativity, and the disappearance of the ability to build social bonds. Scientists agree that limiting interactions with nature during childhood can lead to biophobic behavior in mature life, alienation, lack of understanding of nature, and consequent disappearance of ecological attitudes (Bunting & Cousins 1985, Sobel 1996, Schultz et al. 2004). Contact with nature alleviates the impact of stress and helps children deal with adversities. The more often they are in the wild, the greater the benefits (Wells, Evans 2003). Also worth mentioning are the cultural aspects of building local identities (Gawryszewska 2009), the symbolic references of contemplative gardens (e.g. Zen Gardens), or artistic values (e.g. Little Sparta by Yana Hamilton Finlay), that show the spiritual roles of nature.

The issue of ecosystem services, understood as *"the ingredients of nature that are directly consumed, felt or used to raise human well-being"* (Solon 2008 after Boyd, Banzhaf 2007.) is just beginning to develop. Four main functions may be distinguished, namely, supply (e.g. food, fuel, materials, etc.), regulatory (e.g. climate, pollution) support (e.g. hydrological cycle, elemental cycle) and cultural (e.g. didactics, art, inspiration, recreation) (Kroneneberg 2016). The idea of ecosystem services is a tool that can synthesize ecological links with economics and ultimately unify economic and ecological assessments (Solon 2008, MEA 2005). This allows the economic valuation of services provided by the environment and the valuation of "green" estate. It is a tool for raising public and politicians' awareness concerning the value and importance of green systems (Costanza et al. 1997, Solon 2008).

One of the first Polish researchers dealing with valuation of ecosystem services is prof. Barbara H. Szczepanowska from IGPiM, who published the findings of her research in 2001 (2007), in which she presented the results of American studies on trees in the city. These findings indicate that having calculated the costs of planting, irrigation and maintenance of trees. Planting trees proves a more financially effective way of saving energy and limiting carbon dioxide emission than many conventional ways. The American researches attempted also to use computer simulations in order to calculate the energy and economic savings that could be achieved by planting 100 million trees (3 trees/a family home), which would produce 30 billion kWh and therefore reduce the costs by 2 billion USD (Dwyer et al., 1992).

The values of the plots themselves were also subjected to estimation depending on their degree of greenery planted on them. The estimations showed a 15% -30% increase in the price of the plots that were overgrown with trees (Szczapanowska 2001). A very interesting issue is the ruling of the US court, made more than 30 years ago in Virginia (USA 1981) which maintained that the loss of one old oak from the plot decreased the property value by 15 thousand USD, which amounted to 10% of the plot value.

There are many different methods of valuing services provided by the natural environment. Professor Jakub Kronenberg from the University of Lodz conducted an assessment of ecosystem services based on the perspective of inhabitants by the conditional selection method (Giergiczny, Kronenberg 2012). This assessment in 2012 allowed to estimate the level of Lodz citizens' willingness to incur additional costs of 17 million PLN per year (PLN 28.15 per capita per year) on increasing the number of trees in the streets in the city center. Calculations exceeded the city's two million PLN budget on greenery by more than eight times. Professor J. Kronenberg's research showed a huge underestimation and a divergence between the city's capabilities and social needs.

Researchers from IGPiM, and prof. B. Szczepanowska in co-operation with Warsaw University of Life Science, developed a *"new method of valuing trees in accordance with Polish conditions"* for the valuation of street trees on Praga North district of Warsaw (Szczepanowska at al. 2014). The method allowed to calculate the total annual value of the tree services of the Praga North, the research area being between the streets of Targowa - Zieleniecka - Zamoyski - Jagiellonska - al. Solidarnosci (research period 2011-2013) to be of more than 2 million PLN (2,016,000 PLN). Another set of measurements was made in relation to the services provided by the park trees in Park Praski in Warsaw (2015), and the data were calculated using the i-Tree ECO methodology (US Forest Service) (Jedraszko-Macukow, Sitarski 2017). The results are currently under development. For now, it is known that the leaf area comprises more than six times the size of park surface (6.5 ha of leaves of trees and shrubs per 1 ha of parkland). The trees consume 20.9t of CO_2 in the course of the year and release 44.8t of O_2 into the atmosphere. Economic calculations are not yet available, but in comparison to the survey of the service of street trees, due to the better condition of the park trees, higher values can be expected. I-Tree ECO research has been developed for many cities, including London, where the total annual profit resulting from trees services has been estimated at £ 132.7 million (Valuing London's Urban Forest). In the 1980s Americans began to estimate the services of urban trees showing their value Interesting data were obtained from the study "American Forests" (2000) which estimated the work of trees in the Baltimore-Washington area to reduce surface runoff and to collect water at the value of 4.68 billion USD. Results of equally large and surprising scale have been obtained in Pugeot Sound, which point to savings at the rate of 2.4 billion US dollars obtained as a result of surface runoff reduction (through trees) to billion. (Szczepanowska 2007).

Greenery counteracts the common phenomenon of urban heat island. In Atlanta, by increasing the coverage of the city with green areas (5.000 newly-planted larger trees - the trunk diameter of 30cm and 60 thousand smaller - the trunk diameter of 10cm), the range of heat islands was reduced by 1/3. (Szczepanowska 2007).

Green systems can also be useful in supporting the ventilation process. Openwork tree crowns cause longer aerodynamic shadows when compared to architectural baffles of similar height (Orzeszek-Gajewska 1982). Planting, depending on requirements, can increase or decrease the speed of wind, as well as alter the direction of air mass flow. Aerodynamic studies of urban building development (Zielonko-Jung 2013) indicate the occurrence of places with insufficient ventilation. In such situations, the use of trees, especially their thick and compact crowns, should be avoided and their functions will be taken over by climbers planted on elevations. Trees can also interact inversely by intensifying the wind speed. High trees, situated parallel to the main wind direction, surrounding alleys will increase wind speed by up to 120%. In case the need to reduce only the airflow occurs, openwork plant composition should be used. In order to achieve effective year-round insulation, a multi-layered composition of coniferous trees and shrubs should be designed. The range of impact varies from 2 to 20 times the height of the trees (forming the insulation) – planted from the windward side, to 60 times their height – planted from leeward side. (Lewinska 2000).

The above mentioned examples describing the value of ecosystem services influence imagination, defining greenery as one of the most important assets of the city. Carrying out valuation of ecosystem services allows us to count the value of the natural environment as a "natural green fund" in accordance with the guidelines provided by the European Commission (Szczepanowska et al., 2014). The value of these services grows significantly over time and alongside the maturity of ecosystems as opposed to the value of technical infrastructure (Szczepanowska 2001 after McPherson 1994).

In cities, the natural environment is very much modified by strong urbanization. The problem poses a major challenge for contemporary researchers. Modern (novel) ecosystems and new ecologies are being discussed openly (Hobbs et.al. 2006, 2013).

Due to the efforts made to build a sustainable and resilient urban environment, the research is not only, at the center of attention for scientists themselves, but also designers, such as urban planners, architects and landscape architects are interested in. The results of the studies form the basis for architectural and urban planning activities. The knowledge of the functions of urban nature and their meaning for economics understood through the concept of ecosystem services is becoming an important tool for city development. These results provide arguments for cost-effective economic use of green infrastructure in a market economy, while also contributing to active environmental protection. Green architecture as an element of green infrastructure creates an important link in the natural city environment.

THE GREEN OVER THE GRAY

Green roof technologies are of great importance in this field. Research in Chicago showed that in August on a sunny day the temperature of the green roof was 33-48°C when the conventional roof at that time was heated to 76°C. The air temperature above the green roof was 4°C lower than in case of the conventional roof (Department of Energy 2004). The study suggests that if all Chicago buildings were to be equipped with green roofs (30% of the total land area) in ten years, it would save the costs worth 100 millions USD per year due to reduced cooling requirements (Peck, Kuhn. 2003). In addition, the emission of CO_2 and other harmful compounds arising from the operation of air conditioners and energy production would be reduced.

In Poland, research on green roofs is most advanced in Wroclaw, where the focus is placed on water management (Szajda-Binfeld et.al. 2012, Burszta-Adamiak 2015). As far as other ecological green roofs functions are concerned, we rely on German research by prof. Manfred Kohler (e,g. 2003) and prof. Marco Schmidt, (e.g. 2015), and on guidelines recorded in the FLL publication (Forchungsgesellschaft Lnadschaftsentwicklung Lnadschaftsbau e.V), which have been translated into Polish (2015). In Warsaw, pilot studies were conducted (Baryła, Sedlicka, Kaczmarczyk 2015) into the temperature changes depending on the green roof coverage (garden roof on a building of The Neofilology Faculty - University of Warsaw). The results of the studies show the average temperature difference between surfaces covered in plants and the board surface was 7.23°C. Similar conclusions are presented in the research carried out in Cracow where the temperature of green roof was lower than the temperature of roofs of old town buildings by about 8°C - 10°C. (Walawender 2015).

Similarly, green walls in the form of façades overgrown with vines or developed in the form of vertical gardens insulate in the same ways as green roofs do. Climbing plants growing on elevations reduce the amplitude of daily fluctuations in air temperature filling the space between the leaf layer and the wall by about 4°C. On summer days, the temperature of the air under the vines is about 4°C - 5°C lower. Consequently, energy savings of 15-30% per year (in moderate climate) are achieved. Climbing plants are very efficient in terms of leaf area - they comprise a very high ratio of biologically active surface in relation to the surface needed for growth. Grown vines occupy about 0.5 m^2 of land and the area covered by its leaves in the season is 2600 m^2 (Borowski, Latocha 2014). German research conducted on the building of the Faculty of Physics at Adlershoff in Berlin proves that shading of the southern façade of a building by covering it with generates costs valued at €1300/a, which is more than ten times cheaper than the conventional forms of electric blinds that cost 16,250 €/a. (Shmidt 2015).

In Poland, however, there is no study on the impact of vertical gardens on the climate of the city. Vertical gardens in the Polish climate are a challenging issue. Hydroponic cultivation developed by P. Blanc is impossible due to the cold winters present in Poland. In the years 2012-2015 research was conducted on the development of the Polish Green Wall (Innotech - K1/I1/40/159571/ NCBR/12). The collaboration of the AB System S.C. team with Warsaw University of Life Sciences and the Bioclimatology and Ergonomics Environment Laboratory resulted in a 1:1 model of the substrate form of the extensive and intensive vertical garden. In the intensive system, various plant species, with a high proportion of shrubs, have been tested and are constantly being observed. In the extensive system, the species selection is focused on succulent plants. The first research shows the Polish Green Wall produced 1.5 kg O_2 from a 1m^2 surface during 214 days of vegetation and was 1.6 x more effective in acoustic insulation than a typical screen. The prototype was built in a suburban area near wild land, therefore it is difficult to estimate what its effectiveness in the city center will be (Zinowiec-Cieplik, Guranowska-Gruszecka, Dankiewicz 2015).

Green architecture, understood as a composition of architectural elements and the greenery is becoming a standard over more than the recent 20 years. Slowly, but successively such vegetation begins to "spill itself over" the façades and penetrate deep into the architectural structures. The role of greening in the fight against water shortages has repeatedly been stressed (Schmidt 2015, Solarek, Rynska, Mirecka 2016, Januchy-Szostak 2010, 2011). They also support the water circulation in the city environment, in which water-tight surfaces constitute the majority. Green roofs, vertical gardens or façades with vines become an ersatz for the *"hydrological window"* (Wolski 2013). They may be a part of a wider system associated with retention and infiltration tanks. Since 2013, The Association of Modern Buildings has been conducting expert work on the environment and natural resources within the Working Group (GR7) to develop technical requirements in line with the principles of sustainable development for the construction industry. It is suggested to include in the biologically active area all forms of perennial greenery growing on architectural structures in specialized substrates - also green walls with areas over 10m^2 (Solarek, Rynska, Mirecka 2016). The current design practice favoring the coexistence of plants with architecture in Poland is concentrated mainly on intensive forms of greenery located on the parking roofs. On the scale of urban investment, green roofs as extensive mat systems unfortunately are sporadically encountered.

The external vertical gardens are a unique subject - in Poland, we have only several examples of such solutions in public spaces: in Wroclaw on the façade of the City Hall (Greenarte project and realization

2015), in Warsaw on the Foundation for Polish Science in Dutch substrate technology (project FAAB Architektura Adam Bialobrzeski Adam Figurski 2009-2014), in Gdansk on airport public space (project and realization Greenarte 2014). At present, the green walls are an expensive (between 2000-4000 PLN/m2) and wobbly investment in the view of harsh climatic condition and the lack of legal mechanism to govern it. If the GR7 demands were included in the law, the development of vertical garden could have accelerated.

The development of greenery on architectural forms presents one more positive aspect, namely biodiversity. Green roofs, vines on elevations and vertical gardens are habitats for small animals: small rodents, birds and most commonly insects, including pollinators. The fight for every plant and every insect is gaining in importance in the context of the data confirming the up to 500 times acceleration in the rate at which extinction of species has been occurring in the recent century (MEA 2005). Modern formations are becoming a place for animals. Green roofs can be designed to create a place for rare and endangered species. Swiss and British research on the richness and protection of the diversity in case of green roofs recommend the construction of extensive forms on every newly built flat-roofed building in the city (Nature and Landscape Conservation Act §9; Building and Planning Act §72) (Brenneisen, 2006, Kadas 2006). Researchers notice an effective tool for biodiversity conservation in the extensive green roofs in the context of regional and landscape planning. (Baumann 2006). Greenery on buildings will be serve its functions more efficiently if whole systems of it are created, connected with each other and related to conventional forms of urban greenery such as parks, greenways, or squares.

The development of green roof technology and vertical gardens is very dynamic. Generally, green roofs are divided into extensive, intensive and semi-intensive systems depending on the thickness of the vegetative substrate and the plant species that may be planted on them.

An example of an intensive cultivation (allowing the planting of shrubs and trees) in Warsaw is the roof garden of BUW (Budzynski M., Badowski Z. - architectural project and Bajerska I. - landscape project 1993). This is the largest garden of this type in Poland, covering an area of 2000 m^2. The park has gained significant popularity among citizens. It has become not only a place of resting for students and for inhabitants of the city but also be a touristic attraction. It shows the new features of green roofs garden.

Extensive gardens, on the other hand are the areas with the least load and the smallest (2-20 cm substrate layer) with ecological and insulating roles without user input. One of the largest extensive roofs is built in Warsaw on the Arkadia Shopping Mall (Mariusz Pasek Ogrodownia project 2004). It is created by the succulent mat growing at 10 cm and substrate playing an ecological role.

The research and prototyping of the Polish Green Wall (Innotech-K1/I1/40/159571/NCBR/NCBR/12) has led to similar conclusions as to the types of crops used for covering green roofs. Polish Green Wall comes in an intense form, up to 35 cm thick, and allows to plant shrubs. Extensive form, however, is equipped with a vegetative layer thickness in the range of 15-25cm and is dominated by succulent plants. The advantage of polish vertical gardens in relation to western European solutions is found in their very good level of adaptation to local climate, especially to the conditions characteristic of winter cold zones. Warsaw lies at the border between zones 6b and 6a zone, where permissible winter temperatures can range from -18°C to -23°C. For comparison, Wroclaw, Poznan, and Gdansk - are located in warmer 7a zone, with the lowest winter temperatures ranging from -15°C to-18°C maximum (Heinze, Schreiber 1984). Berlin lies in zone 7b (-15°C to -12°C) and Amsterdam 8a (-9°C to -12°C) and although both cities are located in a latitude similar to Warsaw, the climate and the applied technology cannot be compared. The disadvantage of the substrate system of the polish vertical garden is its weight reaching up to 2.5

kN /m^2 in extensive forms and up to 5.40 kN /m^2 in the intensive form. Creation of such walls requires appropriate construction (Zinowiec-Cieplik, Guranowska-Gruszecka, Dankiewicz 2015).

Polish research concerning the problem of green roofs, as well as the research on climbers is quite advanced. In contrast, studies on vertical gardens have just begun. The studies conducted mainly by botanical/horticultural centers are focused on natural elements. There is no interdisciplinary analysis that would combine natural sciences and technical sciences on sustainable building and architectural technologies. For the development of the concept of the Polish Green Wall and its involvement in the creation of green architecture, it would be important to verify both, the functional possibilities and ecosystem services, as well as to develop such technological solutions in urban environments that would differ from the prototype conditions.

KODnZEB GREENERY RESEARCH PROJECT

Against the background outlined above, the concepts of greenery prepared by the KODnZEB research project at the Warsaw University of Technology (2016-2017) for the building of the Faculty of Building Services, Hydro and Environmental Engineering (FBSHEE), and the Dormitory House "Muszelka" appear as an undertaking of rather insignificant scale, but these projects show the contemporary trends in the formation of green infrastructure. In both cases, researchers attempted to integrate greenery with architecture. It was not always possible. It turns out the space on the façades and roofs of buildings is also a valuable area to be covered with power or ventilation installations. In the case of the FBSHEE, the southern and western façades were occupied by photovoltaic panels and the curtain wall. Place for plant pots in the form of gutters above the ventilation systems was managed only on the eastern and northern side of the building façade. The project envisages planting in pots such forms of vegetation that will not deprive the interior of light, mostly bushes together with clematis with a dense type of cover. For the preservation of rhythmic, it is proposed to limit the selection of plants to several species and apply them interchangeably for individual layer, which means a homogeneous set of species for a particular level. The composition of façade plants will complement the courtyard plant design, enriched with plants climbing on steel links. In case of the modernized the FBSHEE, the courtyard is a vital element of the green project and together with the green façades, they co-create the environmental system of this small interior. In order to increase the area covered with greenery and to ensure the comfort of people staying in this space, it was decided to eliminate the admission of cars and the possibility to park cars. The plant composition has been matched to the characteristic pavement pattern in the form of interpenetrating swaths of ground covered with greenery. It forms the "disappearing" (increasingly scattered) decorative grass which eventually turns into a mixed naturalistic structure with evergreen ferns and covers plants when it reaches the trees crowns. This composition is intended to be open to the emergence of spontaneous vegetation. Evergreen species are a great part of design.

In the case of the Dormitory House "Muszelka", the entire roof surface was taken up by the photovoltaic panels, and green roof had to be eliminated. The space for greenery was left in the inner courtyard covered with valuable and mature tree stand. The green design consisted of arranging the composition so that the central turf was made available to students for resting and recreation purposes. Rhythmic planting was proposed at the edges to allow free access to the lawn. Design introduced a mini green roof on the bicycles pavilion. It was proposed that the southern façades (from Mochnackiego St.) were covered with climbers from the outside. Due to the historic character of the building, the proposal may seem controversial. There are several arguments to support such a decision. The first reason is the south exposure, which is subject to strong summer heating and winter cooling. The application of climbing plants would improve the insulation of the façade and increase air humidity. This façade is, moreover, a heavy, modernist and monotonous frontage, which requires a light break in form towards a more friendly and less overwhelming composition. In addition, the central part of the student housing complex, of which Muszelka is a part, is already overgrown with climbers. As the research shows, historic monuments, buildings of highest historic value, are overgrown with climbing plants, the examples being Wawel, Ksiaz or Malbork. The condition for such plants to be used is proper care and cultivation as they should not dominate the buildings and of course the necessary arrangements with the conservation services (Jackiewicz, Borowski 1998).

CONCLUSION

As shown by the he KODnZEB projects, greenery is not always the most important of elements. Architecture is not only supposed to be environmentally neutral, but it has to support the natural environment. Openness to plant spontaneity, biodiversity development, air purification, leveling and fighting the urban heat island – these are some of the support functions described in the article above.

Figure 1. Project for the modernization of composition of plants close to the building of the Faculty of Building Services, Hydro and Environmental Engineering, Warsaw University of Technology

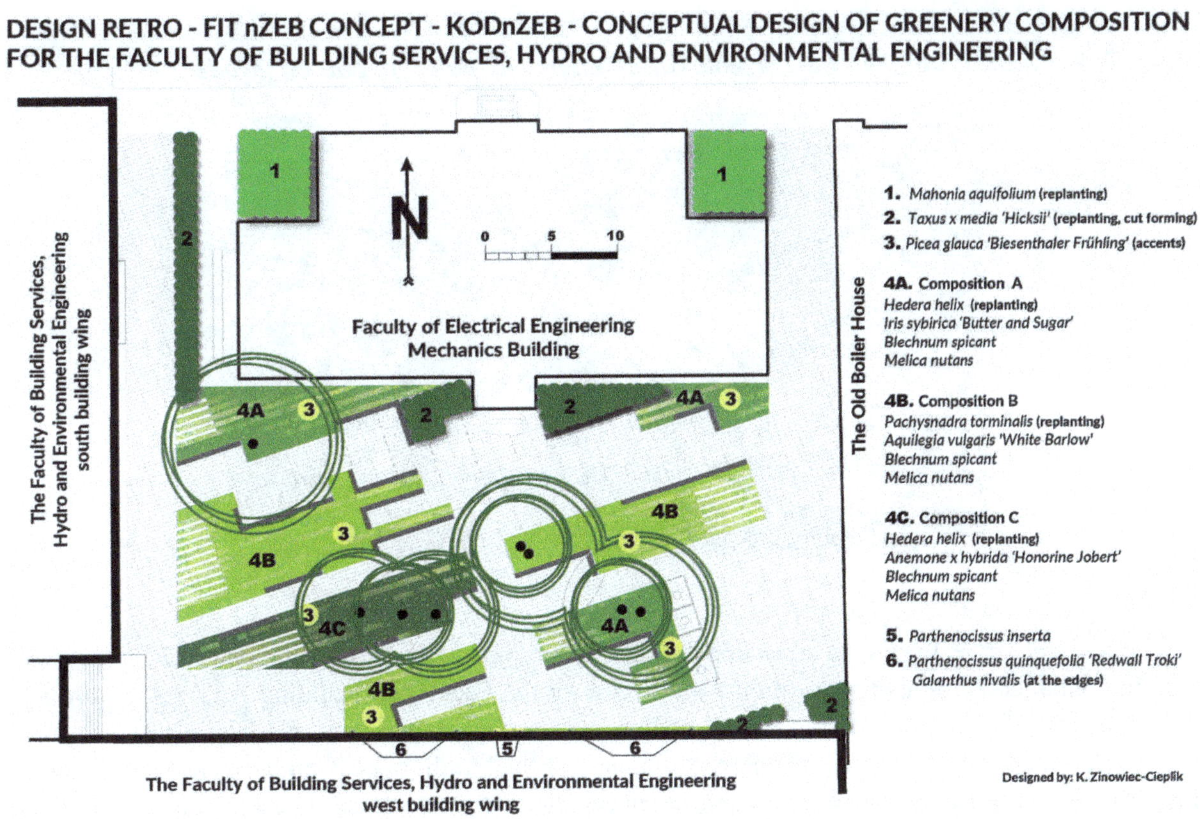

Figure 2. Project for the modernization of composition of plants close to the Student Building "Muszelka", Warsaw University of Technology

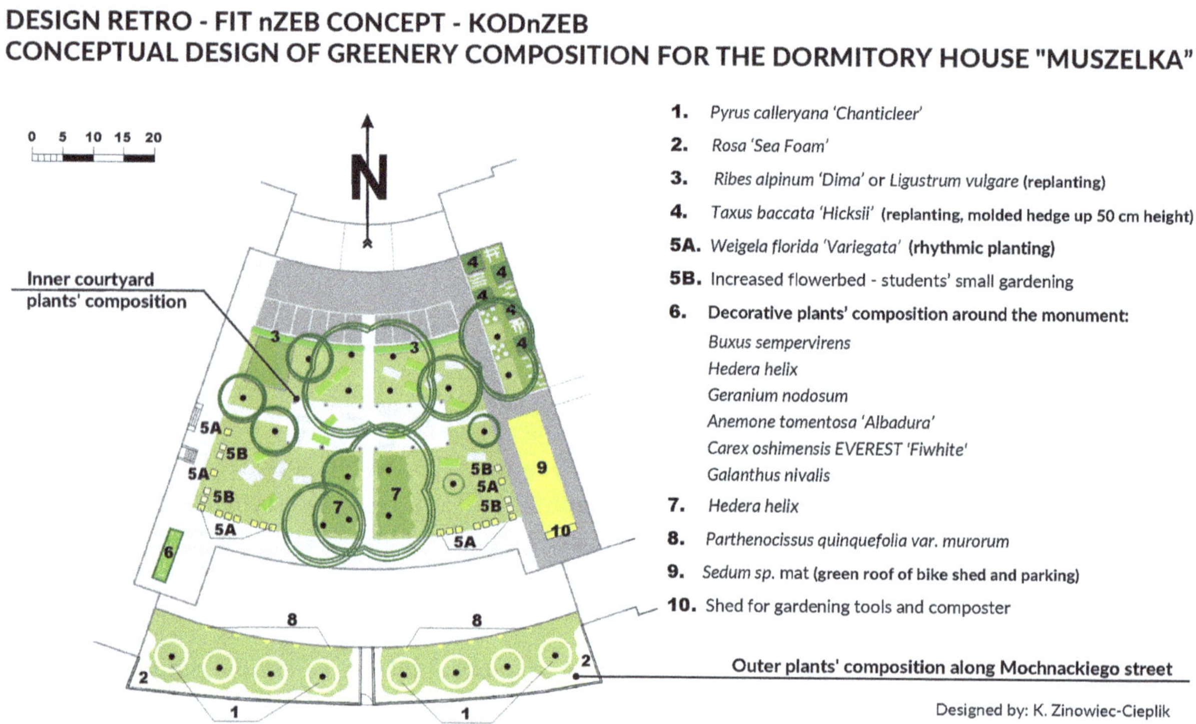

Figure 3. Visualization of the FBSHEE courtyard design with plants in pots on façade

The research project presents an example of a very simple and basic application of plant compositions in containers or vines on elevations. In the case of the Faculty of Building Services, Hydro and Environmental Engineering, the façade vegetation was used as an element of the green area, which, however, does not directly affect the energy efficiency of the building itself. In the case of "Muszelka" dormitory house, is the greenery serves a different function - it works as an insulator against sun radia-

Figure 4. Visualization of the "Muszelka" courtyard design

tion and moisture. It is difficult to assess what could be the efficiency of plants on the elevation in this respect, because no comparative research to rely on has been conducted in the Warsaw region. A possible implementation of the project assumptions will allow to conduct appropriate studies in the future.

The example of KODnZEB project is very modest but also one of the first to undertake an interdisciplinary research project combining technical and natural sciences.

Nowadays, the search for innovative solutions for green architecture goes far beyond traditional forms of greening façades. Designers looking for new solutions assume several directions.

One of the directions to take is the search for materials that, by virtue of their structure and construction, would facilitate the creation of such places in which vegetation would flourish. Examples of this may be the proposal of "biological concrete" developed by scientists from the Department of Construction Engineering of the Barcelona Polytechnical University (Universitat Politeccnica de Catalunya BarcelonaTech). Such concrete forms the basis for the development of lichen and moss (BarcelonaTech 2012). Another example may be the search by the Biota Lab (Bartlett School of Architecture and the University College London), whose scientific team is actively developing knowledge of bioreactive materials like clagareous composites, cleycrete, hydrogels and others.

Interesting prototypes of algae technologies are created by ecoLogicStudio Maria Pasquero and Marco Poletto (2014, 2015a, 2015b). They also prepared project of land and submarine algae farm in region scale for the Swedish Simrishamn Municipality and Marine Center (Pasquero, Poletto 2011). Another interesting and practical example may be the BIG Building in Hamburg (Splitterwerk Architects and Arup 2013) opened four years ago. It is the first prototype form of a 4-storey full-size building with façades of algae bioreactors, which produce heat (clean energy) and also shade the interior against overheating.

Another research led by Ferdinand Ludwig from the University of Stuttgart draws on a living architecture, relying on reinforcing the plant material (mainly willows) by steel structures - Baubotanic. In this case, the plants are the dominant material and the reinforcing elements serve only as an addition. An architect must understand the needs and requirements of plants in order to create such structures using plants that are stable, long lasting and poses aesthetic values. Architecture these days enters the world of nature as it had never before; it clings to nature through the cooperation of designers with specialists in environmental sciences from cellular microbiology to macro scale processes in ecosystems. This requires designers to be particularly sensitive to the natural world, understanding and accepting its rules.

RESEARCH PROGRAM: INNOTECH-K1/I1/40/159571/NCBR/NCBR/12

- *AB System S.C. team landscape architects:*
- MSc Bartosz Dankiewicz, MSc Katarzyna Dankiewicz, Eng. Karolina Adamska, MSc Izabela Główka-Roman;
- *Warsaw University of Life Sciences team:* Prof. Jacek Borowski (dendrology), Prof. Stefan Pietkiewicz (plant physiology), Prof. Tadeusz Łoboda (plant physiology),PhD Wojciech Kwasowski (soil science);
- *Bioclimatologiese team (Bioclimatology and Ergonomics Environment Laboratory):* Prof. Krzysztof Błazejczyk, PhD Jarosław Baranowski, MSc Wojciech Błazejczyk, MSc Anna Błazejczyk
- *Construction and technical team*: ABsystem

REFERENCES

Architects, S. (2013). *The Clever Treefrog 1*. Retrieved from http://algae-architecture.com/?p=269#more-269

BarcelonaTech. (2012). *Researchers at the UPC develop a biological concrete for constructing "living" façades with lichens, mosses and other microorganisms*. Retrieved from http://www.upc.edu/saladepremsa/al-dia/mes-noticies/researchers-at-the-upc-develop-a-biological-concrete-for-constructing-201cliving201d-facades-with-lichens-mosses-and-other-microorganisms

Bauman, N. (2006). Ground-Nesting Birds on Green Roofs in Switzerland: Preliminary Observations. *Urban Habitats. An Electronic Journal on the Biology of Urban Areas Around the World, 4*. Retrieved from http://www.urbanhabitats.org/v04n01/urbanhabitats_v04n01_pdf.pdf

Bednarek, A. (1979). *Wplyw parkowej roslinności drzewiastej, trawnikow i placow na warunki mikroklimatyczne na przykladzie Warszawy. Les (Ljubljana), 27*, 89–107.

Borowski J. (2010). *Ile warte sa drzewa w miescie?* Retrieved from http://www.muratorplus.pl/technika/zrownowazony-rozwoj/ile-warte-sa-drzewa-w-miescie_68688.html

Borowski, J. (2011). *Bezcenne drzewa*. Zwiazek Szkolkarzy Polskich. Retrieved from http://www.zszp.pl/pliki/MPO07-11s.60-61.pdf

Borowski J. (2013). *Dlaczego warto sadzic i pielęgnowac drzewa?* Stowarzyszenie na rzecz Ochrony Krajobrazu Kulturowego Mazur "Sadyba". Retrieved from https://sadybamazury.wordpress.com/2013/04/06/dlaczego-warto-sadzic-i-pielegnowac-drzewa-jacek-borowski/

Borowski, J., & Latocha, P. (2014). *Zastosowanie roslin pnacych i okrywowych w architekturze krajobrazu*. Warszawa: Wydawnictwo SGGW w Warszawie.

Bouchanan, P. (1992). *Emilio Ambasz Inventions: The Reality of the Ideal*. Rizzoli International Publications, Inc.

Boyd, J., & Banzhaf, S. (2007). What are ecosystem services? *Ecological Economics, 63*(2–3), 616–626. doi:10.1016/j.ecolecon.2007.01.002

Brenneisen, S. (2006). Space for Urban Wildlife: Designing Green Roofs as Habitats in Switzerland. *Urban Habitats, 4*(1), 27-33. Retrieved from http://www.urbanhabitats.org/v04n01/urbanhabitats_v04n01_pdf.pdf

Bunting, T. E., & Cousins, L. R. (1985). Environmental dispositions among school-age children. *Environment and Behavior, 17*(6), 725–768. doi:10.1177/0013916585176004

Burszta-Adamiak E. (2015). *Zielone dachy jako element zrownowazonych systemow odwadniajacych na terenach zurbanizowanych.* Wroclaw: Wyd. Uniwersytetu Przyrodniczego we Wroclawiu.

Bytnerowicz, A. (1979). *Pochlanianie niektorych zanieczyszczen srodowiska przez drzewa w Warszawie. In Materialy z konferencji naukowo-technicznej, 28 stycznia 1978 r* (pp. 33–45). Wroclaw, Warszawa: Polska Akademia Nauk, Ogrod Botaniczny, Ossolineum.

Carson, R. L. (1962). *Silent Spring*. London: Hamish Hamilton.

Chelminski, J. (2017). *Zielen w Warszawie. Policzyli i wycenili drzewa na Pradze. Sa warte majatek.* Retrieved from http://warszawa.wyborcza.pl/warszawa/7,54420,21864813,zielen-w-warszawie-ile-warte-sa-nasze-drogie-drzewa.html

Clement, G. (1991). *Le jardin en movement*. Paris: Sens and Tonka.

Clement, G. (2003). *Le Manifeste du Tiers-Paysag*. Paris: Sujet/Objet.

Clement G., Eveno C., (1997). *Le jardin planetaire.* L'Aube/Chateau-Vallon.

Clouston, B. (1977). *Landscape Design with Plants*. Landscape Institute, William Heinemann Ltd.

Costanza, R., d'Arge, R., de Groot, R., Farber, S., Grasso, M., Hannon, B., & van den Belt, M. et al. (1997). The value of the world's ecosystem services and natural capital. *Nature, 387*(6630), 253–260. doi:10.1038/387253a0

Cruz, M. (2015 –2016). *Bio – Receptive Materials – EPSRC Funded Research Project.* Biota Lab-Bartlett School of Architecture and University College London. Retrieved from http://www.biota-lab.com/?page_id=2920

Czarnecki, S., & Stawinska, E. (1984). *Badanie wplywu zieleni na zmniejszenie halasu w aglomeracjach miejskich. In Wplyw zieleni na ksztaltowanie srodowiska miejskiego* (pp. 109–124). Warszawa: Instytut Ksztaltowania Srodowiska.

Dabski, M. (2015). Rola i ksztaltowanie zieleni w otoczeniu szpitali. *Space and Form, 24*(1).

Department of Energy. (2004). Federal Technology Alert: Green Roofs. DOE/EE-0298. Washington, DC: Author.

Dwyer, J. F., McPherson, E. G., Schroeder, H. W., & Rowntree, R. A. (1992). Assessing of the Benefits and Cost of the Urban Forest. *Journal of Arboriculture*, *18*(5), 227–234.

EPBD Recast, Directive 2010/31/EU of the European Parliament and of the Council o 19 May 2010 on the energy performance buildings (recast). (2010). *Official Journal of the European Union.*

Frac, M. (2009). Algi – energia jutra (biomasa, biodiesel). *Acta Agrophysica, 13*(3), 627-638. Retrieved from http://www.old.actaagrophysica.org/artykuly/acta_agrophysica/ActaAgr_168_2009_13_3_627.pdf

Fronty, L. (1978). *Le fugeres poussent meme au plafond. VSD, March 1978 5/26.* Retrieved from http://www.verticalgardenpatrickblanc.com/patrick-blanc/dates-clefs

Gasiorowski, M. (2011). Funkcje roslinności w stawach. *Zielen miejska, 10.* Retrieved from http://www.stawy-kapielowe.com.pl/images/stories/artykuly/zielen_miejska_10_2011.pdf

Gawronski, S., & Szmit, B. J. (2011). *Roslinne oczyszczanie.* Warszawa: Zwiazek Szkolkarzy Polskich. Retrieved from http://www.zszp.pl/pliki/MPO09-Fitoremediacja.pdf

Gawronski, S. W., Gawronska, H., & Rokosza, J. (2000), *Drzewa, krzewy i rosliny zielne w procesie fitoremediacji w terenie zurbanizowanym.* III Forum Architektury Krajobrazu, Warszawa, Poland.

Gawryszewska, B. J. (2009). Revitalisation as a social process – the work with local communities. *Revitalising Built Environments: Requalifying Old Places for New Uses.*

Giergiczny, M., & Kronenberg, J. (2012). *Jak wycenic wartosc przyrody w miescie? Wycena drzew przyulicznych w centrum Lodzi.* Retrieved from http://www.sendzimir.org.pl/images/zrz_3_pl/04_jak_wycenic_wartosc_przyrody_w_miescie.pdf

Heinze, W., & Schreiber, D. (1984). Eine neue Kartierung der Winterhärtezonen für Gehölze in Mitteleuropa. *Mitteilungen der Deutschen Dendrologischen Gesellschaft*, *75*, 11–56.

Hobbs, R. J. (2006). Novel ecosystems: theoretical and management aspects of the new ecological world order. *Global Ecology and Biogeography*, *15*, 1–7. Retrieved from https://www.fs.fed.us/global/iitf/pubs/ja_iitf_2006_hobbs001.pdf

Hobbs, R. J. (2013). *Novel Ecosystems: Intervening in the New Ecological World Order*. John Wiley & Sons, Ltd. doi:10.1002/9781118354186

Hobbs, R. J. (2013). *Novel Ecosystems: Intervening in the New Ecological World Order*. Wiley-Blackwell. doi:10.1002/9781118354186

Hobbs, R. J., Arico, S., Aronson, J., Baron, J. S., Bridgewater, P., Cramer, V. A., & Zobel, M. et al. (2006). Novel ecosystems: Theoretical and management aspects of the new ecological world order. *Global Ecology and Biogeography*, *15*(1), 1–7. doi:10.1111/j.1466-822X.2006.00212.x

Hundertwasser, F. (1980). *The House Should Not Be Measured by Normal Standards.* Hundertwasser at the press conference with Mayor Gratz, Vienna, Austria. Retrieved from http://www.hundertwasser-haus.info/en/blog/2011/07/19/the-house-should-not-be-measured-by-normal-standards/

Jackiewicz, B., & Borowski, J. (1998). Rosliny pnace na budowlach zabytkowych. *Ochrona zabytkow, 51*(203), 402-417. Retrieved from http://bazhum.muzhp.pl/media//files/Ochrona_Zabytkow/Ochrona_Zabytkow-r1998-t51-n4_(203)/Ochrona_Zabytkow-r1998-t51-n4_(203)-s402-417/Ochrona_Zabytkow-r1998-t51-n4_(203)-s402-417.pdf

Januchta-Szostak A. (2010). Miasto w symbiozie z wodą. *Czasopismo Techniczne nr 6-A/1/2010; zeszyt 14, rok 107*. Politechniki Krakowskiej.

Januchta-Szostak, A. (2011). Metody wielofunkcyjnego zagospodarowania miejskich obszarow zieleni towarzyszacych ciekom. Nadrzeczne parki buforowe. *Przeglad Komunalny*, *8/2011*, 58–63.

Jedraszko-Macukow, M., & Sitarski, M. (2017), *Possibilities for estimating ecosystem services of the Praga Park in Warsaw using the i-tree Eco method*. Presentation 16th May 2017 during a Seminar on "Trees in the city - value, benefits, strategy". Warsaw University, University Research Center for Natural Environment and Sustainable Development, Institute of Spatial Management and Housing, Biological and Chemical Warsaw University Research Centre, Seminar organized by the SITO / NOT Science and Technology Association of Horticultural Engineers and Technicians. (in Polish).

Kadas, G. (2006), Rare Invertebrates Colonizing Green Roofs in London. *Habitats*, *4*(1), 66-73. Retrieved from http://www.urbanhabitats.org/v04n01/urbanhabitats_v04n01_pdf.pdf

Kawon, K., & Zmuda, S. (1977). Rola zieleni w ksztaltowaniu srodowiska czlowieka, regionow przemyslowo-miejskich. Studia nad ekonomika regionu, T8, 181-200.

Kohler, M. (2003), Green Roofs as a Contribution to Reduce Urban heat Islands. *RIO 3 - World Climate & Energy Event*, 1-5. Retrieved from http://www.rio12.com/rio3/proceedings/RIO3_493_M_Koehler.pdf

Kowarik, I. (2015). *Integration of biodiversity conservation into urban development*. Presentation 19th of March, 2015, McGill University. Retrieved from http://qcbs.ca/wp-content/uploads/2015/04/Kowarik-Montreal-2015.pdf

Koziel, W., & Wlodarczyk, W. (2011). Glony – produkcja biomasy. *Acta Agrophysica 2011, 17*(1), 105-116. Retrieved from http://www.old.actaagrophysica.org/artykuly/acta_agrophysica/Acta-Agr_188_2011_17_1_105.pdf

Kronenberg, J. (2016). Uslugi ekosystemow – nowe spojrzenie na wartosc srodowiska przyrodniczego. In *EcoMiasto#Srdowisko. Zrownowazony, inteligentny rozwoj miasta*. Lodz: Wydawnictwo Uniwersytetu Lodzkiego. Retrieved from http://repozytorium.uni.lodz.pl:8080/xmlui/bitstream/handle/11089/17980/4-063_088-Kronenberg.pdf?sequence=1&isAllowed=y

Lewinska, J. (2000). *Klimat miasta. Zasoby, zagrozenia, ksztaltowanie*. Krakow: Instytut Gospodarki Przestrzennej i Komunalnej Miasta.

Louv, R. (2014). *Ostatnie dziecko lasu*. Warszawa: Mamania.

Lovelock, J. E., & Margulis, L. (1979). *Gaia: A New Look at Life on Earth*. Oxford University Press.

Maksymiuk, G. (2013). Wplyw zielonej infrastruktury na wartosc ekonomiczna nieruchomosci. Problemy ekologii krajobrazu, 36, 145-153.

Manson, R. (1977). Biological Architecture: A Partnership of Nature and Man. *The Futurist, 3.*

McMahon, E.T. (2000). Green Infrastructure. *Planning Commissioners Journal, 37.*

Millennium Ecosystem Assessment. Ecosystems and Human Well-Being: Synthesis. (2005). Washington, DC: Island Press.

Mumford, L. (1961). *The City in History*. San Diego, CA: Harcourt Inc.

Mumford, L. (1967-70). *The Myth of the Machine.* Harcourt Brace Jovanovich.

Myczkowski, S. (1976). *Człowiek – przyroda - cywilizacja.* Warszawa: PWN.

Orzeszek-Gajewska, B. (1982). *Kształtowanie terenow zieleni w miastach.* Warszawa: PWN.

Pasquero, C., & Poletto, M. (2011). *AlgaeFarm project.* Retrieved from http://www.ecologicstudio.com/v2/project.php?idcat=3&idsubcat=59&idproj=110

Pasquero, C., & Poletto, M. (2014). *Algae Canopy – ecoMachines project for Milano.* Retrieved from http://www.ecologicstudio.com/v2/project.php?idcat=3&idsubcat=59&idproj=137

Pasquero, C., & Poletto, M. (2015a). *Urban Algae Folly project and realization for Braga.* Retrieved from http://www.ecologicstudio.com/v2/project.php?idcat=3&idsubcat=71&idproj=148

Pasquero, C., & Poletto, M. (2015b). *Urban Algae Folly, project and realization for Milano.* Retrieved from http://www.ecologicstudio.com/v2/project.php?idcat=3&idsubcat=71&idproj=147

Peck, S., & Kuhn, M. (2003). *Design guidelines for green roofs.* Ontario Association of Architects, Canada Mortgage and Housing Corporation. Retrieved from https://www.eugene-or.gov/DocumentView.aspx?DID=1049

Philips, A. (2013). *Designing Urban Agriculture: A Complete Guide to the Planning, Design, Construction, Maintenance and Management of Edible Landscapes.* Wiley.

Poradnik TEEB dla miast: uslugi ekosystemow w gospodarce miejskiej, (2011). Krakow: Fundacja Sedzimira. Retrieved from http://uslugiekosystemow.pl/wp content/uploads/2017/07/teeb_poradnik_dla_miast_pl.pdf

Rozanski, S. (1959). *Budowa miasta i jego klimat.* Warszawa: Arkady.

Schmidt M. (2015). *Making the city compact more livable and resilient thanks to the greening.* Presentation 19th May 2015 on Hotel de ville de Paris. Retrieved from https://api-site.paris.fr/images/72760

Schultz, P. W., Shriver, C., Tabanico, J. J., & Khazian, A. M. (2004). Implicit connections with nature. *Journal of Environmental Psychology, 24*(1), 31–42. doi:10.1016/S0272-4944(03)00022-7

Sobel, D. (1996). *Beyond Ecophobia: Reclaiming the Heart of Nature Education.* Great Barrington, MA: The Orion Society.

Solarek, K., Rynska, D., & Mirecka, M. (2016). *Urbanistyka i architektura w zintegrowanym gospodarowaniu wodami.* Warszawa: Oficyna Wydawnicza Politechniki Warszawskiej.

Solon, J. (2008). Koncepcja "Ecosystem Services" i jej zastosowania w badaniach ekologiczno-krajobrazowych. *Problemy Ekologii Krajobrazu, 21*, 25–44.

Sullivan, W., & Kuo, E. (1996). Do trees strengthen urban communities reduce domestic violence? *Arborist News, 5*, 33–34.

Szajda-Binfeld, E., Pływaczyk, A., & Skarżyński, D. (2012). *Zielone dachy. Zrownowazona gospodarka wodna na terenach zurbanizowanych.* Wroclaw: Uniwersytetu Przyrodniczego we Wroclawiu.

Szczepanowska, H. B. (1984). *Wspoldzialanie czynnikow siedliskowych i ich wpływ na stan i rozwoj roslinnosci na terenach zieleni, Wpływ zieleni na ksztaltowanie srodowiska miejskiego red. Szczepanowskiej H.B.* Warszawa: PWN.

Szczepanowska, H. B. (2001). *Drzewa w miescie*. Warszawa: Hortpress.

Szczepanowska, H. B. (2007). Ekologiczne, spoleczne i ekonomiczne korzysci z drzew na terenach zurbanizowanych. *Czlowiek i Srodowisko, 31*(3-4), 5–26. Retrieved from https://www.igpim.pl/publikacje/str07_3-4/Szczepanowska.pdf

Szczepanowska, H. B. (2015). Drzewa w miescie - zielony kapitał wartosci i usług ekosystemowych. *Czlowiek i Srodowisko, 39*(2), 5–28. Retrieved from https://czlowiekisrodowisko.igpim.pl/wp-content/uploads/2016/06/01-Szczepanowska.pdf

Szczepanowska, H. B., Sitarski, M., & Suchocka, M. (2014). *Funkcjonowanie drzew i krzewow w warunkach oddzialywania infrastruktury technicznej miasta.* Artykul powstal na podstawie referatu, wygloszonego podczas konferencji: MIASTO IDEALNE – MIASTO ZROWNOWAZONE Planowanie przestrzenne terenow zurbanizowanych i jego wpływ na ograniczenie skutkow zmian klimatu Warszawa, 24 pazdziernika 2014 r. Retrieved from http://ucbs.uw.edu.pl/wp-content/uploads/SitarskiSzczepanowska-Funkcjonowanie-POPRAW.pdf

Szczurek, M., & Zych, M. (2012). *Dzielo - Dzialka*. Krakow: muzeum Etnograficzne w Krakowie.

Szumacher, I. (2005). Funkcje ekologiczne parkow miejskich. *Prace i Studia Geograficzne, 46*.

Szumacher, I. (2011). Funkcje terenow zieleni miejskiej a swiadczenia ekosystemow. *Prace i Studia Geograficzne, 49*.

Ungers, O. M. (1981). *A Vertical Garden. Design for the Facade of a Department Store* (Vol. 31). Lotus.

Urban Gardening – Back to the Future. (2013). Cieszyn: Zamek Cieszyn. Retrieved from http://www.zamekcieszyn.pl/baza_w_files/1368089604-59katalog_urban-net.pdf

Valuing London's Urban Forest Results of the London i-Tree Eco Project. (2015). Retrieved from https://www.forestry.gov.uk/pdf/LONDONI-TREEECOREPORT151202.pdf/$FILE/LONDONI-TREEECOREPORT151202.pdf

Walawender, J. P. (2015). *Wplyw dachow zielonych na warunki klimatyczne w miescie*. Portal Zielona Architektura. Retrieved from http://zielonainfrastruktura.pl/wplyw-dachow-zielonych-na-warunki-klimatyczne-w-miescie/

Wells, N., & Evans, G. (2003). Nearby Nature: A Buffer of Life Stress Among Rural Children. *Environment and Behavior, 35*(3), 311–330. doi:10.1177/0013916503035003001

Wines, W. (2016). Green Architecture. *Encyclopaedia Britannica online*. Retrieved from http://www.britannica.com/art/green-architecture

Wohlleben, P. (2016). *Sekretne rzycie drzew, wyd.* Kraków: Otwarte.

Wojcicka, I. (1971). *Uciazliwosci klimatu miast i mozliwosci jego poprawy za pomoca roslinnosci.* Warszawa: Inst. Urb. I Arch.

Wolski, P. (2013). Znaczenie okien hygrologicznych. *Problemy Ekologii Krajobrazu, 36,* 129-144. Retrieved from http://agro.icm.edu.pl/agro/element/bwmeta1.element.agro-aee7a572-f4b5-4206-a2a8-2a5153ef6114

Wytyczne do projektowania, wykonania i pielegnacji dachow zielonych – wytyczne dla dachow zielonych. (2015). Opole: DAFA.

Zielonko-Jung, K. (2013). *Ksztaltowanie przestrzenne architektury ekologicznej w strukturze miasta.* Warszawa: Oficyna Wydawnicza Politechniki Warszawskiej.

Zimny, H. (2005). *Ekologia miasta.* Warszawa: Agencja Reklamowo-Wydawnicza Arkadiusz Grzegorczyk.

Zimny, H., & Wysocki, Cz. (1984). *Produktywnosc trawnikow na terenach zieleni miejskiej Warszawy. In Wplyw zieleni na ksztaltowanie srodowiska miejskiego* (pp. 163–180). Warszawa: PWN.

Zinowiec-Cieplik, K., Guranowska-Gruszecka, K., & Dankiewicz, B. (2015). *Polish Green Wall -research program.* Innotech-K1/I1/40/159571/NCBR/12 - poster for 1st European Urban Green Infrastructure Conference 2015, Vienna, Austria.

KEY TERMS AND DEFINITIONS

Garden in Motion: A Gilles Clement idea defining the important role of wasteland in building the environment of the human life, its values (ecological, aesthetic, etc.), and the need for protection; over time, the author has been developed the first idea to the concept of the Planetary Garden (limited and endangered natural resources for which humans are responsible) to receive the final form of the Third Landscape Manifesto, where Gilles Clement postulates the creation of biodiversity reserves open to spontaneous natural succession, especially in heavily built and transformed landscape.

Modern (Novel) Ecosystems: Modified by and as a result of human activities natural systems, not having their counterparts in nature. They are created in anthropogenic areas, modified and contaminated land, post-industrial areas and ruins, agricultural degradation fields, and are resistant to climatic conditions and processed substrates. They are characterized by a large number of alien and invasive species, according to some researchers, affect more than three quarters of the wild nature of our planet.

Polish Green Wall: A vertical garden system corresponding to the polish climate conditions of 6 and 5 planting zones, developed by the polish scientists in co-operation with AB System S.C. within the research grant Innotech-K1/I1/40/159571/NCBR/12.

Chapter 13
Impact of Climbing Plants on Buildings and Their Environment

Jacek Borowski
Warsaw University of Life Sciences, Poland

ABSTRACT

In this chapter, the impact of climbing plants on facades of buildings and their surroundings is presented. Benefits and risks of plant growth on the walls are discussed with respect to their durability. Economic benefits from the presence of vines are shown including energy savings for home heating and cooling. Additionally, the phytoremediation (cleaning up the environment by plants) properties of vines are describe. It should be stated that climbing plants can contribute to damage only in places where facades are damaged, plaster cracked, or where plants are incorrectly planted.

INTRODUCTION

For many years, there has been a total agreement concerning the positive impact of plants on the quality of living in a city. With this position comes the undisputed need to plant them. There is less and less space for animate nature in central districts of modern urban agglomerations. Difficult conditions and lack of space result in the need to seek solutions that allow introducing plants to unavailable places. Building façades are such places.

When we move away from peripheral districts toward city centres, the acreage of built-up areas increases; however, the surface area of vertical walls increases even more (Figure 1).

These spaces can be brought to life and become biologically active only thanks to such specialized plants as climbers. Selected plants must meet the entire set of urban conditions; they should create a large mass of greenery with high aesthetic values. Climbers in natural conditions cling to vertical supports, often beginning their growth in deep shadow, in forest understory, ending in tree canopies, in direct sunlight. For this very reason, climbing plants exhibit a large tolerance to habitat conditions, including unfavourable conditions prevalent in cities. And, more importantly, they occupy little space in built-up conditions (Borowski & Latocha, 2014).

DOI: 10.4018/978-1-5225-4105-9.ch013

Figure 1. The surface area of walls suitable for covering with plants by building development type
Baumann, 1991, as amended.

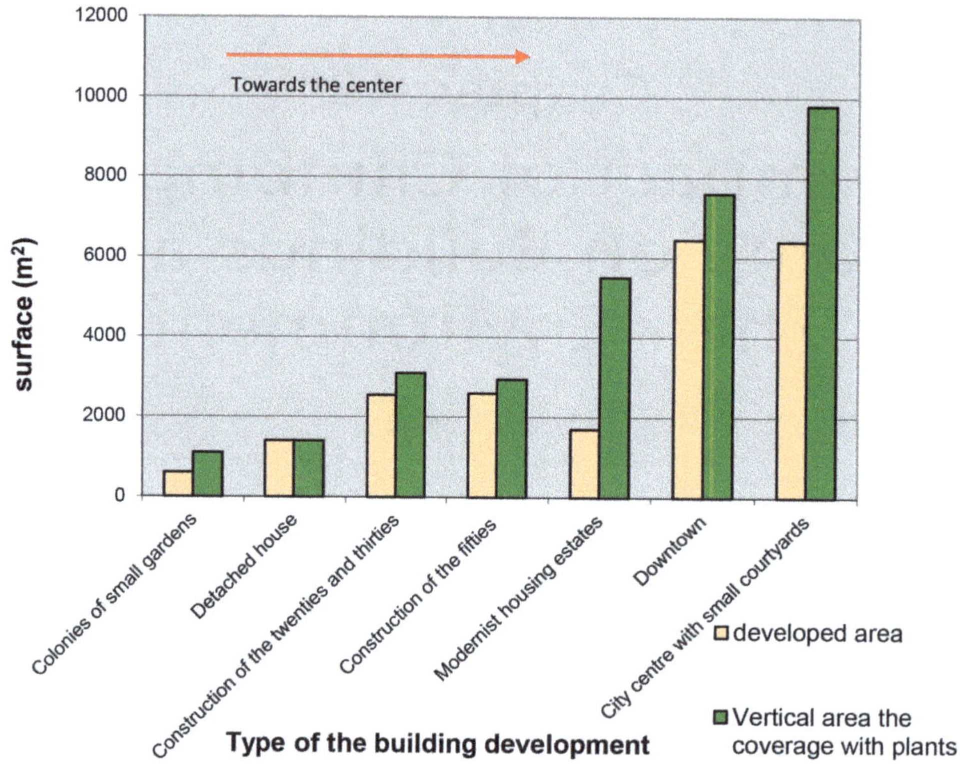

One of the properties that distinguish climbers are very long and flexible sprouts; many species climb very high, up to 30 m, and produce an enormous amount of leaves in a short period. Such plants are ideal for greening walls of large buildings.

BUILDING GREENING METHODS

The most important benefits of building greening systems include: improving the appearance of surroundings, cleaning the air, exhibiting biofiltration capabilities, reducing noise and vibrations, increasing biodiversity, social and psychological effects, intensifying water circulation, reducing the urban heat island effect, affecting temperature inside buildings, protecting façades, increasing the value of real estate ("A guide to Green", 2014).

Plants can be placed on buildings using three basic methods: green roof, green wall, and green façade (Figure 2).

Green roofs are the most expensive to make and keep; they can result in very interesting effects; nevertheless, they require a specially designed building. Green wall systems are easier and cheaper. Again, there are several options.

Figure 2. Vegetation added to the building
A Guide to Green, 2014.

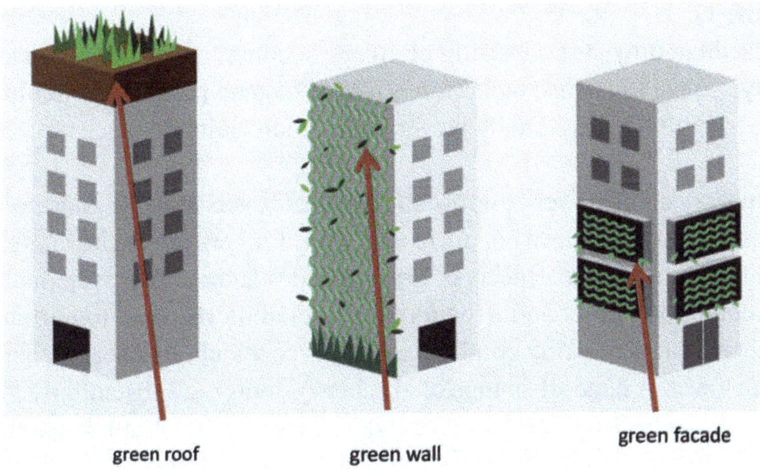

The façade greening method in the form of a vertical garden was invented by Professor Stanley Hart White from the University of Illinois in 1931-38. Despite the passage of years, "green walls" are still rarely found. Their most famous proponent, Patrick Blanc, began his work on "Mur Végétal" in late 1980s (Blanc, 2013).

Unquestionable benefits of green façades, compared to systems made of climbers, include better insulation and larger plant mass affecting its environment. The diversity of applicable species and varieties is also larger. These benefits mostly apply to modular green systems (Table 1).

New artistic aspects of green façades are being continuously discovered. Using a variety of colours and textures of woody plants and perennials can result in interesting effects. It's possible to "paint" pictures that change over time. Increasingly often, green façades are designed to function as temporary advertising banners (Rutgers, 2012).

Table 1. Comparing systems by sustainability

	Modular Green System	**Industrial Felts System**	**Façade Greening (Green Wall)**
Urban agriculture	+	++	--
Biofiltration	++	-	+
Air quality	++	++	+
Noise reduction -	++	+	0
Biodiversity	++	+	0
Social & psychological	++	++	++
Water cycle	++	+	+
Heat island	++	+	+
Heat	+	+	++

A '+' sign means that the benefits of this item are high; a '–' sign means, that the system does not provide a lot of benefit. (Rutgers, 2012 amended)

Industrial felts systems, or systems utilizing a similar hydroponic technology are more artificial and do not foster biodiversity. The substrate is less capable of retaining water, which does not contribute to its biofiltration (Table 1).

Biofiltration is facilitated by systems using climbers in containers. Systems made of climbers, due to the lack of soil layer, have less noise reduction. There is also no possibility to introduce utility plants of so-called urban agriculture. There are, however, many more suitable places to implement urban agriculture in cities.

Nonetheless, climber systems are very useful and efficient in overshadowing and cooling façades (Table 1). New technologies allow us to achieve unique results, and climbers can create interesting artistic forms.

While comparing certain positive aspects of green façades and climber systems seems more favourable to the former, comparing costs and expenditures, including the ever-important water usage, is a completely different tale. Implementing climber systems is considerably cheaper, and complexity of the entire system is much lower. A huge advantage of climber systems is a substantially lower usage of water needed to maintain the plants (Table 2). Climbers typically grow in natural substrates, next to walls or in containers; they also do not require expensive care after planting (Table 2).

The comparison is made on the basis of economic advantages and expediency to vegetate a façade, since the benefits in terms of sustainability are yet to be researched. In terms of weight, a lightweight system was deemed more positive than a heavyweight one, since it will require less construction and less transport, and will thereby cost less (-- = heavy; ++ = light). Vegetation density was judged more positively when more plants grow on a system per square meter, as it is assumed that the aim of a system is to facilitate the growth of plants as much as possible (-- = not dense; ++ = very dense). Maintenance was judged more positively when the system requires less care, since this will be more convenient for the building user, and will cost less after the building process (-- = requires a lot; ++ = requires nothing). Installation was judged more positively when the system can be applied more easily, since this will lead to lower installation costs (-- = difficult; ++ = easy). Complexity was judged as more positive when a system is less complex, since it is assumed that a system can be assembled more easily, and will therefore cost less effort to make (-- = very complex; ++ = very simple). Water use was judged as more positive when the system uses less water, since more water will lead to more costs for the irrigation system and the water itself (-- = uses a lot of water; ++ = water saving).

A large problem of modular green systems is weight, calculated at 244 kg/m^2 of wall; a strong structure is necessary to maintain the system, and maintenance is tedious (Rutgers, 2012). On the other hand, the

Table 2. Practical aspects of the systems

	Modular Green System	Industrial Felts System	Facade Greening (Green Wall)
Weight	-	+	++
Vegetation diversity	++	+	O
Maintenance	--	+	-
Installation	-	--	++
Complexity	O	-	+
Water use	+	--	++

Rutgers, 2012 amended.

weight of climbers is incomparably lower; for example, the weight of *Jasminum* (Jasmine), *Rosa* (Rose) is 6-12 kg/m^2, *Clematis* (Clematis), *Tropaeolum* (Flame Nasturtium) – 3-12, *Vitis ssp.* (Ornamental Grape), *Ampelopsis ssp.* (Porcelain Vine) – 12-26, *Lonicera ssp.* (Honeysuckle), *Actinidia kolomikta* (Arctic kiwi) *Wisteria ssp.* (Wisteria) – 10-26 kg/m^2 ("A guide to Green", 2014). Naturally, climber weight is augmented by the mass of the structure, unless we use climbers that don't require such structures. Thus, *Hedera helix* (English ivy) weighs up to 25 kg/m^2, and it is the only weight that must be supported by a wall (Dunnett, 2008). English Ivy, which is the most often used climber in such systems, can grow 20 to 30 meters high with average speed of 1 to 2 meters per year (Zimmermann, 2011). However, the combined weight of climbers and structures on the wall is low. A steel trellis structure weighs only 1 to 3kg/m^2. The plants add a little weight to this, depending on the size and species. The total is much lighter than the previous two systems (Rutgers, 2012). Building a simple system of self-clinging climbers, growing in the soil at the base of the façade is the least expensive at approx. € 30 - 45/m^2 (Perini & Rosasco, 2013).

RULES OF SELECTING CLIMBER SPECIES

The first basic issue when selecting climber species is choosing between self-clinging climbers, and twining and tendril climbers (Figure 3).

Before choosing a plant, we need to understand their requirements and know what to expect of them, as well as what conditions can we provide for them. Therefore, we have to consider:

- **Display:** There are fundamentally different conditions on walls facing different corners of the world ("displays"). They are completely different on the dry and intensely sunny southern side, reaching several dozen degrees Celsius in the summer, than on the cold, damp northern display, practically devoid of direct sunlight.

Figure 3. On the left: self-clinging climbers, on the right: twining and tendril climbers
A Guide to Green, 2014.

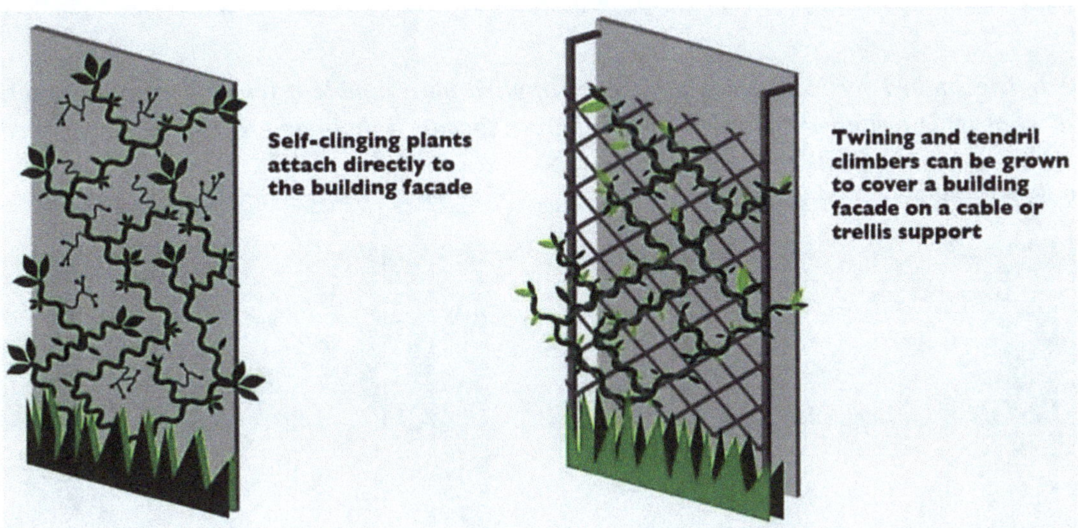

- **Size and Height Reached by the Climber:** This impacts the type of required support and how deep it has to be anchored to the wall, and how will the final effect look.
- **Habitat Requirements, Mainly Soil and Dampness:** Particularly important in urban environments, where we can't always provide the plants with soil of appropriate quality and volume.
- **Tolerance to Low Temperatures:** In practice, you only need to learn about the hardiness zone, which determines where the plants can be cultivated, which is related to the selection of an appropriate habitat.
- **Aesthetic Aspects:** Preferably falling in line with consumer tastes, in harmony with colour and texture of the building's façade.

When using climbers on façades, it is possible to plant them in four ways (Figure 4).

Significance of Climbers for the Environment and Living Comfort

It's difficult to overestimate the significance of climbers for the quality of environment in cities. Although positive impact on the environment is not exclusive to climbers, they require significantly less space than woody plants and often have a large biologically active area.

A large Virginia creeper, covering only about half a square metre of land, can produce approx. 2,600 m^2 of foliage during the vegetative period (Table 3), which is comparable to seven lime trees with a canopy diameter of 10 m, and considerably more than a hundred-year-old beech with a leaf area of approx. 1,600 m^2 (Borowski & Latocha, 2014). The estimated data presented in the table indicates how large climbers get, and how strong their impact on the environment is. In case of evergreen plants, such as English ivy, this impact persists throughout the year (Table 3).

It is not difficult to notice that naked house façades get warmer on sunny summer days, up to as much as 30 degrees Celsius higher than greened façades. The foliage constitutes a kind of mobile plant-based shutters. Following the sun, climber leaves deviate from vertical in the morning, reach maximum opening about noon, and close (return to vertical) in the evening. The tight leaf layer contributes to the creation of convection currents called the "stack effect", resulting in the movement of air, cooling and drying the façade. The scheme of the climber's influence on the wall is presented on the drawing (Figure 5).

Figure 4. The studied vertical greening systems, applied on a double-brick wall with air cavity and mineral wool: 1. Direct green façade, 2. Indirect green façade, 3. Indirect green façade combined with planter boxes, 4. Living wall system
Perini & Rosasco, 2013.

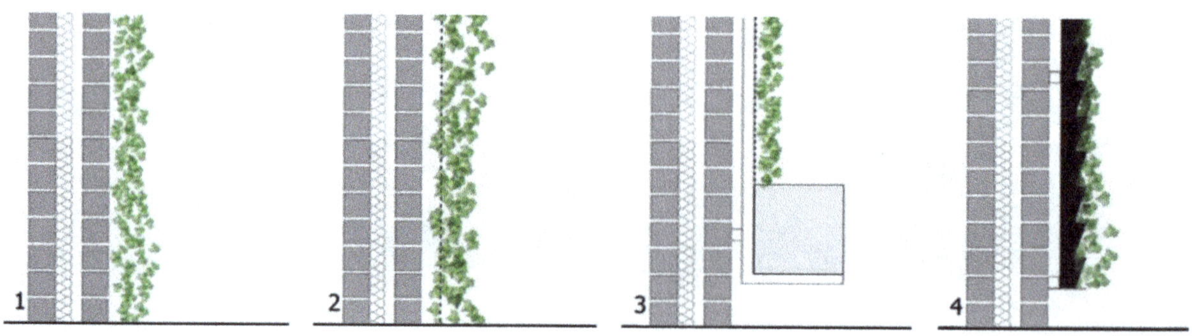

Table 3. Certain dimensions of a large Parthenocissus quiquefolia (Virginia creeper) and its impact on the environment

Virginia Creeper Covering 531 m² of wall	
Measured Parameter	**The Entire Plant (Approximate Values)**
Leaf area	2,600 m^2
Leaf mass	3,000 kg
Transpired water	15,000 l/day
Transpiration cooling	37.5 GJ/day
Oxygen production	250 kg O_2/year
Absorption of CO_2	500 kg/year

Borowski & Latocha, 2014.

Figure 5. Positive impact of climbers on walls
Borowski & Latocha, 2014.

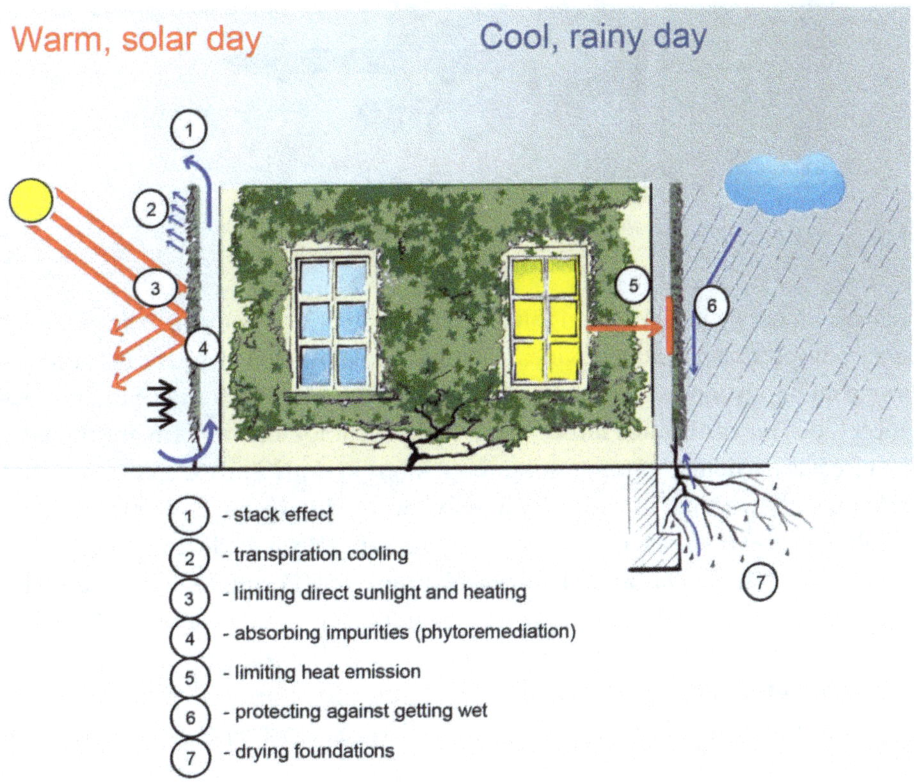

Kohler, on the basis of own research, presented the energetic balance of the facade covered by climbers (Kohler, 2012) (Figure 6).

During summer, climbers naturally regulate the amount of sunlight falling on the building's surface. Vegetation of appropriate species, cared for tight, and almost completely covering the building's façade, constitutes an efficient protection from the sun. Mainly for this reason, energy savings from covering the

Figure 6. Presentation of energy balance of elevation
Kohler, 2012.

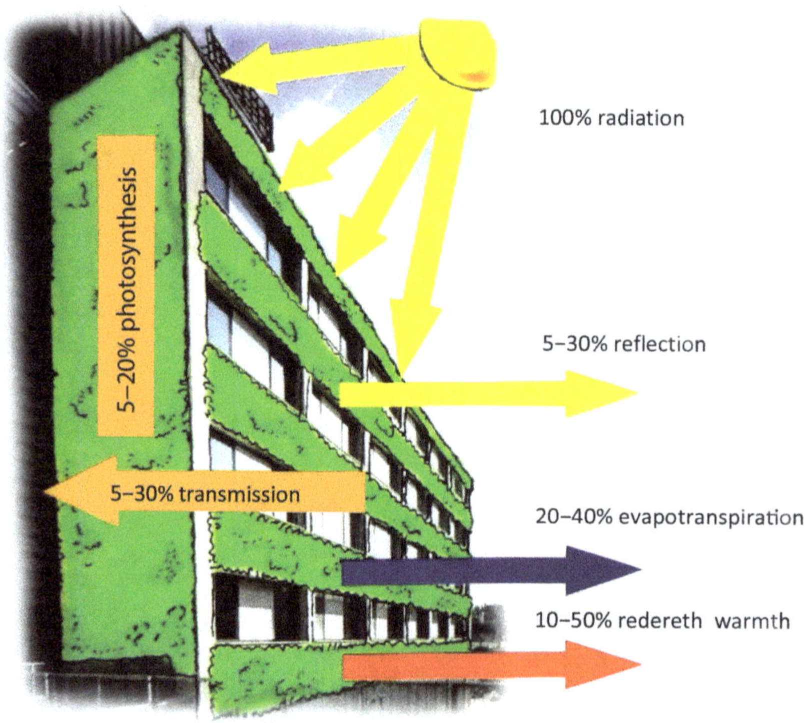

façade using dense climbers with evergreen foliage approach 15-30% per year. The plants' ability to cool the environment by evapotranspiration provides a natural insulation, lowering the building's service costs, at the same time continuously producing oxygen. According to the Australian Stainless Steel Development Association (2009) it constitutes direct social and psychological benefits for the building's users.

The cooling effect is particularly important in very sunny countries. Research conducted by Mazzali, Peron and Scarpa has shown that, in case of a massive wall with 40-cm concrete (without insulation), reduction of cooling energy can reach up to 65.8% (Mazzali, Peron & Scarpa, 2012). Research shows that a foliage layer causes the creation of an air insulation layer (Perini, Ottele, Fraaij, Haas & Raiteri, 2011), which contributes to the reduction of the need for air conditioning energy to 40-60%, for example, in the Mediterranean Sea Basin area (Alexandri & Jones, 2008; Mazzali etal., 2012). In the experiments, large thermal capacity of green façades has been confirmed by a drop in temperature of the external surface of the wall to 14 degrees Celsius in July and in early September (Coma, Perez, Sole, Castell & Cabeza, 2014).

There are also acoustic benefits resulting from the presence of climbers on a building, although – due to the low thickness of leaf layer – they are small (Giachetta & Magliocco, 2007). Additionally, the effect on the sense of hearing caused by leaves rustling, sprouts rubbing against themselves, or bird sounds, can partially negate the discomfort of other sounds (Borowski & Latocha, 2014).

Many studies indicate the importance of greenery in phytoremediation including combating air pollution. This property also applies to climbers, which is best exemplified in the study by Bruse, Thonnessen and Radtke (1999). It indicated that Boston ivy accumulated considerable amounts of dust impurities.

Practically all cadmium and lead particles washed from the surface of the climber's leaves came from traffic pollution of a nearby street (Figure 7).

The climbers' ability to accumulate dust on leaf surfaces was confirmed by studies of another team of researchers (Dzierzanowski, Popek, Gawronska, Sæbo & Gawronski, 2011).

The role of climbers as growth environment for many species of birds and insects is also non-trivial. German research shows that Virginia creeper and English ivy shrubs were inhabited by insects from 19 orders and over 50 families (Kohler, 1988).

Due to the aforementioned reasons, many cities have begun covering their buildings with climbers in late 20th century. One such example is Berlin, where 245,584 square metres of green façades were installed in 1983-1997 (Kohler, 2008).

HAZARDS

It should be reasoned that using vertical greenery provides many benefits; however, there are some doubts. Most times, they concern the possible harmful impact of plants on façades directly covered by them. The concerns pertain to mechanical and chemical (biological) effects, as well as excessive dampness on walls covered with self-clinging plants. However, situations where the destructive effects of plants are proven are very rare. The most popular creepers cannot chemically damage walls because their adhesive discs are biologically active for just a few days. The contact of epidermis with the surface results in the production of plant-based glue that strengthens the connection of adhesive discs with the surface (He, Zhang & Deng, 2011). Contact glues do not damage the surface. Sometimes, at the contact point of Virginia creeper and plaster, very small defects can be noticed; they are caused by biological activity and probably short-term activity of organic acids (Borowski & Latocha, 2014) (Figure 8).

Similarly short is the effect of adventitious roots of ivy, climbing hydrangea, trumpet creeper, or various other creepers on the substrate. The carbonic acid they produce acts very shortly, and the epidermis of roots suberizes quickly. Like adhesive discs, they do not absorb or produce water. Both of them lignify after a short time and their function is limited to fixing the plant to the surface (Althaus, 1987). Very rarely, on old walls, it may happen that root-clinging climbers, planted at the base of an old wall, can penetrate it while still growing, even after being cut at the base (Borowski & Latocha, 2014).

Figure 7. Leaf stripped with Polyvinyl-Butyrals (left), uncleaned leaf surfaces, June (middle) and October (right)
Bruse et al., 1999.

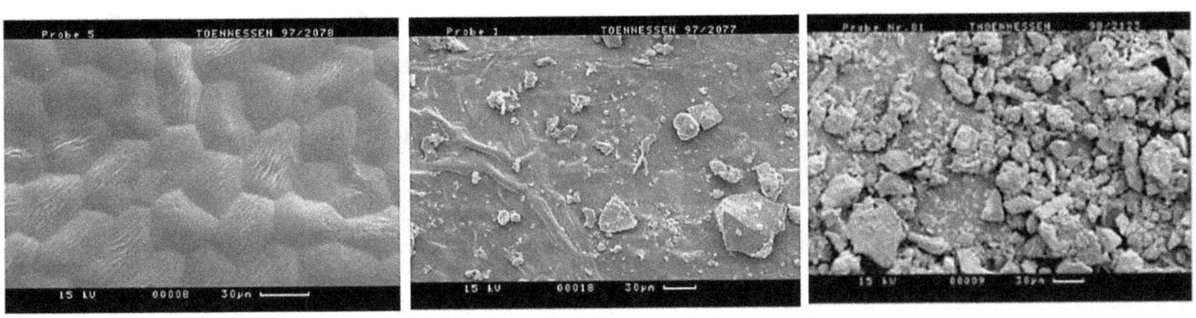

Figure 8. Microstructure of the sample surface of plaster located around living adhesive discs of a Virginia creeper; micro crazing's are visible near the adhesive disc
Borowski & Latocha 2014.

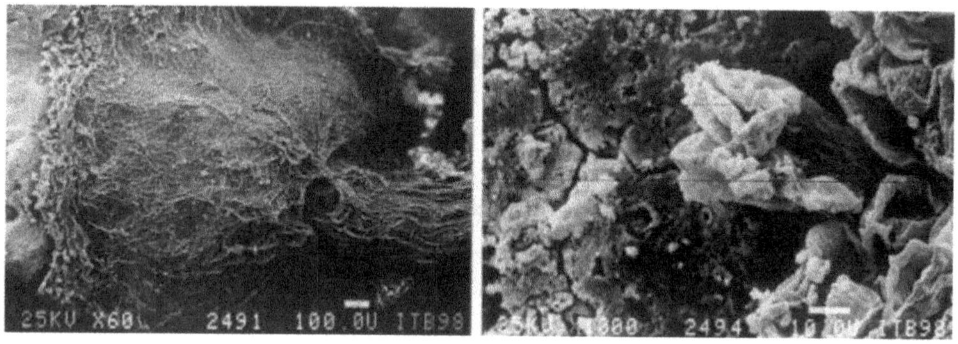

Structural studies of plasters performed at the Building Research Institute indicate that live layers of green protect façades from effects of acid rains. They exhibit a higher quality of plasters for most climber-covered plaster samples for at least thirty years, compared to the control group without climbers. Processes of airing construction materials intensify along with the increase in amplitude of temperature and humidity. Decreased by the plant layer, the amplitude of these fluctuations certainly prevents the erosion of the surface (plaster). Another non-trivial part is the reduced impact of the destructive effect of direct sunlight (Borowski & Latocha, 2014).

It should be emphasized that external parts of building equipment are not adapted to carrying often significant loads. They also cannot bear prying and breaching caused by growing climbers. Plants planted too close to rainwater downpipes can clog and break them off; on the other hand, lightning rods often cannot carry very lush climbers.

In summary, it should be stated that they can contribute to damage only in places where façades are damaged, plasters cracked, or where plants are incorrectly planted. Situations where the plants' destructive effect on buildings is proven are extremely rare.

CLIMBERS ON THE nZEB AND RETROFIT BUILDINGS

There is lack of reports how climbers can be used in nZEB and Retrofit buildings.

In their work, architects are dealing primarily with the construction and parameters of nZEB buildings and Retrofit buildings. Their interest in plants, including their values and their choice, is obviously less important. However, plants can significantly affect the insulation of buildings, and thus reduce the building's energy requirements for heating and cooling.

In the definition of building nZEB (Kurnitski et al., 2011) it is stated, that "energy need for heating is caused by heat losses and is reduced by solar and internal heat gains. Net energy need is the energy need minus heat gains, i.e. thermal energy without any system losses needed to maintain indoor climate conditions".

Energy losses are lower as the thermal insulation of buildings increases, to which the climbers contribute.

As stated by researchers (Azahr & Sulaiman, 2013) the green façade passively influences the building's energy efficiency through shading. It also stabilizes the inside temperature when the air conditioning is off. This contributes to minimizing the energy required for cooling.

Impact of climbers in the urban area of tropical climate countries has been also confirmed by Othman and Sahidin (2016). Their study has proved that wall covering with climbers helps to sustainably design buildings in cities in the tropics. However, the effectiveness of the green cover depends on the structure of the wall and its location to directions of the world.

Climbers are also beneficial as passive systems for energy savings in buildings located in the Mediterranean Continental climate (Coma et al., 2014). Cameron, Taylor and Emmett (2015) report about the thermal insulation provided with climbers under climate conditions of the British Isles. They have tested the most commonly used evergreen ivy in Europe (*Hedera helix*) and concluded that vegetation reduced convective heat loss.

Although there are no robust experiments showing the beneficial effects of climbing plants on the properties of nZEB buildings, the benefits from climbers such as thermal insulation, can certainly improve the performance of each building.

As climbers can climb all types of elevations I do not see any obstacles for them to behave the same way on nZEB buildings as well. It may be very important using climbing plants when upgrading Retrofit buildings. Also in this case, reducing energy consumption can be significantly facilitated in the presence of climbers.

I am absolutely convinced that covering elevations with plants should be applied to nZEB and Retrofit buildings.

REFERENCES

A guide to Green roofs, walls and facades in Melbourne and Victoria, Australia. (2014). Retrieved December 16, 2016, from http://www.growinggreenguide.org/wp-content/uploads/2014/02/growing_green_guide_ebook_130214.pdf

Alexandri, E., & Jones, P. (2008). Temperature decrease in a urban canyon due to greenwalls and green roofs in diverse climates. *Building and Environment*, *43*(4), 480–493. doi:10.1016/j.buildenv.2006.10.055

Althaus, Ch. (1987). *Fassadenbegrünung*. Berlin: Patzer Verlag.

Australian Stainless Steel Development Association. (2009). Vertical landscaping. *Australian Stainless Magazine*, *46*, 1–2.

Azhar, M.K., & Sulaiman, M. (2013). *Application of Green Façade as Passive Cooling Technique for Energy Efficiency Architecture*. Department, Faculty of Engineering and Built Environment, Universiti Kebangsaan Malaysia.

Blanc, P. (2013). *The Vertical Garden: From Nature to the City by Patrick Blanc*. W. W. Norton & Company. Retrieved December 15, 2016, from http://scholar.google.pl/scholar?q=Living+facades

Borowski, J., & Latocha, P. (2014). *Zastosowanie roslin pnacych i okrywowych w architekturze krajobrazu*. Warszawa: Wydawnictwo SGGW.

Bruse, M., Thonnessen, M., & Radtke, U. (1999). Practical and theoretical investigation of the influence of facade greening on the distribution of heavy metals in urban Streets. *Proceedings International Conference on Urban Climatology & International congress of Biometeorology*. Retrieved December 15, 2016, from http://www.envi-met.com/documents/papers/facade1999.pdf

Cameron, R. W. F., Taylor, J., & Emmett, M. (2015). A Hedera green façade - Energy performance and saving under different maritime-temperate, winter weather conditions. *Building and Environment*, *92*, 111–121. doi:10.1016/j.buildenv.2015.04.011

Coma, J., Perez, G., Sole, C., Castell, A., & Cabeza, L. F. (2014). New green facades as passive systems for energy savings on buildings. *Energy Procedia*, *57*, 1851–1859. doi:10.1016/j.egypro.2014.10.049

Dunnett, N. (2008). *Planting green roofs and living walls*. Portland, OR: Timber Press.

Dzierzanowski, K., Popek, R., Gawronska, H., Sæbo, A., & Gawronski, S. (2011). Deposition of Particulate Matter of Different Size Fractions on Leaf Surfaces and in Waxes of Urban Forest Species. *International Journal of Phytoremediation*, *13*(10), 1037–1046. doi:10.1080/15226514.2011.552929 PMID:21972570

Giachetta, A., & Magliocco, A. (2007). *Progettazione sostenibile. Dalla pianificazione territoriale all'ecodesign*. Roma: Carocci.

He, T., Zhang, L., & Deng, W. (2011). Biological adhesion of Parthenocissus tricuspidata. *Archives of Biological Sciences*, *63*(2), 393–398. doi:10.2298/ABS1102393H

Kohler, M. (1988). Besiedlung von Kletterpflanzen durch Insekten und Spinnen in Berlin. *Zeitschrift für Angewandte Zoologie*, *75*, 195–202.

Kohler, M. (2008). Green facades – a view back and some visions. *Urban Ecosystems*, *11*(4), 423–426. doi:10.1007/s11252-008-0063-x

Kohler, M. (2012). *Handbuch Bauwerks-begrünung, Planung-Konstruktion-Aufführung*. Köln: Müller, Rudolf.

Mazzali, U., Peron, F., & Scarpa, M. (2012). Thermo-physical performances of living walls via field measurements and numerical analysis. Eco-architecture IV. Harmonisation between architecture and nature WIT transactions on ecology and the environment, 165, 239-50. doi:10.2495/ARC120231

Othman, A. R., & Sahidin, N. (2016). Vertical Greening Façade as Passive Approach in Sustainable Design. *Procedia: Social and Behavioral Sciences*, *222*, 845–854. doi:10.1016/j.sbspro.2016.05.185

Perini, K., Ottele, M., Fraaij, A. L. A., Haas, E. M., & Raiteri, R. (2011). Vertical greening systems and the effect on air flow and temperature on the building envelope. *Building and Environment*, *46*(11), 2287–2294. doi:10.1016/j.buildenv.2011.05.009

Perini, K., & Rosasco, P. (2013). Cost-benefit analysis for green façades and living wall systems. *Building and Environment*, *70*, 110–121. doi:10.1016/j.buildenv.2013.08.012

Rutgers, R. (2012). *Living façades. A study on the sustainable features of vegetated façade cladding. Masterthesis about the sustainable benefits of plants on a façade.* Faculty of Architecture Delft University of Technology. Retrieved December 15, 2016, from https://issuu.com/rrutgers/docs/living_facades

Zimmermann, A. (2011). *Constructing landscape: materials, techniques, structural components.* Boston: Birkhauser.

KEY TERMS AND DEFINITIONS

Biodiversity: A variety of life forms together with all their variability at the level of genes, species, and ecosystems on an Earth scale or closer biogeographic units.

Biofiltration: The process of cleaning the environment and biologically degrading pollutants by passing them through a biofilter (e.g., plants).

Climbing Plants: A versatile group of plants that require support for normal vertical growth.

Phytoremediation: Purification of environmental degradation using plants.

Self-Clinging Climbers: Plants that climb even over smooth surfaces by adventitious roots, and tendrils with adhesive discs (e.g., Virginia creeper).

Tendril Climbers: Plants climbing using tendrils (e.g., grape vine).

Twining Climbers: Climbers that twine around supports using twining stems, tendrils, or twisting leaf stalks.

Chapter 14
The Impact of Forms of Buildings on the Air Exchange in Their Environment:
Based on the Example of Urban Development in Warsaw

Katarzyna Zielonko-Jung
Warsaw University of Technology, Poland

Marta Poćwierz
Warsaw University of Technology, Poland

ABSTRACT

Possibilities for energy-efficient, natural ventilation of buildings in an urban environment depend on the airflow around them. This chapter deals with the issue of dependence of air exchange in urban spaces on the building forms used in them and on relative position of these buildings. The authors focus on the problem of air stagnation in dense urban development. This phenomenon increases the energy demand of buildings. The purpose of the following study is to present this problem and identify it in the selected example of existing settlements in Warsaw. The existing situation was compared with the revised spatial layout. The conclusions relate to spatial features of those building arrangements that are exposed to the problem of insufficient ventilation.

INTRODUCTION

Currently more than half of human population inhabits cities and this share is constantly rising. It is likely that by the year 2050 the figures describing the population dwelling in cities will have reached up to 70%. Cities tend to be the largest centers of housing, services, culture and communication, as well as being the largest sources of energy consumption and emissions of harmful substances. Urban housing development is therefore among the most important targets for the development of energy-saving strategies, including the modernization of existing buildings that were built in various historical periods.

DOI: 10.4018/978-1-5225-4105-9.ch014

Making the proper use of natural components of the climate is an essential element of building energy concepts for nearly zero energy buildings (nZEB). The city's natural environment is strongly transformed by anthropogenic factors and as such creates a specific meso and micro climatic environment. Studies on the nature of meteorological phenomena taking place in cities reveal the interpretation of the city structure as being a surface strongly deformed in comparison to flat surfaces or surfaces slightly deformed by such factors as terrain shaping. In the latter cases, climatic factors such as sun, wind or rain are not obstructed by complex obstacles, which is the case in cities. In addition, most of the surfaces on which climate components exert influence in urban layouts consist of paved surfaces, walls and roofs of buildings, all of which happen to be heat accumulating surfaces (Harman, 2003). Thus, physical processes related to solar radiation, temperature, airflow or water circulation differ from the ones present in non-urban areas and are frequently far more complex and varied. Moreover, those processes are closely linked to the geometric shaping of the buildings, as well as their mutual arrangement (Ratti, Di Sabatino, Britter, 2006). The subject matter to be discussed in the following chapter is the aerodynamic phenomena that take place around such urban developments in case of which the link is particularly clear, as these phenomena affect the thermal characteristics and the air exchange in the building environment, which in turn in a factor directly influencing their energy economy (Ratti, Baker, Steemers, 2005).

In the case of temperate climate, which is being discussed in the chapter, the indication for wind conditions is that there is a need to protect against cool winds and rapid wind acceleration, both of which are characteristic of particular geometrical configurations of building developments (for instance high rise building environments, narrowings in building developments, long, densely built up streets, corners and creases of the building line). At the same time, proper air exchange in urban areas must be ensured. Air stagnation not only severely impairs the urban areas microclimate, but also negatively affects the energy efficiency of buildings (Che-Ani, Shahmohanadi, Sairi, Mohd-Nor, Zain, Surat, 2009). As a result of air stagnation, natural ventilation of buildings (especially at summer nights) becomes inefficient; hence such buildings display an increased demand for cooling. What is more, the multiplication of the phenomenon of too little local air exchange over a large area leads to exacerbation of the effect of urban heat island. This results in overheating of the city space and the accumulation of air pollutants.

The following chapter focuses on the problems of air stagnation in urban layouts. The phenomenon may generally be considered as having a negative impact on buildings that display low energy demands. It decreases the possibility of employing natural and hybrid ventilation and intensifies the problem of interior overheating in the summer season.

The purpose of the following chapter is to identify the phenomenon in densely built up urban areas. In the first part of the chapter, the issue of airflow around buildings has been presented, together with research methods and the theoretical background that determines the connection between wind parameters and the heat phenomena that appear in buildings. In the second part of the chapter, an analysis of the airflow around the building complex of a selected part of Warsaw has been performed. The example represents the spatial determinants of central inner-city zones of today's large Polish cities in which quarterly buildings with high density are predominant.

For the purpose of the following chapter, the results of experimental research conducted on 1:400 scale building models in the average speed aerodynamic tunnel were used. The aim of the experiment was to investigate the correlation between the phenomenon of air stagnation and the geometry of building development.

In each case, the existing situation was compared with the adjusted variant, the purpose of building development shape adjustment being the minimization of the observed phenomenon. Also, hypotheses concerning the possibility of performing additional activities by means of supplementation of building developments with properly selected greenery were formulated. The conclusions that were reached concern the spatial characteristics of developments that are especially likely to be exposed to the problem of insufficient ventilation. The conclusions relate also to the possibility for the discussed solutions to be applied in the process of city planning in Poland.

BACKGROUND

The knowledge concerning architectural and technological solutions applied to buildings with low energy requirements has been undergoing rapid development over the recent 50 years. While the scale of such buildings, as well as their components has been relatively well identified, only a few authors refer to the phenomena occurring in the vicinity of buildings that affect airflow between a given buildings and its surroundings.

Aerodynamics that was initially used in case of energy needs, has been subject to extensive development over the last 50 years (Blocken, 2014). While the scale of the building is relatively well diagnosed, few authors regard the phenomena occurring in the vicinity of buildings that influence the flow of energy between buildings and their environment. The phenomena in the field of airflow around buildings are among the most complex ones and as such are difficult to recognize (Zielonko-Jung, 2013). The field of research that allows the study of those phenomena is wind engineering (Flaga, 2008). The branch originates in the needs of certain industries, such as the airplane, automotive or sport industry. Gradually though, its application has been expanded to new fields, that is including architecture and urban planning.

Wind engineering is a field of research that has established itself over the recent decades. With regard to buildings, it can be divided into fields related to: construction, ventilation of buildings and the environment surrounding building developments, the latter of which consist the subject matter to be discussed in this chapter. Being directly related to the issue of sustainable development, these specific fields have become the scope of particular interest for researchers. Commonly, the simplest form of geometry, such as detached solids (Köose, Dick, 2010), regular lattices of solids (Blocken, Carmeliet, 2004) or spaces between two walls that represent street models (Kovar-Panskus, et al.,, 2002) constitute the subject of research. Publications on more complex geometrical layouts that incorporate the environment, and therefore correspond to the actual problems encountered at architectural design are relatively rare (Blocken, Persoon, 2009, Yoshie, et al., 2007, Gumowski, Olszewski, Pocwierz, Zielonko-Jung, 2015). Obtaining full and reliable research results in such cases is a much more difficult task.

The contemporary researches are noticeably being directed to the development of simulation research methods that allow for obtaining the most accurate data possible (ANSYS, 2011). In the near future, those methods may popularize the possibilities of optimizing such shapes of building developments that will ensure wind comfort for each unique situation. A number of outstanding global research centers including the University of Technology in Eindhoven, the Architectural Institute of Japan in Tokyo, Massachusetts Institute of Technology in Cambridge, USA engage in the topic. Yet, there is a noticeable lack of publications that attempt to link wind comfort issues to the issues urban and architectural design and urban green design. Faced with the increasing problems of the decline in air quality observable in large cities, attempts to explore this subject deeply are justified. The issue of insufficient urban winds discussed in this chapter addresses the problem of the urban heat island, the phenomenon identified and named as early as in the nineteenth century. The issue has currently become one of the most important aspects of urban climate balancing (Mills, 2012).

In the hitherto conducted studies, a lack of publications that link the matters concerning airflow around buildings to the problems encountered in urban architecture practice, including building design and low energy requirements issues, is observable.

WIND VELOCITY AS A PARAMETER INFLUENCING ENERGY CONSUMPTION OF THE BUILDING

The energy demands for a particular building depends on certain essential elements (Ratti, et al., 2005):

- Climatic conditions of the building localization,
- Urban layout geometry that results from the interaction of the building and its environment
- Architectural design of the building,
- Installation systems efficiency,
- Behaviour of users.

The above factors affect each other in the way shown schematically in Figure 1, which presents the building nZEB in a much broader context than its own scale only.

When the framework of a building design is concerned, the possible variation in building's energy requirements may amount to 2.5 times (this means that a more energy-saving project may lead to reduction in energy consumption by 2.5 times). Variation at the rate of x2 can be attributed to such factors as level of installation systems efficiency and behavior of users, whereas the influence exerted by the geometry of the urban layout is the factor most difficult to estimate. It is assumed that the cumulative effect of all the factors shown in the diagram above amounts to the variation x10, while in practice this rate can actually reach up to 20 (Ratti, et al., 2005). The conclusion that follows is that in this case the urban context plays a significant role.

Wind is a climate component having a direct effect on the buildings performance understood as the total of physical processes that take place between the building indoor environment and its surroundings. Wind speed is essential for the purposes of natural ventilation, especially when attempting to introduce it in case of such urban buildings whose use and size impose the need to apply mechanical systems. Imposing the introduction of the so-called cross ventilation requires such forms of building, as well as the interior layout and the facade solution that are well considered (Marchwinski, Zielonko-Jung 2012). Optimization of these elements is possible if the distribution of local wind speeds around a particular building is known. For large scale buildings, especially the high-rise ones, the wind speeds may reach

Figure 1. Factors that affect energy consumption in buildings
Ratti, et al., 2005.

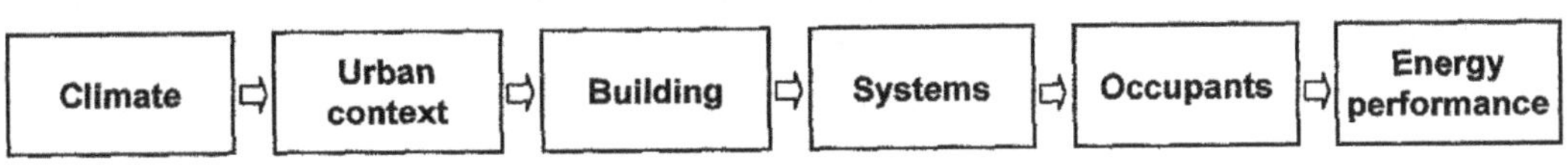

various values that are subjected to dynamic changes over time. The design process in this case requires conducting detailed aerodynamic studies, which take into account the building's geometric relation to neighboring buildings (Daniels, 1997).

Local wind speed constitutes also a parameter that influences thermal phenomena in the building. It determines the actual value of the convective heat transfer coefficient for external surfaces (Blocken, et al., 2006, *Environmental design - CIBSE Guide Book*, 2006, Mirsadegh, Cóstola, Blocken, Hensen, 2013). This coefficient is a significant factor when assessing heat loss for buildings. It depends on a number parameters, some of which include solid geometry, roughness of wall surface, temperature differences between facade and ambient air and local air flow conditions. In the urbanized area, local wind conditions around a building result from the location and the geometry of neighboring buildings. Terrain shaping proves to be an important factor as well, as it affects the average wind speed profile and turbulence intensity (Blocken, et al., 2006).

A number of research have been conducted the point of which was to determine the interdependence of the convective heat transfer coefficient for external surfaces and wind velocity. The model created by MacAdams, based on research by Jurges (Mirsadegh, et al., 2013) remains one of the most popular models applied to this type of calculation. The equation he gives has been linearized, while the undisturbed flow velocity has been changed by the local wind speed. A great deal of empirical relationships that determine this coefficient exists, the description of which can be found in (Blocken, et al., 2006). An equation that determines the h_c coefficient used in the Building Energy Simulation (BES) programs is presented below.

$$\backslash h_c = \begin{cases} 3.8V_{loc} + 7.4 & \textit{for rough surfaces} \\ 3.6V_{loc} + 6.5 & \textit{for smooth surfaces} \end{cases}$$

The lack of consistency in the results obtained by various authors may stem from limited knowledge concerning the actual value and variability of local wind speed V_{lok} for various building geometries and for their mutual location.

THE NATURE OF AIRFLOW FOR COMPACT BUILDING DEVELOPMENT

Determining the most desirable wind flow conditions that meet the demand of the criteria for wind comfort depends on such features as the type of urban space and its intended use (for instance car parks, main roads, assembly areas, entrances, open spaces), the time of year, type of activity expected of people and other factors. The generally assumed rule is that the wind speed in pedestrian areas should not exceed 5m/s for over 20% of the annual time. If the rate of wind speed amounts to less than 1m/s for 90% of the time, the given area is unlikely to be receiving sufficient ventilation (Lewińska, 2000). Determining the intensity of air movement comfort ranges requires investigating the correlation between wind speed and temperature. The need for protection against wind demonstrates itself when the temperature drops below -5 ° C with the wind speed exceeding 1 m/s. The problem of insufficient ventilation, on the other hand, occurs with temperatures reaching above 25 ° C with the wind speeds below 3 m/s (Klemm, 2011).

The airflow in spaces between buildings depends mostly on the size and shape of buildings, as well as their mutual location. Hereafter, the typical phenomena that occur in two basic types of urban spaces, the street and the square, are described.

The Street

The dependence of airflow on the building development geometry in the so-called street canyons is usually described using the H/S parameter, that is, the relation of height of buildings to the distance between them. The most problematic situation occurs when the wind blows perpendicularly to the street, in which case buildings are obstructing the wind flow. The space between them may be divided into two circulation zones: recirculation zone and ventilation zone (Harman, 2003). The first zone is located at the leeward wall of the building, in the vacuum zone. Within this area, the air swirls downwards (Figure 2a). The ventilation zone, on the other hand, is located at the windward wall of the opposite building. Air from the upper zones enters the street space to subsequently rise up. As the distance between buildings decreases, the ventilation area declines (Figure 2b). In case of narrow streets, a complete lack of ventilation is observable, while the air confined to the area is subject to a closed rotary motion (Figure 2c).

Figure 2. Recirculation zones (in gray) and ventilation zones (tintless) within the spaces around buildings
Harman, 2003.

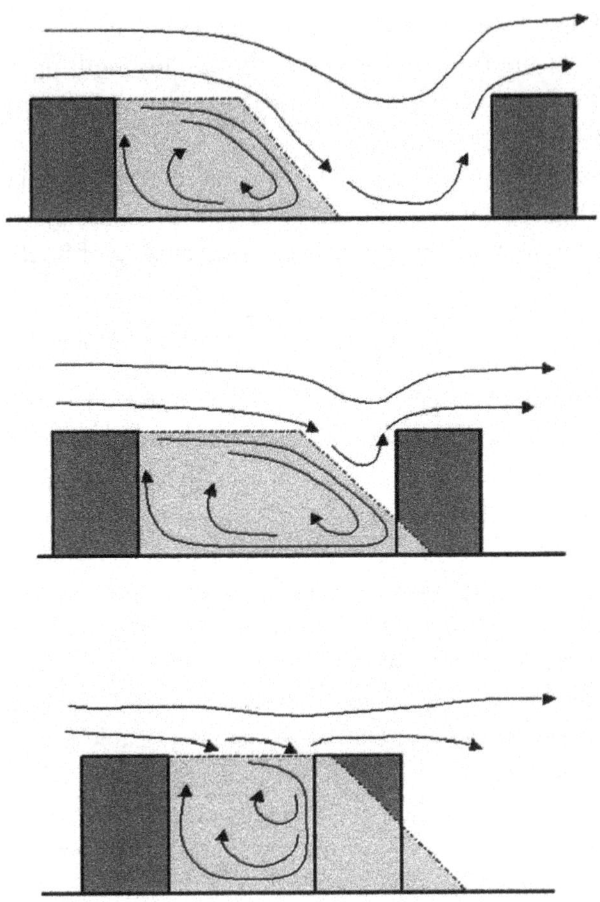

The section of recirculation area is trapezoidal, while its proportions in relation to the street profile provide the research subject (Figure 2). Apart from the buildings geometry, significant factors include wind speed profile that is associated with the roughness of neighboring area's terrain and the roof form. Oke suggests that the ratio of the lower base of the trapezoid and the height of the windward amounts to 2-3, Castro and Robins claim that in case of flat roof buildings, the ratio equals 2, Harman assumes the value of 3, while for Okamoto the value is 3.5 (Harman, 2003). If the distance between buildings is less than the value of the lower base of the trapezoid recirculation zone, street ventilation does not reach the ground and, as such, may prove insufficient. For streets whose width is equal to or less than the upper base of the trapezoid (in Harman's assumption, it totals half the value of the lower base), the ventilation ceases completely. Assuming that the lower base of the trapezoid recirculation zone equals 2.7 of the building height (an average of the values given above), it can be stated that with proportion coefficient H/S of less than 0.37, an intense air exchange with winds perpendicular to the street may occur, whereas for the coefficient greater than 0.74 the air is likely not be exchanged at all.

As mentioned above, it is important to strike a balance between protection of streets against winds and their ventilation. As suggested by Bottema, this goal is possible for the canyon ratio H/S of the value of 0.1 - 0.5 (Bottema, 1993). Oke maintains that the ideal parameter of H/S ratio equals 0.65 (Ratti, et al., 2006). The evident discrepancies between the results of scientific research prove how complex the problem is. The values given should be considered as approximations only, as an order of magnitude, as specific recommendations for particular situations can only be obtained through individually performed research or computer simulations.

Another feature of building development, on which the intensity of air exchange within a street is dependent is the continuity of the front elevation. Within tightly built-up streets, wind is powerfully directed, even if the direction in which it blows is not fully compatible with the layout of the street and its speed tends to increase (Flaga, 2008).

Winding of street spaces can also be significantly increased by the difference in heights of building developments. This applies high buildings scattered among lower building layouts. The existence of such buildings causes vertical air movement and intensifies the exchange of it (Blocken, Carmeliet, 2004).

Square

The issues related to aerodynamics of central system urban interiors, such as squares or courtyards rarely find themselves to be investigated by researchers. Commonly, they are treated as a special case of a street canyon that reaches low values of H/S parameter. In fact, the squares geometry is more complex than that of streets. Being restricted on three or four, rather than two sides, squares can assume various shapes. Given large size of space, the problem of lack of air exchange characteristic for deep street canyons does not occur. On the contrary, such spaces may be exposed to winds overly intense. Bottema addresses this problem with regard to Danish cities on which urban spaces, even of devoid of elevated buildings, happen to be extremely windy (Bottema, 1993). Based on the research conducted, he proposes the maximum dimensions of squares to be expressed as multiples of building development heights, which allows for protection against excessive winding (Table 1). The juxtaposition presented below refers not only to the dimensions of squares, but takes into account also the type of activity of the square users (long stay, short stay, stroll, fast march), as well as the surroundings of the square (water, countryside, suburbs, small town, big city). In order to provide proper protection against intense winds,

Table 1. Widths of squares providing protection against intense winds for various user activities and conditions

Surroundings	User Activity			
	Short Stay	Long Stay	Stroll	Fast March
Large water area (distant by 10-20km)	S<7H	S<9H	S<9H	S<9H
Rural area	S<9H	S<14H	S<17H	
Small suburbs (radius of about.2km)	S<10H	S<17H	-	-
Large suburbs (radius of > 10-20km)	S<15H	-	-	-
Small town (radius of about 2km)	S<15H	-	-	-
Large city (radius of > 10-20km)	S<19H	-	-	-

Bottema, 1993.

the squares adjacent to open water containers should be considerably smaller (S <7H, thus H/S> 0.14) than the ones found in a large city (S <19H, thus H/S> 0.05).

Courtyards of smaller surface area built in quarter of building development can generally be considered a special case of a square. In case of such squares, the problems that may arise are similar to those encountered in narrow streets, but they may be additionally worsened by the "closure" of space that occurs not in one only, but in both directions. Openings in the quarter of building developments, as well as gates play a vital role in the intensification of ventilation of such squares (Laskowski, 1987). The important aspects of such openings include their size, number and location.

Striving to make even more intensive use of land in cities leads to the creation of quarters of building development and inner courtyards that are increasingly tight and small in size. In Poland, this trend is supported by the regulations in force that do not account for wind condition. Therefore, conducting research into the relationship between the geometry of courtyards and their microclimatic comfort seems particularly necessary. This idea has become an incentive to embark on a research of our own, the results of which are presented later in this chapter.

RESEARCH METHODS

Two basic research methods are employed when analyzing airflows around buildings: experimental research in the wind tunnel and numerical simulations. Over the past two decades, a considerable development of the latter has been noticed. However, the reliability of these methods does not match the reliability of experimental methods, especially if extensive, complex building systems are concerned. Currently, the combination of the two methods yields the most reliable results. This is the case, since experimental results make a general identification of phenomena possible; allow isolation of problem areas and provide the basis for determining parameters that are later introduced into computer programs.

Experimental methods can be divided into point methods and field methods. The former methods yield quantitative information concerning a specific physical quantity, for instance speed or pressure, but only within a discrete points set. The latter, however, provide continuous quality information concerning a certain volume within a given area. Experimental methods include visualization and erosion techniques of all kind. Hereinafter, two methods that prove significant for the study of the issue covered by this chapter are discussed.

Oil streak technique serves as an example of visualization techniques. The model of building development is attached to the base, the surface of which is covered with a thin layer of oil tinted with appropriate pigment. In the process of an aerodynamic tunnel experiment, the oil previously spread on the surface is blown off areas of high flow velocity and is accumulated in areas of much less intense flow velocity. The results of such an experiment yield an image representing the average direction of the moving air, while the arrangement of streaks corresponds to the flow turbulence. In the progress of the experience photographs are taken. Analysis of the photographs sequences facilities the understanding and the interpretation of flow phenomena. If the method fails to provide quantitative data, it provides a starting point for further, more detailed research.

Erosion techniques - including scour technique (detailed description of which can be found in (Blocken, Carmeliet, 2004) indicate the extent to which wind speed changes in the vicinity of buildings. Saltation, by definition, means the transportation of rock crumbs that perform low leaps over the sediment surface on ballistic trajectory. The sand used for the experiment is characterized with standardized diameter and very low humidity coefficient. A board of uniform roughness, to which models of buildings are attached, is needed for the experiment. Scour technique makes it possible to observe the changes in flow velocity values that occur as a consequence of the introduction of an obstacle.

Measurement using scour technique comprises two stages, the first one of which involves setting a reference speed at which the applied sand layer is almost completely blown off the board. The speed is measured in undisturbed area at a certain distance from the ground using a Prandtl tube (the choice of place of measure depends on the cross-section of the aerodynamic tunnel in which the tests are performed). The second step involves the measurement proper, which is performed for each wind direction under consideration. In this case, the building models are placed on a rough board, with the sand spread evenly over the empty spaces between of the buildings. The test is performed in a tunnel for which the flow velocity is gradually increased by adding ΔV to subsequently obtained values. For each change of velocity, several moments are needed for the flow to stabilize and for the sand to cease erosion. After that time, a photograph of the area under measurement should be taken.

Photos for successive speeds are processed and superimposed. As a result of this procedure, a colorful map presenting the local amplification factor α is obtained. In order to determine it, both the value of reference speed and the value of speed being currently set up are needed (a detailed description of the ways to determine the local amplification factor may be found in (Blocken, Carmeliet, 2004). In zones where $\alpha > 1$, air stream accelerates resulting from the presence of buildings. These places are characterized with good ventilation, but are less comfortable for pedestrians. In areas where $\alpha < 1$, the flow slows down and in extreme cases, stagnation zones known also as air stagnant areas are formed.

Knowing the value of wind speed at a given moment and the value of the local amplification factor, the approximate flow rate in a given area can be determined. The values obtained should obviously be treated as estimates, as they are used in order to identify places of poor or good ventilation and may specify areas of severe discomfort in case of very high wind speeds.

The speed value obtained resulting from the erosion process does not serve exactly as a parameter of the local wind velocity V_{loc} appearing in the equation above, as it illustrates the velocity present in the lower parts of the building. With the right approximation, this value can serve as a local value for low buildings (up to 12 m). In case of higher buildings, additional studies, such as velocity measurements along a height at a certain distance from the building wall (as described in (Blocken, et al., 2006) and

(Mirsadegh, et al., 2013), or computer simulation should be carried out to define the wind speed profile as a function of altitude. The speed obtained in the process of saltation in the immediate vicinity of a building displays a highly significant property - it takes into account the geometry of the building under consideration, as well as the shape and location of buildings in its immediate vicinity. Even if this value cannot be applied directly for the heat permeability coefficient formulas for external surfaces, it does provide a comparative basis and a starting parameter for future research.

EXPERIMENTAL RESEARCH OF EXAMPLES OF URBAN BUILDINGS DEVELOPMENT

Purpose, Object, and Research Method

The aim of the experimental research was to identify the phenomenon of insufficient ventilation in the existing urban developments in Warsaw and to observe the dependence of this phenomenon on the geometrical parameters of the building developments. Three examples of buildings developments were selected, in case of which these issues were likely to occur. The basic criterion taken into consideration was the ratio of building height to the distance between buildings. The selected examples of building developments were chosen due to their compact character, where distance between buildings was less than twice their height. (Figure3) For one of the selected cases, built in the layout of a four cornered quarter (Figure 3 below), an experiment aerodynamic tunnel was performed. This is the building quarter situated between the street Dobra, Lipowa, Leszczynska and Wybrzeze Koscuiszkowskie (the neighborhood of Warsaw University Library BUW). The height of buildings varies between 3-7 stories and the distance between buildings is 1-2 building height.

The existing situation was tested and the geometrical layout was corrected. The aim of the correction was an attempt to minimize the phenomenon of insufficient ventilation. The corrected variant underwent tests as well and then the results for both variants were compared.

The oil streak technique was chosen as it proved sufficient for a visual assessment of the phenomena occurring in the area selected for test purposes. A model in scale of a 1: 400 which depicted the building development layout in a radius of 200 m was constructed. The visualization was performed for two wind directions occurring most frequently in Warsaw, namely the western and northern direction. The fore field of the tunnel was formed in such a way as to convey the profile of wind velocity that is characteristic of the city. The outcome image was interpreted according to a sequence of photographs taken every 15 seconds.

Airflow Around Building Development Analysis: The Existing Situation

Western Wind Direction

The air enters the urban planning system under consideration from the spaces between buildings numbered 3 and 11, 10 and 9, 9 and 8 (Figure 4). Behind all of the listed buildings, there appear leaning-edge air vortices of different shapes in which air undergoes low speed rotation. Air hits the wall of building 1 and bounces off it. In between the eastern-side wall of building 8 and western-side wall of 1G, an

Figure 3. The examples of building development existing in Warsaw, in case of which problems of insufficient ventilation are likely to occur. The bottom image presents the situation chosen for experimental research (site plan and model plan).

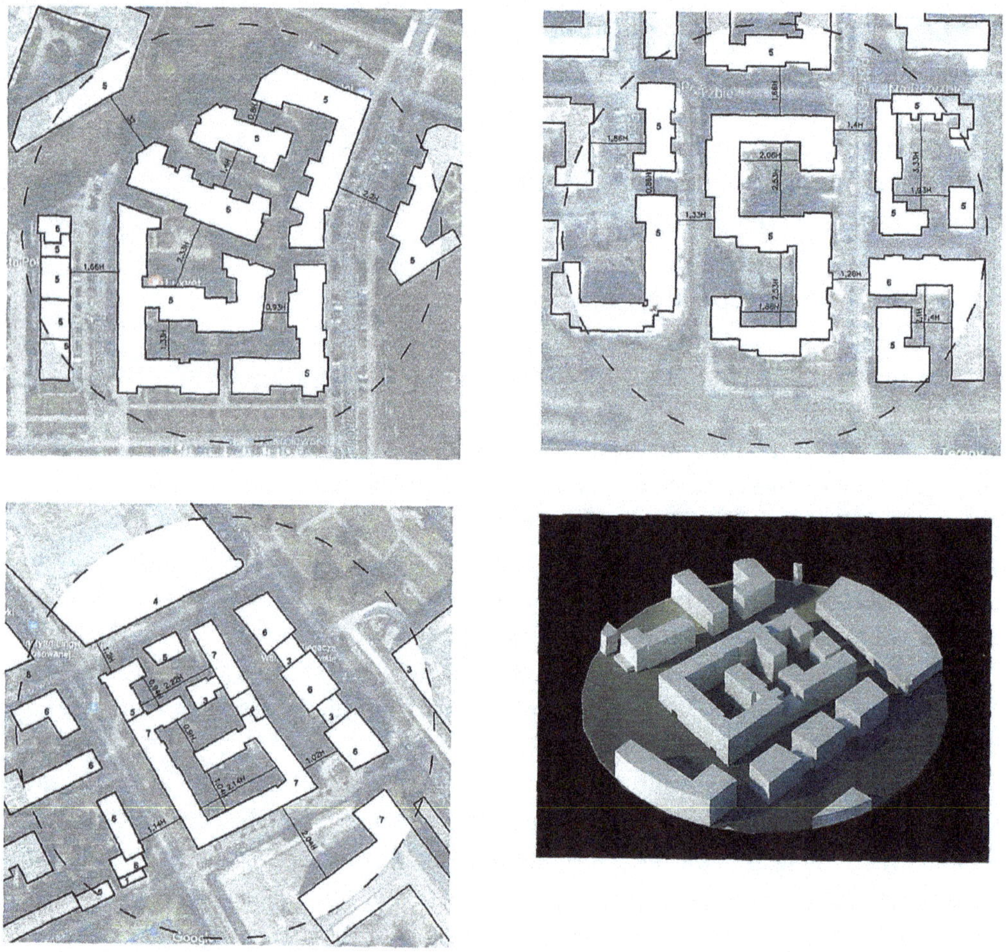

extensive, but slightly distorted area in which the so-called horseshoe vortex operates can be observed. The air bounced off the building moves in the direction opposite to the main stream.

The building assembly 1, with all its wings, poses a vast barrier for the airflow to move freely. In the courtyard restricted by wings marked in the picture1-H, B, 1C and D, the air is trapped and rotates very slowly. The only space where ventilation occurs is the wing area 1A. The air bounces off it and moves downwards along the wall and then permeates into the spaces between building wings 1H and 1B. However, it is slowed down and consequently held back in the area surrounding wing 1D.

An indistinctive movement of the air bounced off 1F wall and flowing towards wing 1D is observed in the courtyard restricted by walls 1- D, E, F, G, I.

The space between buildings 1 (eastern walls 1A, 1C, 1E) and 4 (western walls 4A, 4B, 4C, and 4D) remains unventilated.

Figure 4. The oil visualization of the existing building development layout (western wind), together with airflow markers. The hatching indicates the areas where air exchange is likely to be limited.

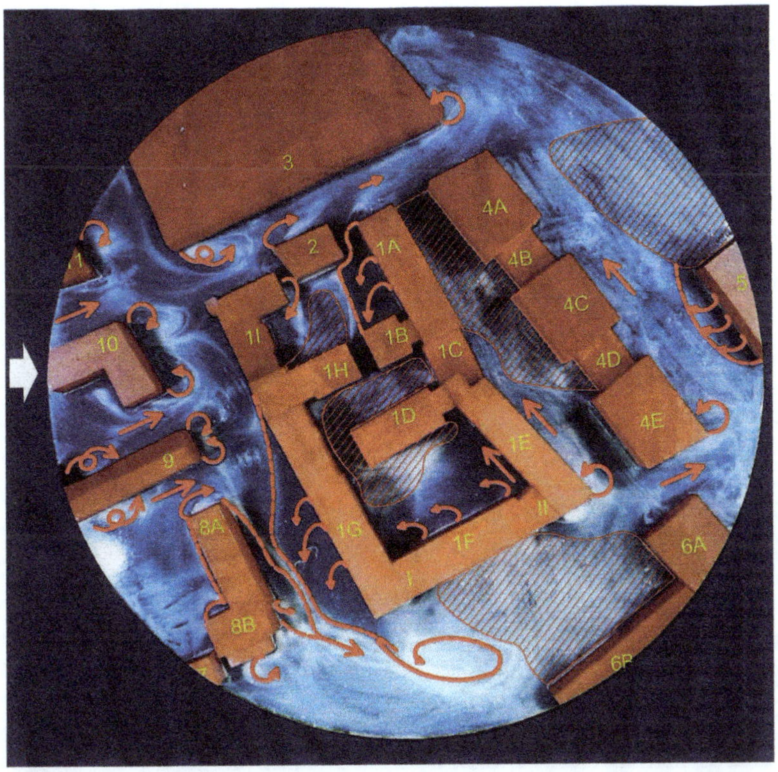

Northern Wind Direction

The main stream of air flows along the street adjacent to the building 1 -I, G, as well as the street stretching along the western walls of the building 4 (Figure 5). Certain amount of the air is sucked into the space between buildings 3 and 4 – A. It also enters the street running along buildings 1 - A, C, E and 4 - A, B, C, D and moves quite freely towards building 6A.

In the courtyards formed by numerous walls of building 1, only negligible airflow occurs. The air that gets through the gap between building 2 and building 1A is sucked to behind building 2 and performs slow rotational movement. Some of the air entering the courtyard from the top bounces off the northern facade of the 1H building, and performs a backwards movement. In the courtyards restricted by buildings 1-, G, H, B, C and D, as well as 1 -G, F, E and D the air remains still. Only minor airflow in the corner of building 1H at its junction with 1 G wing is observable. The courtyard, therefore, prove to be very poorly ventilated.

An extensive stagnation zone can also be observed in the southern part of the tested area, behind the 1F wing. The zone is adjacent to the horseshoe vortex formed by the air bounced off the walls of building 6.

Figure 5. Oil visualization of the existing building development (northern wind) together with airflow markers. The hatching indicates the areas where air exchange is likely to be limited.

Problem Identification, Geometry of Building Development Correction

The main problem observed is the tendency for air stagnation within the courtyards located inside the building developments in the shape of the quarter. (1A-I, 2). The purpose of the building development layout geometry correction was to minimize the occurrence of this phenomenon. A decision was made to remove the selected elements of the building development located inside the building quarter (the courtyard was initially divided into three parts, after the changes were introduced, it is divided into two parts), as well as to additionally open the courtyard (hence the broadening of the development gap and the introduction of gate in the places where air stagnation is the strongest).

Four changes were introduced to do so:

1. Shortening of building 1I wing in order to increase the distance between the wing and building 2,
2. Introduction of the gate opening in building 1G, adjacent to building1D,
3. Removal of segment 1B,
4. Removal of segment1D.

Airflow Around Building Development Analysis: The Improved Situation

Western Wind Direction

Compared to the case presented in the previous research, the airflow conditions in the courtyard restricted by wings A, I, H of building 1 and building 2 have improved significantly (Figure 6). Distinct air movement id observable, as proved by the heavily washed oil. Having removed segments B and D of building 1, the air was enabled to enter rather freely into the courtyard restricted by facades 1 - C, E, F, G and H. It is then sucked behind the wall of building 1H, where it performs a slow rotational movement within a certain area. The air entering form the top of the building quarter created by building 1 bounces off the windward walls C and E and moves backwards in relation to the main stream of airflow. However, a considerable part of the area in the courtyard under consideration (the area stretching along 1G wall towards facade 1E) still remains poorly ventilated.

The air flow outside the building quarter formed by the walls of building 1 remained largely unchanged, the location of vortex or stagnant areas remains rather constant. The images may not be identical, but in fact no experiment is replicable in 100%.

Figure 6. Oil visualization of the improved building development (northern wind) together with airflow markers. The hatching indicates the areas where air exchange is likely to be limited.

Northern Wind Direction

Having introduced the changes and removed buildings 1B and 1D, no significant improvement of ventilation in the building quarter restricted by the remaining walls of building 1 has been observed (Figure 7). Local air movement of increased value can only be seen in the immediate vicinity of building 1H - an outline of the horseshoe vortex on the north side is visible, which denotes the presence of airflow bounced off facade and heading in the direction opposite to the mainstream airflow. Slow vortex movement of airflow on the south side is also observable. Undoubtedly, such a poor level of ventilation of the building quarter in case of northern wind in both cases described in the study is due to the presence of large building 3, which poses a significant obstacle to the air flow.

CONCLUSION OF THE CONDUCTED RESEARCH INTO THE EXAMPLE OF BUILDING DEVELOPMENT

Research has shown that insufficient air exchange in the building development layout under consideration is likely to occur. The reason for this phenomenon to occur is provided by the existence of small dis-

Figure 7. The oil visualization of the improved building development layout (northern wind), together with airflow markers. The hatching indicates the areas where air exchange is likely to be limited.

tances between buildings, as compared to their heights. Another reason for the phenomenon is "closed" geometric layout, namely the building quarter. Stagnant air zones are formed in the inner courtyards and on parts of streets that surround the development. The aim of the introduced changes was to improve air circulation within the building quarter.

In the case of the western wind, the geometric correction of the building development resulted in the reduction of air stagnation zones that exist within the building quarter. This was the case especially in the smaller courtyard located on the north side. The improvement was, however, insignificant. In addition, following the introduction of the changes in the building development layout, there appeared a disadvantageous increase of stagnation zones east of building 1.

With the northern wind direction, an unfavorable expansion of air stagnation zone was observed both within the building development and the smaller courtyard from the north in case of the improved layout.

Other minor changes in airflow seem unimportant for the improvement of wind comfort within the analyzed building developments.

To conclude, the wind situation in case of the analyzed building development failed to improve significantly as a result of the introduced changes. Presumably, a much more critical correction that would result in a greater reduction in building cubature or height would be required to note a greater improvement.

SOLUTIONS AND RECOMMENDATIONS

City Denseness: Environmental Risks

The case study that was presented above addresses the complex problem of urban environment shaping. The undertaken study, both theoretical and experimental, indicate that urban building developments fashioned in accordance with the modern trends towards spatial planning are prone to suffer from insufficient air exchange in spaces around buildings. The occurrence of this problem results in a decrease in the quality of urban space microclimate and creates unfavorable conditions that hamper the operation of buildings with reduced energy demand, including nZEB. This is an aspect of great importance, particularly in terms of retrofit of the buildings existing in cities, as the highest concentration of most energy-consuming buildings that require retrofit is found in urban areas. Urban areas are currently undergoing the process of intensified densification, which is in line with the postulate of protection against extensive urban sprawl. Post-industrial areas, together with areas formerly remaining under the management of communication are also being developed. The process is dominated by economic factor, which prioritizes the utmost employment of usable space that can be extracted from a given area.

The regulations in force governing distances between buildings in Poland do not apply to the ventilation conditions of urban spaces. They concern rather such issues as the distance of the building from the plot boundaries, as well as mutual obscuring of buildings and the amount of sunlight access to the interior. The latter only applies to rooms intended for accommodating people in flats, nurseries, kindergartens and schools. In order for them to meet the minimum sun exposure requirements, the distance between buildings should vary from 1.5 to 3 of their height (H/S = 0.33-67), depending on the street orientation. In case of developments situated in central inner-city zones the requirement may be reduced to half of its value, whereas in certain cases (one-room apartments) is not applicable at all. Another issue beyond regulations concerns the conditions of sunlight access for buildings, whose purposes is other than residential or educational. Consequently, by using all the deviations from regulations concerning required sunshine access for living spaces in urban areas, as well as owing to the lack of any requirements concerning services commonly located on ground floors of buildings, it is only possible to design such courtyards whose geometrical proportions prevent them from being effectively ventilated. This problem concerns especially courtyards within narrow quarters of building developments and narrow, long, densely built-up streets.

The above statement is confirmed by the research discussed. As it follows from the research, a slight correction in geometric layout of a too tight building may only reduce the problem rather than eliminate it. Presumably, introducing door openings and breaks in order to open narrow quarters of buildings can only be effective where the interior of the building quarter is combined with other well ventilated spaces. On the other hand, connecting them to closed, windproof interiors will probably contribute to the intensification of ventilation to a lesser degree. (Similar conclusions follow from the research described in (Zielonko-Jung, 2014).

Therefore, improving the conditions of poor microclimate quality of building environment that result from excessive density of building developments in existing situations poses difficulty, as it may require considerable demolition of the existing structures. This creates an obvious problem, in both economic and technical terms. However, it is also an urban problem, since as a result of the demolition the established compositional system can lose its clarity, while the logic of space division may be disrupted.

Possibilities for Improving Problem Situations, Importance of Greenery

In view of the above difficulties, it is important to look for opportunities to improve local microclimates around buildings in the existing tight building developments, whose layout obstructs air exchange. The analysis of source research point to the significance of greenery and unpaved water permeable surfaces in creating microclimatic conditions.

Greenery is known to increase air humidity. Moreover, it is reported to demonstrate the ability to decrease air temperature, which increases air circulation. Greenery is responsible for the production of oxygen and the retention of various gaseous and particulate matter contained in air and water. Keeping the amount of hardened surfaces to the minimum in favor of greenery reduces the accumulation of heat resulting in overheating of built-up areas in hot seasons. In case of green surfaces, the temperature amounts to 1-2 ° C below the ambient temperature, whereas differences in heating of lawns and asphalt roads reach up to 20 ° C (Yeang, 2008).

In the case described in the chapter, the introduction of the largest possible area of unpaved surfaces within the courtyard area and covering it with low greenery seems a beneficial solution. Due to the likely poor sun exposure of this area, it is necessary to select the right plant species.

In similar situations, application of green walls and roofs can be considered. In summer, the temperature of a traditional roof may reach the temperature of up to 80 ° C, while the roofs covered with greenery reach the temperature of only about 25 ° C (Yeang, 2008). In the light of a ground surface deficit, it seems justified to increase the biologically active space by means of introducing green roofs and walls.

In turn, it is unadvisable to introduce high greenery inside and in the immediate vicinity of courtyards, as by doing so the airflow will be restricted even further. High plants may lead to alleviations of possibly adverse rapid acceleration and variations in airflow patterns that also occur in complex building developments. In the examined case, however, such instances were not observed.

Tools and Research Methods

The issues concerning airflow around buildings are very complex and as such are yet to be sufficiently investigated in terms of theoretical issues. Currently, airflow testing is only required as a part of the designing process in special cases, namely highrise buildings and building developments located in the corridors that aerify cities. Such research proves expensive and time-consuming, as well as involving professionals. Nevertheless, data concerning aerodynamic phenomena, including the distribution of the above-mentioned local V_{loc} wind speed values, may prove applicable in number of other situations, including modernization of existing buildings in such as way as to reduce their energy needs. Obtaining this sort of data may pose difficulties. Issues such as credibility and comparability of the data to different variants of designing decisions are crucial. Another issue of importance is the cost of obtaining data, which may affect the decision whether to include such studies in the design process

FUTURE RESEARCH DIRECTIONS

The issue discussed in the chapter fails to be fully understood, both in theoretical and practical terms. This is the case since the issue is of complex nature and it attempts to combine activities of various fields not integrated with one another strongly enough. However, a number of especially vital directions for future research in this field may be identified.

The first of those directions is the development of research methods, especially of simulation methods that are capable of providing support to urban and architectural designing, both in the field of building design, as well as building modernization. A good example of such a tool is provided by a digital city model updated on a regular basis that would be used in order to simulate climatic and energy-related phenomena at diverse detail levels. Such a model would enable researchers to check the environmental impacts caused by spatial decisions, to be used for instance while making local development plans, creating master plans, or finally designing buildings themselves. Even though the current level of development of simulation methods makes the creation of such a multilevel tool impossible, this goal may be achieved in the future. Tools of such kind could also provide a source of accurate, reliable data to be used in case of nZEB building designing, as well as for retrofit of the existing buildings.

Theoretical studies, which could provide the basis for design guidelines, indicate good practice, as well as support design intuition are vital. The development of knowledge concerning geometrical parameters of building developments are of special importance, as such parameters may cause problems related to airflow, including the problem of insufficient ventilation. It is also necessary to determine boundary parameters to describe density of urban developments, as well as to establish guidelines for combining urban interiors with each other in such a way as to promote fluent ventilation. It is, moreover, particularly important to develop research into the ability of greenery to influence climatic conditions around buildings and the potential of greenery to modify aerodynamic phenomena. It seems that these two issues are not linked in scientific research to a sufficient extend.

The possibility of implementing airflow optimization issues around buildings as a common, valid element of the designing process appears to be yet another matter of importance. At present, this problem is hardly taken into consideration. The implementation of those issues depends largely on technological development in the scope of research methods, the credibility of such research, their ability to be efficiently integrated into the design process at its subsequent stages and is certainly conditioned by the financial aspect. Yet, development of legal and administrative instruments that would allow the above problems to be analyzed in the process of spatial planning and urban and architectural design is just as crucial.

CONCLUSION

The urban environment produces a specific microclimatic environment for buildings, which is a vital issue to be considered in the processes of designing new buildings and retrofit of the existing buildings with low energy needs. Airflow around intensive building development is an issue of complex nature, as it leads to the situation in which a considerable variation in wind direction and speed occurs simultaneously in a relatively small area. For the sake of identification and optimization of energy- exchange related phenomena, it is crucial to obtain reliable data concerning local wind speed V_{loc} in the close vicinity of buildings. This was made possible thanks to the introduction of wind flow simulation methods. Currently, research that combines experimental and numerical methods shows great potential.

The performed analysis of geometric parameters for urban building developments that are characteristic for large Polish cities indicates the risk of local air stagnation within closed, dense urban indoors. The research presented in the chapter above provides a description of such a situation. The tendency to condense urban development poses a risk of proliferation of the phenomenon, which may add to the intensification of adverse urban heat island effect. All the issues mentioned make it necessary to consider retrofit of urban building developments into nZEB type of buildings in a broader context. The problem engages with the logic concerning spatial shaping of buildings in larger areas, which should be oriented towards the creation of spaces with proper ventilation conditions, as it would create most favorable environment for nZEB type of buildings.

REFERENCES

ANSYS. (2011). *Ansys Fluent Theory Guide, version 14.0.* Canonsburg: ANSYS.

Blocken, B. (2014). 50 years of Computational Wind Engineering. Past, present and future. *Journal of Wind Engineering and Industrial Aerodynamics*, *129*, 69–102. doi:10.1016/j.jweia.2014.03.008

Blocken, B., & Carmeliet, J. (2004). Pedestrian wind environment around building: Literature review and practical examples. *Journal of Thermal Envelope and Building Science*, *28*(2), 107–159. doi:10.1177/1097196304044396

Blocken, B., Janssen, H., Roels, S., Derome, D., & Carmeliet, J. (2006). Local wind speed near exterior building surfaces for convective transfer coefficients. *Research in Building Physics and Building Engineering: Proceedings of 3rd International Building Physics Conference.*

Blocken, B., & Persoon, J. (2009). Pedestrian wind comfort around a large football stadium in an urban environment: CFD simulation, validation and application of the new Dutch wind nuisance standard. *Journal of Wind Engineering and Industrial Aerodynamics*, *97*(5-6), 255–270. doi:10.1016/j.jweia.2009.06.007

Bottema, M. (1993). *Wind Climate and Urban Geometry* (Unpublished doctoral dissertation). Eindhoven University of Technology.

Che-Ani, A. I., Shahmohanadi, P., Sairi, A., Mohd-Nor, M. F. I., Zain, M. F. M., & Surat, M. (2009). Mitigating the Urban Heat Island Effect: Some Points without Altering Existing City Planing. *European Journal of Scientific Research*, *2*(35), 204–2016.

Daniels, K. (1997). *The Technology of Ecological Building*. Berlin: Birkhauser.

Flaga, A. (2008). *Inzynieria wiatrowa*. Warszawa: Arkady.

Gumowski, K., Olszewski, O., Pocwierz, M., & Zielonko-Jung, K. (2015). Comparative analysis of numerical and experimental studies of the airflow around the sample of urban development. *Bulletin of the Polish Academy of Sciences. Technical Sciences*, *63*(3), 729–737. doi:10.1515/bpasts-2015-0084

Harman, I. N. (2003). *The Energy Balance of Urban Areas* (Unpublished doctoral dissertation). University of Reading, Reading, UK.

Klemm, K. (2011). *Kompleksowa ocena warunkow mikroklimatu w luznych i zwartych strukturach urbanistycznych*. Warszawa: KILiW PAN.

Kose, D. A., & Dick, E. (2010). Prediction of the pressure distribution on a cubical building with implicit LES. *Journal of Wind Engineering and Industrial Aerodynamics*, *98*(10-11), 628–649. doi:10.1016/j.jweia.2010.06.004

Kovar-Panskus, A., Louka, P., Sini, J. F., Savory, E., Czech, M., Abdelqari, A., & Toy, N. et al. (2002). Influence of geometry on the mean flow within urban street canyons - a comparison of wind tunnel experiments and numerical simulation. *Water, Air, and Soil Pollution*, *2*(5-6), 365–380. doi:10.1023/A:1021308022939

Laskowski, L. (1987). *Wybrane zagadnienia fizyki miasta*. Retrieved from http://www.yego.pl/x_C_I__P_63511184__PZTA_01.html

Lewinska, J. (2000). *Klimat miasta, zasoby, zagrozenia, ksztaltowanie*. Krakow: Instytut Geografii i Gospodarki Przestrzennej.

Marchwinski, J., & Zielonko-Jung, K. (2012). *Wspolczesna architektura proekologiczna*. Warszawa: PWN.

Mills, G. (2012). Luke Howard, Tim Oke and the Study of Urban Climate. *American Meteorogical Society Journal*. Retrieved October 12, 2012, from www.ams.confex.com/ams/pdfpapers/144519.pdf

Mirsadegh, M., Cóstola, D., Blocken, B., & Hensen, J. (2013). Review of external convective heat transfer coefficient models in building energy simulation programs: Implementation and uncertainty. *Applied Thermal Engineering*, *56*(1-2), 134–151. doi:10.1016/j.applthermaleng.2013.03.003

Ratti, C., Baker, N., & Steemers, K. (2005). Energy Consumption and Urban Texture. *Energy and Building*, *37*(7), 762–776. doi:10.1016/j.enbuild.2004.10.010

Ratti, C., Di Sabatino, S., & Britter, R. (2006). Urban Texture Analysis with Image Processing Techniques: Winds and Dispersion. *Theoretical and Applied Climatology*, *84*(1-3), 77–90. doi:10.1007/s00704-005-0146-z

Warszawa, Centralny Osrodek Informacji Budownictwa. (n.d.). Retrieved from http://www.yego.pl/x_C_I__P_63511184__PZTA_01.html

Yeang, K. (2008). *Eco Design. A Manual for Ecological Design*. Chichester, UK: John Wiley & Sons.

Yoshie, R., Mochida, A., Tominga, Y., Kataoka, H., Harimoto, K., Nozu, T., & Shirosawa, T. (2007). Cooperative project for CFD prediction of pedestrian wind environment in the Architectural Institute of Japan. *Journal of Wind Engineering and Industrial Aerodynamics*, *95*(9-11), 1551–1578. doi:10.1016/j.jweia.2007.02.023

Zielonko-Jung, K. (2013). *Kształtowanie przestrzenne architektury ekologicznej w strukturze miasta*. Academic Press.

Zielonko-Jung, K. (2014). *Kompas Lewis & Clark Authentic Models*. Retrieved from http://www.yego.pl/x_C_I__P_63511184__PZTA_01.html

KEY TERMS AND DEFINITIONS

Convective Heat Transfer Coefficient h_c for External Walls: A coefficient required to estimate heat loss from buildings and dependent largely on the local wind conditions that surround the building.

Flow Reinforcement Coefficient: A parameter that describes the ratio of airflow velocity at the ground in the absence of buildings to airflow velocity at the ground in vicinity of building. The coefficient provides the basis for determining spots at which airflow is likely to slow or accelerate resulting from the presence of objects.

Horseshoe Vortex: The zone of still horseshoe shaped vortex that is formed in front of windward wall of the building and results from the collision of two air streams, namely the incoming one and the bounced off the building one.

Recirculation Zone: The zone of reverted airflow that appears commonly on leeward side of buildings.

Stagnation Zone: Area in which air remains stagnant, or undergoes a very slow rotary motion.

Urban Building Quarter: An assembly of buildings that constitute building development of the area restricted by a set of streets surrounding it.

Urban Heat Island Effect: Meteorological phenomenon common for urban areas. It is characterized by increased accumulation of solar radiation and temperature increase in cities, as compared to surrounding areas.

Ventilation Zone: A well-ventilated area that appears in between of buildings.

Compilation of References

A guide to Green roofs, walls and facades in Melbourne and Victoria, Australia. (2014). Retrieved December 16, 2016, from http://www.growinggreenguide.org/wp-content/uploads/2014/02/growing_green_guide_ebook_130214.pdf

Addis, B. (2006). *Building with reclaimed Components and Materials.* London: Earthscan.

Aelenei, L., Petran, H., Tarrés, J., Riva, G., Ferreira, A., Camelo, S., & Gonçalves, H. et al. (2015). New Challenge of the Public Buildings: nZEB Findings from IEE RePublic_ZEB Project. *Energy Procedia, 78,* 2016–2021. doi:10.1016/j.egypro.2015.11.195

Ahn, K. U., Kim, Y. J., Park, C. S., Kim, I., & Lee, K. (2014). BIM interface for full vs. semi-automated building energy simulation. *Energy and Building, 68,* 671–678. doi:10.1016/j.enbuild.2013.08.063

Akhsassi, M., El Fathi, A., Erraissi, N., Aarich, N., Bennouna, A., Raoufi, M., & Outzourhit, A. (in press). Experimental investigation and modeling of the thermal behavior of a solar PV module. *Solar Energy Materials and Solar Cells.*

Akram, M., Rahman, I.A., & Memon, I. (2014). A Review on Expert System and its Applications in Civil Engineering. *International Journal of Civil Engineering and Built Environment, 1*(1).

Alexandri, E., & Jones, P. (2008). Temperature decrease in a urban canyon due to greenwalls and green roofs in diverse climates. *Building and Environment, 43*(4), 480–493. doi:10.1016/j.buildenv.2006.10.055

Almeida, M., Ferreira, M., Rodrigues, A., & Bragança, L. (2014). Cost Optimality and nZEB target in building renovation of Portuguese residential buildings. *Proceeding of iiSBE Net Zero Built Environment 2014, 17th Rinker International Conference.*

Alreshidi, E., Mourshed, M., & Rezgui, Y. (2017). Factors for effective BIM governance. *Journal of Building Engineering, 10,* 89–101. doi:10.1016/j.jobe.2017.02.006

Althaus, Ch. (1987). *Fassadenbegrünung.* Berlin: Patzer Verlag.

ANSYS. (2011). *Ansys Fluent Theory Guide, version 14.0.* Canonsburg: ANSYS.

Aparicio S., Alonso J. (2012). Reestructuracion de la vivienda social en Holanda: Caso Europarei, Uithoorn. *Revista de Architectura, 27.*

Architects, S. (2013). *The Clever Treefrog 1.* Retrieved from http://algae-architecture.com/?p=269#more-269

Ashby, M. F. (2013). *Materials and the Environment. Eco-Informed Material Choice.* Oxford, UK: Elsevier Inc.

ASHRAE. (2010). *ANSI/ASHRAE Standard 55-2010.* Atlanta, GA: American Society of Heating, Refrigerating and Air-Conditioning Engineers, Inc.

ASHRAE. (2013). *ASHRAE Handbook - Fundamentals (SI Edition).* Atlanta, GA: ASHRAE.

Australian Stainless Steel Development Association. (2009). Vertical landscaping. *Australian Stainless Magazine*, *46*, 1–2.

Azhar, M.K., & Sulaiman, M. (2013). *Application of Green Façade as Passive Cooling Technique for Energy Efficiency Architecture*. Department, Faculty of Engineering and Built Environment, Universiti Kebangsaan Malaysia.

Balaras, C.A. (2004). Decision support software for sustainable building refurbishment. *Transactions-American Society of Heating Refrigeration and Air Conditioning Engineers, 110*(1), 592-601.

Bank Gospodarstwa Krajowego. (2016). *Figures of the Thermomodernisation and Renovation Fund [Dane liczbowe Funduszu Termomodernizacji i Remontów]*. Warsaw: National Economy Bank. (in Polish)

Barbolini, F., Cappellacci, P., & Guardigli, L. (2017). A design strategy to reach nZEB standards integrating energy efficiency measures and passive energy use. *Energy Procedia*, *111*, 205–214. doi:10.1016/j.egypro.2017.03.022

BarcelonaTech. (2012). *Researchers at the UPC develop a biological concrete for constructing "living" façades with lichens, mosses and other microorganisms*. Retrieved from http://www.upc.edu/saladepremsa/al-dia/mes-noticies/researchers-at-the-upc-develop-a-biological-concrete-for-constructing-201cliving201d-facades-with-lichens-mosses-and-other-microorganisms

Basista, A. (2001). *Betonowe dziedzictwo. Architektura w Polsce w czasach komunizmu*. Warszawa, Poland: Wydawnictwo Naukowe PWN.

Bassanino, M., Fernando, T., Wu, K.-C., Ghazimirsaeid, G., Klobut, K., Mäkeläinen, T., & Hukkalainen, M. (2016). A collaborative environment for energy-efficient buildings within the context of a neighborhood. In *Proceedings, eWork and eBusiness in Architecture, Engineering and Construction*. CRC Press.

Bassanino, M., Fernando, T., & Wu, K.-C. (2013). Can virtual workspaces enhance team communication and collaboration in design review meetings? *Architectural Engineering and Design Management*, *10*(3-4), 200–217. doi:10.1080/17452007.2013.775102

Bauman, N. (2006). Ground-Nesting Birds on Green Roofs in Switzerland: Preliminary Observations. *Urban Habitats. An Electronic Journal on the Biology of Urban Areas Around the World, 4*. Retrieved from http://www.urbanhabitats.org/v04n01/urbanhabitats_v04n01_pdf.pdf

Beccali, M., Galatioto, A., Leone, G., & Longo, S. (2013). Is the NZEB benchmarking approach suitable for assessing energy retrofit design? *Applied Mechanics and Materials*, *361-363*, 402–407. doi:10.4028/www.scientific.net/AMM.361-363.402

Becchio, C., Bottero, M., Corgnati, S., & Ghiglione, C. (2015). nZEB design: Challenging between energy and economic targets. *Energy Procedia*, *78*, 2070–2075. doi:10.1016/j.egypro.2015.11.226

Becchio, C., Dabbene, P., Fabrizio, E., Monetti, V., & Filippi, M. (2015). Cost optimality assessment of a single family house: Building and technical systems solutions for the nZEB target. *Energy and Building*, *90*, 173–187. doi:10.1016/j.enbuild.2014.12.050

Beckman, W. A., Klein, S. A., & Duffie, J. A. (1977). *Solar Heating Design by the f-chart Method*. New York: Wiley-Interscience.

Bednarek, A. (1979). *Wpływ parkowej roslinności drzewiastej, trawnikow i placow na warunki mikroklimatyczne na przykladzie Warszawy. Les (Ljubljana)*, *27*, 89–107.

Behling & Behling. (1996). *Sol Power: The evolution of Solar Architecture*. Munich: Prestel Publishing.

Bernier, P., Fenner, R. A., & Ainger, C. (2010) Assessing the sustainability merits of retrofitting existing homes. *Proceedings of the Institution of Civil Engineers - Engineering Sustainability, 163*, 197-207. doi:10.1680/ensu.2010.163.4.197

Biovalco. (n.d.). *Biomass Fuels*. Retrieved September 3, 2017, from http://www.biovalco.com/energy/biomass-fuels/

Blanc, P. (2013). *The Vertical Garden: From Nature to the City by Patrick Blanc*. W. W. Norton & Company. Retrieved December 15, 2016, from http://scholar.google.pl/scholar?q=Living+facades

Blocken, B. (2014). 50 years of Computational Wind Engineering. Past, present and future. *Journal of Wind Engineering and Industrial Aerodynamics, 129*, 69–102. doi:10.1016/j.jweia.2014.03.008

Blocken, B., & Carmeliet, J. (2004). Pedestrian wind environment around building: Literature review and practical examples. *Journal of Thermal Envelope and Building Science, 28*(2), 107–159. doi:10.1177/1097196304044396

Blocken, B., Janssen, H., Roels, S., Derome, D., & Carmeliet, J. (2006). Local wind speed near exterior building surfaces for convective transfer coefficients. *Research in Building Physics and Building Engineering: Proceedings of 3rd International Building Physics Conference*.

Blocken, B., & Persoon, J. (2009). Pedestrian wind comfort around a large football stadium in an urban environment: CFD simulation, validation and application of the new Dutch wind nuisance standard. *Journal of Wind Engineering and Industrial Aerodynamics, 97*(5-6), 255–270. doi:10.1016/j.jweia.2009.06.007

Bogusz, A., & Polakowski, L. (2012). *Green Public Procurement*. Warsaw: Public Tender Department.

Bonacina, C. F., Masera, G., & Pavan, A. (2015). Investment Grade Energy Audit: A Financial Tool for the Cost-effective Renovation of Residential Buildings. *Energy Procedia, 70*, 709–718. doi:10.1016/j.egypro.2015.02.180

Borowski J. (2010). *Ile warte sa drzewa w miescie?* Retrieved from http://www.muratorplus.pl/technika/zrownowazony-rozwoj/ile-warte-sa-drzewa-w-miescie_68688.html

Borowski J. (2013). *Dlaczego warto sadzic i pielęgnowac drzewa?* Stowarzyszenie na rzecz Ochrony Krajobrazu Kulturowego Mazur "Sadyba". Retrieved from https://sadybamazury.wordpress.com/2013/04/06/dlaczego-warto-sadzic-i-pielegnowac-drzewa-jacek-borowski/

Borowski, J. (2011). *Bezcenne drzewa*. Zwiazek Szkolkarzy Polskich. Retrieved from http://www.zszp.pl/pliki/MPO07-11s.60-61.pdf

Borowski, J., & Latocha, P. (2014). *Zastosowanie roslin pnacych i okrywowych w architekturze krajobrazu*. Warszawa: Wydawnictwo SGGW w Warszawie.

Bose, J. E., Parker, J. D., & McQuiston, F. C. (1985). *Design/data manual for closed-loop ground-coupled heat pump systems*. ASHRE.

Bottema, M. (1993). *Wind Climate and Urban Geometry* (Unpublished doctoral dissertation). Eindhoven University of Technology.

Bouchanan, P. (1992). *Emilio Ambasz Inventions: The Reality of the Ideal*. Rizzoli International Publications, Inc.

Boyd, J., & Banzhaf, S. (2007). What are ecosystem services? *Ecological Economics, 63*(2–3), 616–626. doi:10.1016/j.ecolecon.2007.01.002

BPIE (2011). *Principles for nearly zero-energy buildings. Paving the way for effective implementation of policy requirements*. Brussels: Buildings Performance Institute Europe.

BPIE. (2014). *Renovation strategies of selected eu countries. A status report on compliance with article 4 of the energy efficiency directive*. Brussels: Buildings Performance Institute Europe.

BPIE. (2015). *Nearly Zero Energy Buildings Definitions Across Europe*. BPIE.

Braungart, M., & McDonough, W. (2002). *Cradle to cradle: Remaking the way we make things*. London: Vintage.

BRE. (2017). *BRE Environmental Assessment Method*. Retrieved from www.breeam.com

BREAAM In-Use International. (2017). Retrieved 22.08.2017 from http://www.breeam.com/in-use

Bregier, T., & Kronenberg, J. (Eds.). (2010). *Wyzwania zrownowazonego rozwoju w Polsce*. Kraków: Fundacja Sendzimira.

Brenneisen, S. (2006). Space for Urban Wildlife: Designing Green Roofs as Habitats in Switzerland. *Urban Habitats, 4*(1), 27-33. Retrieved from http://www.urbanhabitats.org/v04n01/urbanhabitats_v04n01_pdf.pdf

Brown, B. (2016). *Shade Losses in PV Systems, and Techniques to Mitigate Them*. Aurora Solar Inc. Retrieved September 3, 2017, from http://blog.aurorasolar.com/shading-losses-for-pv-systems-and-techniques-to-mitigate-them/

Bruse, M., Thonnessen, M., & Radtke, U. (1999). Practical and theoretical investigation of the influence of facade greening on the distribution of heavy metals in urban Streets. *Proceedings International Conference on Urban Climatology & International congress of Biometeorology*. Retrieved December 15, 2016, from http://www.envi-met.com/documents/papers/facade1999.pdf

Build Up Skills. (2012). *The professional qualifications and staff for the green building*. Warszawa: NAPE.

BuildDesk. (2012). *REPORT: Energy performance of buildings in Poland [RAPORT: Stan energetyczny budynków w Polsce]*. BuildDesk Polska. (in Polish)

Building Code. (2013). *Regulation of the Minister of Transport, Construction and Maritime Economy dated 5 July 2013r. amending the regulation on technical conditions to be met by buildings and their location* [Rozporządzenie Ministra Transportu, Budownictwa i Gospodarki Morskiej z dnia 5 lipca 2013r. zmieniające rozporządzenie w sprawie warunków technicznych, jakim powinny odpowiadać budynki i ich usytuowanie]. Academic Press.

Buildings Performance Institute Europe. (2011). *Principles for nearly zero-energy buildings - Paving the way for effective implementation of policy requirements*. Author.

Buildings Performance Institute Europe. (2012). *Implementing nearly zero-energy buildings (nZEB) in Poland – Towards a definition and Roadmap*. Author.

Bujak, P. (2013). *Modeling building information in a modern architect's workshop - perspective of application, orginal* [Modelowanie informacji o budynku we współczesnym warsztacie architekta – perspektywy zastosowania] (PhD thesis). Warsaw University of Technology.

Bulgaria. (2013). *Outline national plan to increase the number of nearly zeroenergy buildings in accordance with article 9 of directive 2010/31/eu on the energy performance of buildings*. Ministry of Regional Development and Public Works.

Bunting, T. E., & Cousins, L. R. (1985). Environmental dispositions among school-age children. *Environment and Behavior, 17*(6), 725–768. doi:10.1177/0013916585176004

Burszta-Adamiak E. (2015). *Zielone dachy jako element zrownowazonych systemow odwadniajacych na terenach zurbanizowanych*. Wroclaw: Wyd. Uniwersytetu Przyrodniczego we Wroclawiu.

Business Case for Green Building. (2013). *A Review of the Costs and Benefits for Developers, Investors and Occupants*. World Green Building Council.

Bytnerowicz, A. (1979). *Pochlanianie niektorych zanieczyszczen srodowiska przez drzewa w Warszawie. In Materialy z konferencji naukowo-technicznej, 28 stycznia 1978 r* (pp. 33–45). Wroclaw, Warszawa: Polska Akademia Nauk, Ogrod Botaniczny, Ossolineum.

Cameron, R. W. F., Taylor, J., & Emmett, M. (2015). A Hedera green façade - Energy performance and saving under different maritime-temperate, winter weather conditions. *Building and Environment*, *92*, 111–121. doi:10.1016/j.buildenv.2015.04.011

Carson, R. L. (1962). *Silent Spring*. London: Hamish Hamilton.

Cemesova, A., Hopfe, C. J., & Mcleod, R. S. (2015). PassivBIM: Enhancing interoperability between BIM and low energy design software. *Automation in Construction*, *57*, 17–32. doi:10.1016/j.autcon.2015.04.014

Central Statistical Office. (2013). *Inhabited buildings, the National Census of Population and Housing 2011*. Warsaw: Central Statistical Office.

Central Statistical Office. (2014). *Energy consumption in households in 2012*. Warsaw: Central Statistical Office.

Chang, K. K. (2015). Modeling of PV performance without using equivalent circuits. *Solar Energy*, *115*, 419–429. doi:10.1016/j.solener.2015.01.025

Chang, W.-J., Ku, C.-C., & Chang, W. (2009). Fuzzy control with passivity synthesis for continuous affine Takagi-Sugeno fuzzy systems. *International Journal of Intelligent Computing and Cybernetics*, *2*(2), 386–408. doi:10.1108/17563780910959947

Chastas, P., Theodosiou, T., Bikas, D., Kontoleon, K. (2017). Embodied energy and nearly zero energy buildings: A review in residential buildings. *Procedia Environmental Sciences, 38*, 554-561. doi:10.1016/j.proenv.2017.03.123

Che-Ani, A. I., Shahmohanadi, P., Sairi, A., Mohd-Nor, M. F. I., Zain, M. F. M., & Surat, M. (2009). Mitigating the Urban Heat Island Effect: Some Points without Altering Existing City Planing. *European Journal of Scientific Research*, *2*(35), 204–2016.

Chelminski, J. (2017). *Zielen w Warszawie. Policzyli i wycenili drzewa na Pradze. Sa warte majatek*. Retrieved from http://warszawa.wyborcza.pl/warszawa/7,54420,21864813,zielen-w-warszawie-ile-warte-sa-nasze-drogie-drzewa.html

Christoforidis, G., Melandri, D., Branciforti, V., Nousdilis, A., Papagiannis, G., & Penalvo, E. (2016). Meeting of energy skills: An educational framework for energy professionals in NZEB. In *EEEIC 2016 - International Conference on Environment and Electrical Engineering*. Institute of Electrical and Electronics Engineers Inc.

Cios, J. K., & Sztandera, L. M. (1992). Continuous ID3 algorithm with fuzzy entropy measures. *Proc. IEEE Internat. Conf. on Fuzzy Systems*, 469-476. doi:10.1109/FUZZY.1992.258659

Clement G., Eveno C., (1997). *Le jardin planetaire*. L'Aube/Chateau-Vallon.

Clement, G. (1991). *Le jardin en movement*. Paris: Sens and Tonka.

Clement, G. (2003). *Le Manifeste du Tiers-Paysag*. Paris: Sujet/Objet.

Clouston, B. (1977). *Landscape Design with Plants*. Landscape Institute, William Heinemann Ltd.

COHERENO. (2013). *NZEB criteria for typical single-family home renovations in various countries*. Retrieved from http://www.cohereno.eu/fileadmin/media/Dateien/D2_1_BPIE_WP2_12092013_3_5_-final.pdf

Coma, J., Perez, G., Sole, C., Castell, A., & Cabeza, L. F. (2014). New green facades as passive systems for energy savings on buildings. *Energy Procedia*, *57*, 1851–1859. doi:10.1016/j.egypro.2014.10.049

Communication from the Commission to the European Parliament, The Council, The European Economic and Social Committee and the Committee of Regions. (2008). *Public Procurement for better environment*. Author.

Costa, A., Keane, M. M., Raftery, P., & O'Donnell, J. (2012). Key factors methodology — A novel support to the decision making process of the building energy manager in defining optimal operation strategies. *Energy and Building*, *49*, 158–163. doi:10.1016/j.enbuild.2012.02.001

Costanza, R., d'Arge, R., de Groot, R., Farber, S., Grasso, M., Hannon, B., & van den Belt, M. et al. (1997). The value of the world's ecosystem services and natural capital. *Nature*, *387*(6630), 253–260. doi:10.1038/387253a0

Cruz, M. (2015 –2016). *Bio – Receptive Materials – EPSRC Funded Research Project*. Biota LabBartlett School of Architecture and University College London. Retrieved from http://www.biota-lab.com/?page_id=2920

Cuce, E., Cuce, P. M., Karakas, I. H., & Bali, T. (2017). An accurate model for photovoltaic (PV) modules to determine electrical characteristics and thermodynamic performance parameters. *Energy Conversion and Management*, *146*, 205–216. doi:10.1016/j.enconman.2017.05.022

Czarnecki, S., & Stawinska, E. (1984). *Badanie wplywu zieleni na zmniejszenie halasu w aglomeracjach miejskich. In Wplyw zieleni na ksztaltowanie srodowiska miejskiego* (pp. 109–124). Warszawa: Instytut Ksztaltowania Srodowiska.

Dabski, M. (2015). Rola i ksztaltowanie zieleni w otoczeniu szpitali. *Space and Form*, *24*(1).

Dalla Mora, T., Cappelletti, F., Peron, F., Romagnoni, P., & Bauman, F. (2015). Retrofit of an historical building toward NZEB. *Energy Procedia*, *78*, 1359–1364. doi:10.1016/j.egypro.2015.11.154

Dalla Mora, T., Cappelletti, F., Peron, F., Romagnoni, P., & Bauman, F. (2015). Towards nZEB buildings: A historical building case study. *REHVA Journal*, *52*, 16–20.

Daniels, K. (1997). *The Technology of Ecological Building*. Berlin: Birkhauser.

de Dear, R., & Brager, G. (1998). Developing an adaptive model of thermal comfort and preference. *ASHRAE Transactions*, *104*(1), 145–67.

Department of Energy. (2004). Federal Technology Alert: Green Roofs. DOE/EE-0298. Washington, DC: Author.

DI. (2017). *dateandtime.info*. Retrieved September 2, 2017, from http://dateandtime.info/

Diakaki, C., Grigoroudis, E., Kabelis, N., Kolokotsa, D., Kalaitzakis, K., & Stavrakakis, G. (2010). A multi-objective decision model for the improvement of energy efficiency in buildings. *Energy*, *35*(12), 5483–5496. doi:10.1016/j.energy.2010.05.012

Directive 2001/77/EC of the European Parliament and of the Council of 27 September 2001 on the promotion of electricity produced from renewable energy sources in the internal electricity market. (2001). *Official Journal of the European Communities*, *44*(283), 33-40

Directive 2002/91/EC of the European Parliament and of the Council of 16 December 2002 on the energy performance of buildings. (2002). *Official Journal of the European Communities*, *4*(1), 65-71.

Directive 2003/54/EC of the European Parliament and of the Council of 26 June 2003 concerning common rules for the internal market in electricity and repealing Directive 96/92/EC. (2003). *Official Journal of the European Union*, *46*(176), 37-56.

Directive 2009/28/EC of the European Parliament and of the Council of 23 April 2009 on the promotion of the use of energy from renewable sources and amending and subsequently repealing Directives 2001/77/EC and 2003/30/EC (Text with EEA relevance). (2009). *Official Journal of the European Union*, *52*(140), 39-85.

Directive 2010/31/EU of the European Parliament and of the Council of 19 May 2010 on the energy performance of buildings (recast). (2010). *Official Journal of the European Union, 3,* 124-146.

Directive 2010/31/EU of the European Parliament and of the Council of 19 May 2010 on the energy performance of buildings. (2010). *Official Journal of the European Union, 53*(153), 124-146.

DMID. (2015). *Decree of the Minister of Infrastructure and Development on the methodology for calculating the energy performance of buildings and dwelling or building which is part of the whole technical self-utility. and the preparation and presentation of certificates of energy performance from 27th February 2015.* (in Polish).

Duffie, J. A., & Beckman, W. A. (2013). *Solar engineering of thermal processes.* Hoboken, NJ: John Wiley & Sons. Inc. doi:10.1002/9781118671603

Dunnett, N. (2008). *Planting green roofs and living walls.* Portland, OR: Timber Press.

Dwyer, J. F., McPherson, E. G., Schroeder, H. W., & Rowntree, R. A. (1992). Assessing of the Benefits and Cost of the Urban Forest. *Journal of Arboriculture, 18*(5), 227–234.

Dzierzanowski, K., Popek, R., Gawronska, H., Sæbo, A., & Gawronski, S. (2011). Deposition of Particulate Matter of Different Size Fractions on Leaf Surfaces and in Waxes of Urban Forest Species. *International Journal of Phytoremediation, 13*(10), 1037–1046. doi:10.1080/15226514.2011.552929 PMID:21972570

Eadie, R., Browne, M., Odeyinka, H., McKeown, C., & McNiff, S. (2013). BIM implementation throughout the UK construction project lifecycle: An analysis. *Automation in Construction, 36,* 145–151. doi:10.1016/j.autcon.2013.09.001

EED. (2012). *Directive 2012/27/EU of the European Parliament and of the Council of 25 October 2012 on energy efficiency, amending Directives 2009/125/EC and 2010/30/EU and repealing Directives 2004/8/EC and 2006/32/EC Text with EEA relevance.* Brussels: European Parliament, Council of the European Union. EEA relevance, COD 2011/0172.

EN ISO 15251 Indoor environmental input parameters for design and assessment of energy performance of buildings addressing indoor air quality, thermal environment, lighting and acoustics.

EN ISO 15831:2006 Clothing - Physiological Effects - Measurement Of Thermal Insulation By Means Of A Thermal Manikin.

EN ISO 7726. Ergonomics of the thermal environment - Instrument for measuring physical quantities. (1998). Geneva, Switzerland: International Organization for Standardization.

EN ISO 9920:2009 Ergonomics of the thermal environment — Estimation of thermal insulation and water vapour resistance of a clothing ensemble.

ENERGY.GOV. (n.d.). *Geothermal Basics.* Retrieved September 3, 2017, from https://energy.gov/eere/forge/geothermal-basics

EnergyPlus. (n.d.). Retrieved September 2, 2017, from https://energyplus.net/

EPBD Recast, Directive 2010/31/EU of the European Parliament and of the Council o 19 May 2010 on the energy performace buildings (recast). (2010). *Official Journal of the European Union.*

EPBD Recast, Directive 2010/31/EU of the European Parliament and of the Council o 19 May 2010 on the energy performance buildings (recast). (2010). *Official Journal of the European Union.*

EPBD. (2010). *Directive 2010/31/EU of the European Parliament and of the Council of 19 May 2010 on the energy performance of buildings (recast).* Brussels: European Parliament, Council of the European Union. COD 2008/0223.

EPBD. (2015). *Implementing the Energy Performance of Buildings Directive (EPBD). Featuring Country Reports.* Lisboa: ADENE.

EPBD. (2016). *Proposal for a Directive of the European Parliament and of the Council amending Directive 2010/31/EU on the energy performance of buildings.* Brussels: European Commission. Com(2016) 765 Final, 2016/0381 (Cod).

EPBD. (n.d.). Retrieved from https://issuu.com/glassforeurope/docs/the_energy_performance_of_buildings/14

Erhorn, H., & Erhorn-Kluttig, H. (2014). *Selected examples of Nearly Zero-Energy Buildings. Detailed Report.* Retrieved September 2, 2017, from http://www.epbd-ca.eu/wp-content/uploads/2011/05/CT5_Report_Selected_examples_of_NZEBs-final.pdf)

ESP-r. (n.d.). Retrieved September 2, 2017, from http://www.esru.strath.ac.uk/Programs/ESP-r.htm

European Commission. (2006). *EC Directorate general for energy and transport. Intelligence energy – Europe Programme. Multiplying success in buildings: 21 innovative projects for an energy-intelligent Europe. Energy performance assessment for existing non-residential buildings.* Brussels: EPA-NR.

European Commission. (2010). *Directive 2010/31/EU of the European Parliament and of the Council of 19 May 2010 on the energy performance of buildings.* Retrieved on 30th May 2017 at http://eur-lex.europa.eu/legal-content/EN/ALL/?uri=CELEX:32010L0031

European Commission. (2012). *Directive 2012/27/EU of the European Parliament and of the Council of 25 October 2012 on energy efficiency.* Retrieved on 30th May 2017 at: http://eur-lex.europa.eu/legal-content/EN/TXT/?qid=1399375464230&uri=CELEX%3A32012L0027

European Commission. (2014). *Technical guidance: Financing the energy renovation of buildings with Cohesion Policy funding, Final report: A study prepared for the European Commission DG Energy.* Author.

European Parliament and of the Council. (2010). Directive 2010/31/EU of the European Parliament and of the Council of 19 May 2010 on the energy performance of buildings (recast). Author.

European Parliament's Committee on Industry, Research and Energy (ITRE). (2015). *Boosting Building Renovation: What potential and value for Europe? EU 2050 Roadmap for a low-carbon economy.* Author.

EUROSTAT. (n.d.). *Heating Degree Days (HDD).* Retrieved September 2, 2017, from http://ec.europa.eu/eurostat/web/energy/data

EUROSTAT. (n.d.). *Renewable energy statistics.* Retrieved September 3, 2017, from http://ec.europa.eu/eurostat/statistics-explained/index.php/Renewable_energy_statistics

Fanger, P. O. (1970). *Analysis and Applications in Environmental Engineering.* New York: McGraw-Hill Book Company.

Fara, L., & Craciunescu, D. (2017). Output Analysis of Stand-Alone PV Systems: Modeling, Simulation and Control. *Energy Procedia, 112*, 595–605. doi:10.1016/j.egypro.2017.03.1125

Fenves, S. J., Mahler, M. L., & Siriam, D. (1985). Knowledge-based expert systems in civil engineering. *IABSE Periodica, 4*, 63.

Finland. (2012). *Finland's national plan to increase the number of nearly zero-energy buildings. Report to the European Commission pursuant to Article 9 of Directive 2010/31/EU on the energy performance of buildings.*

Firlag, S., & Rucińska, J. (2007). *Simplified method of designing an air-ground heat exchange.* CESB 2007 PRAGUE International Conference - Central Europe Towards Sustainable Building 1.

Firląg, S., & Rucińska, J. (2007). Simplified method of designing an air-ground heat exchange. *CESB 2007 PRAGUE International Conference - Central Europe Towards Sustainable Building.*

Firląg, S. (2015). How to meet the minimum energy performance requirements of Technical Conditions in year 2021? *Procedia Engineering, 111*, 202–208. doi:10.1016/j.proeng.2015.07.077

Flaga, A. (2008). *Inzynieria wiatrowa.* Warszawa: Arkady.

Frac, M. (2009). Algi – energia jutra (biomasa, biodiesel). *Acta Agrophysica, 13*(3), 627-638. Retrieved from http://www.old.actaagrophysica.org/artykuly/acta_agrophysica/ActaAgr_168_2009_13_3_627.pdf

France. (2013). *Action plan for increasing the number of nearly zero-energy buildings.*

France: ш Ref. Ares (2013) 54186 - 16/01/2013.

Fronty, L. (1978). *Le fugeres poussent meme au plafond. VSD, March 1978 5/26.* Retrieved from http://www.verticalgardenpatrickblanc.com/patrick-blanc/dates-clefs

FutureBuilt. (2010-2020). Retrieved from www.futurebuilt.no

Gajewski R. R., Węglarz, A., & Żmijewski, K. (2007). *Poly-Optimal Method of Designing Environmentally Friendly Buildings.* IOS Press BV.

Gasiorowski, M. (2011). Funkcje roslinności w stawach. *Zielen miejska, 10.* Retrieved from http://www.stawy-kapielowe.com.pl/images/stories/artykuly/zielen_miejska_10_2011.pdf

Gawronski, S. W., Gawronska, H., & Rokosza, J. (2000), *Drzewa, krzewy i rosliny zielne w procesie fitoremediacji w terenie zurbanizowanym.* III Forum Architektury Krajobrazu, Warszawa, Poland.

Gawronski, S., & Szmit, B. J. (2011). *Roslinne oczyszczanie.* Warszawa: Zwiazek Szkolkarzy Polskich. Retrieved from http://www.zszp.pl/pliki/MPO09-Fitoremediacja.pdf

Gawryszewska, B. J. (2009). Revitalisation as a social process – the work with local communities. *Revitalising Built Environments: Requalifying Old Places for New Uses.*

GBPN. (2013). *What is a deep renovation definition?* Technical report. GBPN.

GbXML.org. (2016). *Software.* Retrieved on 16th March 2016 at http://gbxml.org/software.php

Geerts, G. (2011). A design science research methodology and its application to accounting information systems research. *International Journal of Accounting Information Systems, 12*(2), 142–151. doi:10.1016/j.accinf.2011.02.004

Gero, J. S., D'Cruz, N., & Radford, A. D. (1983). Energy in context: A multicriteria model for building design. *Building and Environment, 18*(3), 99–107. doi:10.1016/0360-1323(83)90001-X

Giachetta, A., & Magliocco, A. (2007). *Progettazione sostenibile. Dalla pianificazione territoriale all'ecodesign.* Roma: Carocci.

Giergiczny, M., & Kronenberg, J. (2012). *Jak wycenic wartosc przyrody w miescie? Wycena drzew przyulicznych w centrum Lodzi.* Retrieved from http://www.sendzimir.org.pl/images/zrz_3_pl/04_jak_wycenic_wartosc_przyrody_w_miescie.pdf

Giurcaa, I., Aúchileanb, I., Naghiua, G. S., & Badea, G. (2016). *Selecting the Technical Solutions for Thermal and Energy Rehabilitation and Modernization of Buildings. Procedia Technology, 22*, 789–796.

Goswami, D. Y. (2015). *Principles of Solar Engineering.* Boca Raton, FL: Taylor & Francis Group.

Gumowski, K., Olszewski, O., Pocwierz, M., & Zielonko-Jung, K. (2015). Comparative analysis of numerical and experimental studies of the airflow around the sample of urban development. *Bulletin of the Polish Academy of Sciences. Technical Sciences*, *63*(3), 729–737. doi:10.1515/bpasts-2015-0084

Guzzetta, A., Meyers, G., & Purse, A. (2007). *Types of wind turbines and associated advantages*. Retrieved September 3, 2017, from http://me1065.wikidot.com/types-of-wind-turbines-and-associated-advantages

Hamdy, M., Hasan, A., & Siren, K. (2013). A multi-stage optimization method for cost-optimal and nearly-zero-energy building solutions in line with the EPBD-recast 2010. *Energy and Building*, *56*, 189–203. doi:10.1016/j.enbuild.2012.08.023

Harman, I. N. (2003). *The Energy Balance of Urban Areas* (Unpublished doctoral dissertation). University of Reading, Reading, UK.

Hartmann, T., & Fischer, M. (2007). Applications of BIM and Hurdles for Widespread Adoption of BIM Event Report. AISC-ACCL eConstruction Roundtable, New Orleans, LA.

Heinze, W., & Schreiber, D. (1984). Eine neue Kartierung der Winterhärtezonen für Gehölze in Mitteleuropa. *Mitteilungen der Deutschen Dendrologischen Gesellschaft*, *75*, 11–56.

Hermelink, A. (2006a). A Retrofit for Sustainability: Meeting Occupants' Needs Within Environmental Limits. *Proceedings of the 14th ACEEE Summer Study on Energy Efficiency in Buildings*.

Hermelink, A. (2006b). *Reality Check: The Example. SOLANOVA*.

He, T., Zhang, L., & Deng, W. (2011). Biological adhesion of Parthenocissus tricuspidata. *Archives of Biological Sciences*, *63*(2), 393–398. doi:10.2298/ABS1102393H

High Quality Environment. (2017). Retrieved from www.hqegbc.org/accueil

Hobbs, R. J. (2006). Novel ecosystems: theoretical and management aspects of the new ecological world order. *Global Ecology and Biogeography*, *15*, 1–7. Retrieved from https://www.fs.fed.us/global/iitf/pubs/ja_iitf_2006_hobbs001.pdf

Hobbs, R. J. (2013). *Novel Ecosystems: Intervening in the New Ecological World Order*. John Wiley & Sons, Ltd. doi:10.1002/9781118354186

Hobbs, R. J., Arico, S., Aronson, J., Baron, J. S., Bridgewater, P., Cramer, V. A., & Zobel, M. et al. (2006). Novel ecosystems: Theoretical and management aspects of the new ecological world order. *Global Ecology and Biogeography*, *15*(1), 1–7. doi:10.1111/j.1466-822X.2006.00212.x

Horyński, M. (2015). *Energy-efficient automated energy management systems in residential buildings, Wydwanictwo Politechniki*. Lubelskiej.

Huld, T., & Irene Pinedo-Pascua, I. (2012). *Photovoltaic Solar Electricity Potential in European Countries*. PVGIS ©. *European Union*, 2001–2012. Retrieved from http://re.jrc.ec.europa.eu/pvgis/cmaps/eur.htm

Hundertwasser, F. (1980). *The House Should Not Be Measured by Normal Standards*. Hundertwasser at the press conference with Mayor Gratz, Vienna, Austria. Retrieved from http://www.hundertwasser-haus.info/en/blog/2011/07/19/the-house-should-not-be-measured-by-normal-standards/

Hyun, S., Marjanovic-Halburd, L., & Raslan, R. (2015). Investigation Into Informational Compatibility Of Building Information Modelling And Building Performance Analysis Software Solutions. *WIT Transactions on the Built Environment*, *149*, 543–553. doi:10.2495/BIM150441

Hyvärinen, J., Mäkeläinen, T., Peura, J., & Rönty, J. (2013). Strategies, Guidelines and Project Level Leadership as Methods for IDDS Practices in Transition. *Proceedings of the 19th CIB World Building Congress 2013: Construction and Society*.

Idem, R. (2010). *Uspolecznianie procesu zrownowazonego projektowania architektonicznego*, Gdańsk Wydawnictwo Politechniki Gdańskiej Knerer und Lang Architekten. Retrieved February 3,2016 from http://www.knererlang.de/

IEA. (n.d.a). *Final energy consumption by sector and fuel*. Retrieved September 3, 2017, from https://www.eea.europa.eu/data-and-maps/indicators/final-energy-consumption-by-sector-9/assessment-1

IEA. (n.d.b). *Geothermal energy*. Retrieved September 3, 2017, from http://www.iea.org/topics/renewables/subtopics/geothermal

IEA. (n.d.c). *Unit Converter*. Retrieved September 3, 2017, from http://www.iea.org/statistics/resources/unitconverter/

IEC 61215-1:2016. Terrestrial photovoltaic (PV) modules - Design qualification and type approval - Part 1: Test requirements.

IEE. (2014). *Energy Efficiency in Poland 2013 Review, Single family houses energy efficiency and air quality*. Kraków, Poland: Institute of Environmental Economics.

Integrated Design. (2012-2014). Retrieved from www.integrateddesign.eu

International Organization for Standardization. (2006). *Ergonomics of the thermal environment - Analytical determination and interpretation of thermal comfort using calculation of the PMV and PPD indices and local thermal comfort criteria* (Standard No. 7730). Retrieved from https://www.iso.org/obp/ui/#iso:std:iso:7730:ed-3:v1:en

International Organization of Standardization. (2007). *Indoor environmental input parameters for design and assessment of energy performance of buildings addressing indoor air quality, thermal environment, lighting and acoustics* (ISO Standard No. 15251). Retrieved from www.futurebuilt.no

Ireland. (2012). *Towards nearly zero energy buildings in Ireland, planning for 2020 and beyond*. Department of the Environment, Community and Local Government.

IRENA. (2012). International Renewable Energy Agency. *Renewable Energy Technologies: Cost Analysis Series, 1*(5).

ISO. (2006). *ISO 10077-1:2006 Thermal performance of windows, doors and shutters — Calculation of thermal transmittance — Part 1: General*. Retrieved from https://www.iso.org/standard/40360.html

ISO. (2007a). *ISO 6946:2007 Building components and building elements -- Thermal resistance and thermal transmittance -- Calculation method*. Retrieved from https://www.iso.org/standard/40968.html

ISO. (2007b). *ISO 10211:2007 Thermal bridges in building construction -- Heat flows and surface temperatures -- Detailed calculations*. Retrieved from https://www.iso.org/standard/40967.html

ISO. (2007c). *ISO 12524:2000 Building materials and products -- Hygrothermal properties -- Tabulated design values*. Retrieved from https://www.iso.org/obp/ui/#iso:std:iso:10456:ed-3:v1:en

ISO. (2007d). *ISO 13789:2007 Transmission and ventilation heat transfer coefficients—Calculation method*. Retrieved from https://www.iso.org/standard/40894.html

ISO. (2007e). *ISO 14683:2007 Thermal bridges in building construction -- Linear thermal transmittance -- Simplified methods and default values*. Retrieved from https://www.iso.org/standard/40964.html

ISO. (2008). *ISO 13790:2008 Energy performance of buildings -- Calculation of energy use for space heating and cooling*. Retrieved from https://www.iso.org/standard/41974.html

ISO. (2012). *ISO 13788:2012 Hygrothermal performance of building components and building elements -- Internal surface temperature to avoid critical surface humidity and interstitial condensation -- Calculation methods.* Retrieved from https://www.iso.org/standard/51615.html

ISO. (2015). *ISO 9972:2015 Thermal performance of buildings -- Determination of air permeability of buildings -- Fan pressurization method.* Retrieved from https://www.iso.org/standard/55718.html

ISO. (2017). *ISO 13786:2017 Thermal performance of building components -- Dynamic thermal characteristics -- Calculation methods.* Retrieved from https://www.iso.org/standard/65711.html

ISO/FDIS 7730:2005, International Standard, Ergonomics of the thermal environment — Analytical determination and interpretation of thermal comfort using calculation of the PMV and PPD indices and local thermal comfort criteria.

Italy. (2013). *National plan for increasing the number of "nearly-zero energy buildings" Guidelines and development framework.*

Jackiewicz, B., & Borowski, J. (1998). Rosliny pnace na budowlach zabytkowych. *Ochrona zabytkow, 51*(203), 402-417. Retrieved from http://bazhum.muzhp.pl/media//files/Ochrona_Zabytkow/Ochrona_Zabytkow-r1998-t51-n4_(203)/Ochrona_Zabytkow-r1998-t51-n4_(203)-s402-417/Ochrona_Zabytkow-r1998-t51-n4_(203)-s402-417.pdf

Jagemar, L., Schmidt, M., Allard, F., Heiselberg, P., & Kurnitski, J. (2011). Towards nZEB - Some Examples of National Requirements and Roadmaps. *REHVA Journal, 48*(3), 14–17.

Jaggs, M., & Palmer, J. (2000). Energy performance indoor environmental quality retrofit—a European diagnosis and decision making method for building refurbishment. *Energy and Buildings, 31*(2), 97-101.

Januchta-Szostak A. (2010). Miasto w symbiozie z wodą. *Czasopismo Techniczne nr 6-A/1/2010; zeszyt 14, rok 107.* Politechniki Krakowskiej.

Januchta-Szostak, A. (2011). Metody wielofunkcyjnego zagospodarowania miejskich obszarow zieleni towarzyszacych ciekom. Nadrzeczne parki buforowe. *Przeglad Komunalny, 8/2011*, 58–63.

Jarzemska, M., & Węglarz, A. (2009). *Technical conditions of improving energy efficiency in the construction sector, Department of Civil Engineering.* Warsaw: University of Technology.

Jedraszko-Macukow, M., & Sitarski, M. (2017), *Possibilities for estimating ecosystem services of the Praga Park in Warsaw using the i-tree Eco method.* Presentation 16th May 2017 during a Seminar on "Trees in the city - value, benefits, strategy". Warsaw University, University Research Center for Natural Environment and Sustainable Development, Institute of Spatial Management and Housing, Biological and Chemical Warsaw University Research Centre, Seminar organized by the SITO / NOT Science and Technology Association of Horticultural Engineers and Technicians. (in Polish).

Joint work under Praca zbiorowa pod redakcją Dariusza Koca. (2001). *Guide for individual investors on Energy Efficient heating of single family dwellings.* Krajowa Agencja Poszanowania Energii S.A.

Kadas, G. (2006), Rare Invertebrates Colonizing Green Roofs in London. *Habitats, 4*(1), 66-73. Retrieved from http://www.urbanhabitats.org/v04n01/urbanhabitats_v04n01_pdf.pdf

Kaldellis, J. K. (2010). *Stand-Alone and Hybrid Wind Energy Systems - Technology, Energy Storage and Applications.* Woodhead Publishing. doi:10.1533/9781845699628

Kaliszuk-Wietecka, A. (2017). *Sustainable Building Selected issues of building physics* [Budownictwo Zrównoważone Wybrane zagadnienia fizyki budowli]. PWN Polish Scientific Publishers.

Kaliszuk-Wietecka, A. (2017). *Sustainable Building - Selected issues in building physics*. Wydawnictwo Naukowe PWN SA. (in Polish)

Kang, H. (2015). Development of a decision support tool for nZEB (Nearly zero emission building) at the early design stage. *Journal of Asian Architecture and Building Engineering*, *14*(2), 475–481. doi:10.3130/jaabe.14.475

Kantola, M., & Saari, A. (2016). Identifying and managing risks involved in the transition to the EU nZEB decree. *Facilities*, *34*(5/6), 339–349. doi:10.1108/F-03-2014-0032

Kantola, M., & Saari, A. (2016). Project delivery systems for nZEB projects. *Facilities*, *34*(1-2), 85–100. doi:10.1108/F-03-2014-0025

Karlessi, T., Kampelis, N., Kolokotsa, D., Santamouris, M., Standardi, L., Isidori, D., & Cristalli, C. (2017). The Concept of Smart and NZEB Buildings and the Integrated Design Approach. *Procedia Engineering*, *180*, 1316–1325. doi:10.1016/j.proeng.2017.04.294

Kawon, K., & Zmuda, S. (1977). Rola zieleni w ksztaltowaniu srodowiska czlowieka, regionow przemyslowo-miejskich. Studia nad ekonomika regionu, T8, 181-200.

Kazi, S., & Koivuniemi, A. (2006). *Sharing through Social Interaction: The Case of YIT Construction Ltd. Real-Life Knowledge Management, Lessons from the Field*. KnowledgeBoard in collaboration with VTT – Technical Research Centre of Finland.

Kerosuo, H., Mäki, T., & Korpela, J. (2013). Knotworking - A novel BIM-based collaboration practice in building design projects. *Proceedings of the 5th International Conference on Construction Engineering and Project Management ICCEPM*, 9-11.

Kim, K., Kumar, P., Szulejko, J., Adelodun, A., Junaid, M., Uchimiya, M., & Chambers, S. (2017). Toward a better understanding of the impact of mass transit air pollutants on human health. *Chemosphere*, (174): 268–279. PMID:28178609

Klemm, K. (2011). *Kompleksowa ocena warunkow mikroklimatu w luznych i zwartych strukturach urbanistycznych*. Warszawa: KILiW PAN.

Klimczuk, K. (2017). *Analysis of the cost during the lifetime of a building after a deep energy efficiency modernization* (Thesis). Warsaw University of Technology.

Klucher, T. M. (1979). Evaluation of Models to Predict Insolation on Tilted Surfaces. *Solar Energy*, *23*(2), 111–114. doi:10.1016/0038-092X(79)90110-5

KODnZEB. (2015-2017). *KodnZEB project, supported by Finance Mechanism EOG 2009-2014, within the framework of cooperation Fund Operation Program PL04 Efficient Energy and promotion of alternative energy sources*. Project realized with the cooperation of NTNU Trondheim.

KodnZeB. (2015-2017). *Unpublished grant outcomes and analysis from KodnZEB project, supported by Finance Mechanism EOG 2009-2014, within the framework of cooperation Fund Operation Program PL04 Efficient Energy and promotion of alternative energy sources. Project realized with the cooperation of NTNU Trondheim Ministry of Climate and Environment*. Retrieved from https://www.regjeringen.no/en/topics/climate-and-environment

KodnZeB. (2015-2017). *Unpublished grant outcomes and analysis from KodnZEB project, supported by Finance Mechanism EOG 2009-2014, within the framework of cooperation Fund Operation Program PL04 Efficient Energy and promotion of alternative energy sources. Project realized with the cooperation of NTNU Trondheim*. Author.

KODnZEB. (n.d.). *Program PL04 Saving energy and promoting renewable energy source within the framework of the project KODnZEB Design retro-fit nZEB concept for two buildings*. Grant number 315/2015/Wn07/OA-XN-04/D no. SAP 511/00023. Retrieved September 3, 2017, from http://kodnzeb.pl/

Kohler, M. (2003), Green Roofs as a Contribution to Reduce Urban heat Islands. *RIO 3 - World Climate & Energy Event*, 1-5. Retrieved from http://www.rio12.com/rio3/proceedings/RIO3_493_M_Koehler.pdf

Kohler, M. (1988). Besiedlung von Kletterpflanzen durch Insekten und Spinnen in Berlin. *Zeitschrift für Angewandte Zoologie*, *75*, 195–202.

Kohler, M. (2008). Green facades – a view back and some visions. *Urban Ecosystems*, *11*(4), 423–426. doi:10.1007/s11252-008-0063-x

Kohler, M. (2012). *Handbuch Bauwerks-begrünung, Planung-Konstruktion-Aufführung*. Köln: Müller, Rudolf.

Komerska, A., Kwiatkowski, J., & Rucińska, J. (2015). Integrated Evaluation of CO2eq Emission and Thermal Dynamic Simulation for Different Facade Solutions for a Typical Office Building. Energy Procedia, 78, 3216-3221.

Korsgaard, V. (1949). Necessity of using a directional mean radiant temperature. *Heating, Piping and Air*, 117-120.

Kose, D. A., & Dick, E. (2010). Prediction of the pressure distribution on a cubical building with implicit LES. *Journal of Wind Engineering and Industrial Aerodynamics*, *98*(10-11), 628–649. doi:10.1016/j.jweia.2010.06.004

Kost, Ch., Mayer, J. N., Thomsen, J., Hartmann, N., Senkpiel, Ch., Philipps, S., & Schlegl, T. et al. (2013). *Levelized Cost of Electricity Renewable Energy Technologies*. Fraunhofer Institute For Solar Energy Systems ISE.

Kovar-Panskus, A., Louka, P., Sini, J. F., Savory, E., Czech, M., Abdelqari, A., & Toy, N. et al. (2002). Influence of geometry on the mean flow within urban street canyons - a comparison of wind tunnel experiments and numerical simulation. *Water, Air, and Soil Pollution*, *2*(5-6), 365–380. doi:10.1023/A:1021308022939

Kowarik, I. (2015). *Integration of biodiversity conservation into urban development*. Presentation 19th of March, 2015, McGill University. Retrieved from http://qcbs.ca/wp-content/uploads/2015/04/Kowarik-Montreal-2015.pdf

Koziel, W., & Wlodarczyk, W. (2011). Glony – produkcja biomasy. *Acta Agrophysica 2011, 17*(1), 105-116. Retrieved from http://www.old.actaagrophysica.org/artykuly/acta_agrophysica/ActaAgr_188_2011_17_1_105.pdf

Kronenberg, J. (2016). Uslugi ekosystemow – nowe spojrzenie na wartosc srodowiska przyrodniczego. In *EcoMiasto#Srdowisko. Zrownowazony, inteligentny rozwoj miasta.* Lodz: Wydawnictwo Uniwersytetu Lodzkiego. Retrieved from http://repozytorium.uni.lodz.pl:8080/xmlui/bitstream/handle/11089/17980/4-063_088-Kronenberg.pdf?sequence=1&isAllowed=y

Kumar, A., Baredar, P., & Qureshi, U. (2015). Historical and recent development of photovoltaic thermal (PVT) technologies. *Renewable & Sustainable Energy Reviews*, *42*, 1428–1436. doi:10.1016/j.rser.2014.11.044

Kumbaroğlu, G., & Madlener, R. (2012). Evaluation of economically optimal retrofit investment options for energy savings in build. *Energy and Building*, *49*, 327–334. doi:10.1016/j.enbuild.2012.02.022

Kurnitski, J. (2013). *REHVA nZEB technical definition and system boundaries for nearly zero energy buildings*. Federation of European Heating, Ventilation and Air-Conditioning Associations, Report No 4. Brussels: REHVA.

Kurnitski, J., Allard, F., Braham, D., Goeders, G., Heiselberg, P., Jagemar, L., … Railio, J. (2012). How to define nearly net zero energy buildings nZEB. *REHVA Journal*, 6-12.

Kurnitski, J., Buso, T., Corgnati, S. P., Derjanecz, A., & Litiu, A. (2014). nZEB definitions in Europe. *The REHVA European HVAC Journal, 51*(2), 6–9. Retrieved from http://www.rehva.eu/fileadmin/REHVA_Journal/REHVA_Journal_2014/RJ_issue_2/P.06/06_09_nZEB_definitions_in_Europe_RJ1402_WEB.pdf%0Ahttp://www.rehva.eu/publications-and-resources/rehva-journal/2014/022014/nzeb-definitions-in-europe.html

Kurnitski, J., Feldmann, C., Heiselberg, P., Massarella, L., Sartori, I., Voss, K., & Wahlström, Å. (2013). *Present Energy Performance Requirements and nZEB Targets in Some Selected Countries. In Cost Optimal and Nearly Zero-Energy Buildings (nZEB)* (pp. 31–46). Definitions, Calculation Principles and Case Studies. doi:10.1007/978-1-4471-5610-9_3

Kurnitski, J., Saari, A., Vuolle, M., Niemela, J., & Tark, T. (2011). Cost optimal and nearly zero (nZEB) energy performance implementations. *Energy and Building*, *43*(11), 3279–3288. doi:10.1016/j.enbuild.2011.08.033

Kwiatkowski, J., Rucińska, J., & Panek, A. (2014). Analysis of different shading strategies on energy demand and operating cost of office building. *Proceedings from World Sustainable Building Conference SB14*, 5, 512-521.

Laskowski, L. (1987). *Wybrane zagadnienia fizyki miasta.* Retrieved from http://www.yego.pl/x_C_I__P_63511184__PZTA_01.html

Laskowski, L. (2008). *Thermal protection and energy performance of the building*. Warszawa: OWPF. (in Polish)

LBNL. (N.D.) Therm. Lawrence Berkeley National Laboratory. Therm 7.5. Retrieved September 2, 2017, from windows.lbl.gov/software/therm/therm.html

Lee, S. H., Hong, T., Piette, M. A., & Taylor-Lange, S. C. (2015). Energy retrofit analysis toolkits for commercial buildings: A review. *Energy, 89*, 1087-1100.

LEED Green Building Certification Scheme. (2017). Retrieved from www.usgbc.org/leed

LEED. (2014). *LEED 2009 for Existing Buildings Operations and Maintenance*. USGBC U.S. Green Building Council. Retrieved on 22.08.2017 at: https://www.usgbc.org/sites/default/files/LEED%202009%20RS_EBOM_04.02.2014_current.pdf

Lenardic, D. (2001-2017). *PVRESOURCES*. Retrieved September 3, 2017, from http://www.pvresources.com/en/introduction/introduction.php

Lewinska, J. (2000). *Klimat miasta, zasoby, zagrozenia, ksztaltowanie*. Krakow: Instytut Geografii i Gospodarki Przestrzennej.

Lewinska, J. (2000). *Klimat miasta. Zasoby, zagrozenia, ksztaltowanie*. Krakow: Instytut Gospodarki Przestrzennej i Komunalnej Miasta.

Lindkvist, C., Karlsson, A., Sørnes, K., & Wyckmans, A. (2014). Barriers and Challenges in nZEB Projects in Sweden and Norway. *Energy Procedia, 58*, 199–206. doi:10.1016/j.egypro.2014.10.429

Lithuania. (2012). *National plan for increasing the number of nearly-zero energy buildings*. Ministry of the Environment of the Republic of Lithuania.

Liu, B. Y. H., & Jordan, R. C. (1963). The Long-term Average Performance of Flat-Plate Solar Energy Collectors. *Solar Energy*, *7*(2), 53–74. doi:10.1016/0038-092X(63)90006-9

Louv, R. (2014). *Ostatnie dziecko lasu*. Warszawa: Mamania.

Lovelock, J. E., & Margulis, L. (1979). *Gaia: A New Look at Life on Earth*. Oxford University Press.

Luque, A., & Hegedus, S. (2011). *Handbook of Photovoltaic Science and Engineering*. Hoboken, NJ: John Wiley & Sons.

Luxembourg. (2013). *National Plan for Luxembourg to increase the number of nearly zero-energy buildings.* Luxembourg: Le Gouvernement du Grand-Duche de Luxembourg, Ministere de l'Economie et du Commerce exterieur.

Mäkeläinen, T. et al. (2016). *Deliverable 2.2a Visionary scenarios of holistic energy efficiency design.* Design4Energy project.

Mäkeläinen, T. et al. (2017). *Deliverable 2.3: Guidelines for optimal building design: Holistic design of buildings matching with neighbourhood energy systems.* Design4Energy project.

Maksymiuk, G. (2013). Wplyw zielonej infrastruktury na wartosc ekonomiczna nieruchomosci. Problemy ekologii krajobrazu, 36, 145-153.

Maleki, S. A. M., Hizam, H., & Gomes, Ch. (2017). Estimation of Hourly. Daily and Monthly Global Solar Radiation on Inclined Surfaces: Models Re-Visited. *Energies*, *10*(1), 1–28.

Mańkowski, S., & Szczechowiak, E. (2012). *Strategic research project entitled "Integrated system for reducing operating energy consumption in buildings" Research Task No. 2 Volume I, Part A: Conditions of transformations in construction.* Academic Press.

Manson, R. (1977). Biological Architecture: A Partnership of Nature and Man. *The Futurist, 3.*

Marchwiński, J., & Zielonko-Jung, K. (2012). *Wspolczesna architektura proekologiczna.* Warszawa: Wydawnictwo Naukowe PWN.

Marshal of the Sejm of the Republic of Poland. (2008). Act on the promotion of energy retrofit and renovations. *Journal of Laws item 130.* (in Polish)

Marshal's Office of the Pomeranian Voivodship. (2015). *Detailed Description of the Priority Axes of the Regional Operational Program of the Pomeranian Voivodship for years 2014-2020.* Author. (in Polish)

Ma, Z., Cooper, P., Daly, D., & Ledo, L. (2012). Existing building retrofits: Methodology and state-of-the-art. *Energy and Building*, *55*(0), 889–902. doi:10.1016/j.enbuild.2012.08.018

Mazzali, U., Peron, F., & Scarpa, M. (2012). Thermo-physical performances of living walls via field measurements and numerical analysis. Eco-architecture IV. Harmonisation between architecture and nature WIT transactions on ecology and the environment, 165, 239-50. doi:10.2495/ARC120231

McAllister, S., Chen, J.-Y., & Fernandez-Pello, A. C. (2011). *Fundamentals of Combustion Processes.* Cham: Springer International Publishing AG. doi:10.1007/978-1-4419-7943-8

McEvoy, A., Markvart, T., & Castañer, L. (2012). *Practical Handbook of Photovoltaics - Fundamentals and Applications.* Amsterdam: Elsevier.

McMahon, E.T. (2000). Green Infrastructure. *Planning Commissioners Journal, 37.*

Memon, Z. A. (2008). The ACPROM Model: An Expert System for Evaluating the Construction Progress. *Proceedings of First International Conference on Construction in Developing Countries (ICCIDC–I).*

Meteo AHRAE. (2001). *Meteorological data for EnergyPlus, Copywrite: American Society of Heating, Refrigerating and Air-Conditioning Engineers.* Atlanta, GA: ASHRAE.

Miettinen, R., & Paavola, S. (2014). Beyond the BIM utopia: Approaches to the development and implementation of building information modeling. *Automation in Construction*, *43*, 84–91. doi:10.1016/j.autcon.2014.03.009

Mignone, G., Hosseini, M. R., Chileshe, N., & Arashpour, M. (2016). Enhancing collaboration in BIM-based construction networks through organisational discontinuity theory: A case study of the new Royal Adelaide Hospital. *Architectural Engineering and Design Management, 12*(5), 333–352. doi:10.1080/17452007.2016.1169987

Mijakowski, M., Narowski, P., & Sowa, J. (2015). Definicja budynku nZEB na potrzeby projektu KODnZEB. Warszawa: Academic Press.

Mijakowski, M., Rucińska, J., Sowa, J., & Narowski, P. (2017). *The concept of improving the internal environment in a utility building modernized to the nZEB standard* [Koncepcja poprawy środowiska wewnętrznego w przykładowym budynku użyteczności publicznej modernizowanym do standardu nZEB]. *Building Physics in Theory and Practice.* Retrieved from www.powerhouse.no/prosjekter/kjorbo/

Millennium Ecosystem Assessment. Ecosystems and Human Well-Being: Synthesis. (2005). Washington, DC: Island Press.

Miller, C., Thomas, D., Irigoyen, S. D., Hersberger, C., Nagy, Z., Rossi, D., & Schlueter, A. (2014). BIM-extracted Energy-Plus model calibration for retrofit analysis of a historically listed building in Switzerland. Proceedings of SimBuild, 2014.

Mills, G. (2012). Luke Howard, Tim Oke and the Study of Urban Climate. *American Meteorogical Society Journal.* Retrieved October 12, 2012, from www.ams.confex.com/ams/pdfpapers/144519.pdf

Ministry of Infrastructure and Construction. (n.d.). *Energy efficiency of buildings.* Retrieved from http://mib.gov.pl/2-567ab849c4abe.htm (in Polish)

Ministry of Infrastructure and Development. (2014a). *Programming the Financial Perspective 2014-2020 - Partnership Agreement.* Author. (in Polish)

Ministry of Infrastructure and Development. (2014b). *Financing the energy renovation of buildings with Cohesion Policy funding.* Author. (in Polish)

Ministry of Infrastructure and Development. (2015). *Ordinance of the Minister of Infrastructure and Development of 27 February 2015 on the methodology of determining a building's or part of a building's energy characteristic and energy performance certificates* (Journal of Laws of 2015, item 376). Author. (in Polish).

Ministry of Infrastructure. (2002). *Ordinance on technical conditions to be met by buildings and their location of 12 April 2002* (Journal of Laws of 2002, No. 75, item 690, consolidated text). Author. (in Polish)

Ministry of Infrastructure. (2015). *Act 91 of the Council of Ministers of 22 June 2015 on accepting the "National plan aimed at increasing the number of buildings with low energy use".* Author. (in Polish)

Ministry of Transport, Construction and Maritime Economy. (2012). *Ordinance of 25 April 2012 on the detailed scope and form of construction design* (Journal of Laws of 2012, item 462). Author. (in Polish)

Mirsadegh, M., Cóstola, D., Blocken, B., & Hensen, J. (2013). Review of external convective heat transfer coefficient models in building energy simulation programs: Implementation and uncertainty. *Applied Thermal Engineering, 56*(1-2), 134–151. doi:10.1016/j.applthermaleng.2013.03.003

Morelli, M., Harrestrup, M., & Svendsen, S. (2014). Method for a component-based economic optimisation in design of whole building renovation versus demolishing and rebuilding. *Energy Policy, 65*, 305–314. doi:10.1016/j.enpol.2013.09.068

Mulawka, J. (1997). Expert systems [in Polish]. Warsaw: Scientific-Technical Publishing House.

Mumford, L. (1961). *The City in History.* San Diego, CA: Harcourt Inc.

Mumford, L. (1967-70). *The Myth of the Machine.* Harcourt Brace Jovanovich.

Myczkowski, S. (1976). *Człowiek – przyroda - cywilizacja*. Warszawa: PWN.

Narowski, P., Mijakowski, M., Panek, A. D., Rucińska, J., & Sowa, J. (2010). Integrated 6R1C Energy Simulation Method – Principles, Verification and Application. *Proceedings CLIMA 2010, 11th REHVA World Congress "Sustainable Energy Use in Buildings"*.

Nordic Innovation. (2015). *30 Sustainable Nordic Buildings. Best practice examples based on the Charter principles*. Nordic Innovation.

NREL. (2017). *Best Research-Cell Efficiencies*. National Renewable Energy Laboratory. Retrieved September 3, 2017, from https://www.nrel.gov/pv/assets/images/efficiency-chart.png

Ó'Riain, M., & Harrison, , J. (2016). Cost-optimal passive versus active nZEB. How cost-optimal calculations for retrofit may change nZEB best practice in Ireland. *Architectural Science Review*, *59*(5), 358–369. doi:10.1080/00038628.2016.1184130

Ojanen, T., Nykänen, E., & Hemmilä, K. (2017). Guidebook for Structural Energy Efficiency in Renovation [Rakenteellinen energiatehokkuus korjausrakentamisessa – Opas]. Ministry of the Environment, the Confederation of Finnish Construction Industries RT and the Federation of Finnish Woodworking Industries.

Orzeszek-Gajewska, B. (1982). *Kształtowanie terenow zieleni w miastach*. Warszawa: PWN.

Othman, A. R., & Sahidin, N. (2016). Vertical Greening Façade as Passive Approach in Sustainable Design. *Procedia: Social and Behavioral Sciences*, *222*, 845–854. doi:10.1016/j.sbspro.2016.05.185

Paiho, S., Abdurafikov, R., Hoang, H., & Kuusisto, J. (2015). An analysis of different business models for energy efficient renovation of residential districts in Russian cold regions. *Sustainable Cities and Society*, *14*, 31–42. doi:10.1016/j.scs.2014.07.008

Pando, A.R., Capestro, I., Colarusso, A., Solari, M., Saba, A., Morando, M., Reccardo, D., Bogacz, A., Raco, F., Uriarte, A., Tellado, B. (2014). *Technologies and strategies for Public Building and District Retrofitting*. Project Funded by the European Community in the 7th Framework Programme, A2PBEER GA nº: 609060 D2.2 Technologies and strategies for Public Building and District Retrofitting.

Pasquero, C., & Poletto, M. (2011). *AlgaeFarm project*. Retrieved from http://www.ecologicstudio.com/v2/project.php?idcat=3&idsubcat=59&idproj=110

Pasquero, C., & Poletto, M. (2014). *Algae Canopy – ecoMachines project for Milano*. Retrieved from http://www.ecologicstudio.com/v2/project.php?idcat=3&idsubcat=59&idproj=137

Pasquero, C., & Poletto, M. (2015a). *Urban Algae Folly project and realization for Braga*. Retrieved from http://www.ecologicstudio.com/v2/project.php?idcat=3&idsubcat=71&idproj=148

Pasquero, C., & Poletto, M. (2015b). *Urban Algae Folly, project and realization for Milano*. Retrieved from http://www.ecologicstudio.com/v2/project.php?idcat=3&idsubcat=71&idproj=147

Peck, S., & Kuhn, M. (2003). *Design guidelines for green roofs*. Ontario Association of Architects, Canada Mortgage and Housing Corporation. Retrieved from https://www.eugene-or.gov/DocumentView.aspx?DID=1049

Perez, R., Ineichen, P., Seals, R., Michalsky, J., & Stewart, R. (1990). Modelling Daylight Availability and Irradiance Components from Direct and Global Irradiance. *Solar Energy*, *44*(2), 99–108. doi:10.1016/0038-092X(90)90071-J

Perini, K., Ottele, M., Fraaij, A. L. A., Haas, E. M., & Raiteri, R. (2011). Vertical greening systems and the effect on air flow and temperature on the building envelope. *Building and Environment*, *46*(11), 2287–2294. doi:10.1016/j.buildenv.2011.05.009

Perini, K., & Rosasco, P. (2013). Cost-benefit analysis for green façades and living wall systems. *Building and Environment*, *70*, 110–121. doi:10.1016/j.buildenv.2013.08.012

Peterson, K., Torcellini, P., & Grant, R. (2015). *A Common Definition for Zero Energy Buildings*. Prepared for the U.S. Department of Energy by The National Institute of Building Sciences, US Department of Energy.

Petzet, M. (2012). *Reduce, Reuse, Recycle: Rethink Architecture. German Pavilion*. Ostfildern: Hatje Cantz.

PHI. (2010). *Quality-Approved Energy Retrofit with PH Components - Criteria for residential-use refurbished buildings*. Darmstadt: Passive House Institute.

Philips, A. (2013). *Designing Urban Agriculture: A Complete Guide to the Planning, Design, Construction, Maintenance and Management of Edible Landscapes*. Wiley.

Pierzchalska, D. & Węglarz, A. (2016). *Praktyczna realizacja idei głębokiej termomodernizacji.* Materiały Budowlane 1/2016 str. 23-25.

Pikas, E., Kurnitski, J., Liias, R., & Thalfeldt, M. (2015). Quantification of economic benefits of renovation of apartment buildings as a basis for cost optimal 2030 energy efficiency strategies. *Energy and Building*, *86*, 151–160. doi:10.1016/j.enbuild.2014.10.004

Poradnik TEEB dla miast: uslugi ekosystemow w gospodarce miejskiej, (2011). Krakow: Fundacja Sedzimira. Retrieved from http://uslugiekosystemow.pl/wp content/uploads/2017/07/teeb_poradnik_dla_miast_pl.pdf

Portugal. (2013). *National plan for increasing the number of nearly zero-energy buildings in Portugal.* Direcao Geral de Energia e Geologia.

Power.Technology.com. (n.d.). *The 10 biggest hydroelectric power plants in the world.* Retrieved September 3, 2017, from http://www.power-technology.com/features/feature-the-10-biggest-hydroelectric-power-plants-in-the-world/

Ratti, C., Baker, N., & Steemers, K. (2005). Energy Consumption and Urban Texture. *Energy and Building*, *37*(7), 762–776. doi:10.1016/j.enbuild.2004.10.010

Ratti, C., Di Sabatino, S., & Britter, R. (2006). Urban Texture Analysis with Image Processing Techniques: Winds and Dispersion. *Theoretical and Applied Climatology*, *84*(1-3), 77–90. doi:10.1007/s00704-005-0146-z

Rehak, D. R., & Fenves, S. J. (1984). *Expert systems in civil engineering, construction and construction robotics.* Retrieved from http://citeseerx.ist.psu.edu/viewdoc/download?doi=10.1.1.823.781&rep=rep1&type=pdf

Reindl, D. T., Beckman, W. A., & Duffie, J. A. (1990a). Diffuse Fraction Correlations. *Solar Energy*, *45*(1), 1–7. doi:10.1016/0038-092X(90)90060-P

Reindl, D. T., Beckman, W. A., & Duffie, J. A. (1990b). Evaluation of Hourly Tilted Surface Radiation Models. *Solar Energy*, *45*(1), 9–17. doi:10.1016/0038-092X(90)90061-G

Riva, G., Foppapedretti, E., Carolis, C., Giakoumelos, E., Malamatenios, C., Signanini, P., Giancarlo, C., Di Fazio, M., Gajdoš, J., Ručinský, R. (2012). *Handbook on Renewable Energy Sources*. Retrieved from http://docplayer.net/16926980-Handbook-on-renewable-energy-sources.html

RMI. (2002). Regulation of the Minister of Infrastructure of 12 April 2002 on technical specifications for buildings and their location. *Journal of Laws 2015*, Item 1422, consolidated text (in Polish).

RMI. (2008). Regulation of the Minister of Infrastructure of 6 November 2008 On the methodology for calculating the energy performance of the building and dwelling or part of the building and the technical-utilitarian independent parts and the preparation and design of energy performance certificates. *Journal of Laws 2014*, Item 888 (in Polish).

Roberts, J. (2005). *Redux. Designs that reuse, recycle and reveal.* Layton, Gibbs Smith.

Rośczak, P. (2003). *Sztuczna inteligencja*. Retrieved from http://www.rosczak.com/mlp/ai.html

Rozanski, S. (1959). *Budowa miasta i jego klimat*. Warszawa: Arkady.

Rubik, M. (2006). *Heat pumps. Guide*. Ośrodek Informacji, Technika instalacyjna w budownictwie. (in Polish)

Rutgers, R. (2012). *Living façades. A study on the sustainable features of vegetated façade cladding. Masterthesis about the sustainable benefits of plants on a façade*. Faculty of Architecture Delft University of Technology. Retrieved December 15, 2016, from https://issuu.com/rrutgers/docs/living_facades

Rynska, E. D. (2008). Rehabilitation and adaptive reuse of historic buildings in Poland. In *Harmonisation between Architecture and Nature. Eco-Architecture II*. WIT Press.

Rynska, E.D., Kozminska, U., & Oniszk-Popławska, A., Szubert, Klinowska, D, & Tofiluk, A. (2017). Sustainable interdisciplinary transformation of Warsaw University of Technology Building. KodnZeb case study. *International Journal of Sustainable Development and Planning*, *12*(4).

Saini, V., & Mhaske, S. (2013). BIM an emerging technology in AEC industry for time optimization. *International Journal of Structural and Civil Engineering Research*, *2*(4), 196–200.

Salvesen, F. (2015). *Powerhouse Kjobro-Renovation to plus energy standard*. Asplen Viak AS, Company presentation for KODnZEB project.

Sartori, I., Merlet, S., Thorud, B., Haug, T., & Andresen, I. (2016). *Zero Villager Bergen. Aggregated loads and PV generation profiles*. ZEB Project report 28-2016. SINTEF Academic Press and Norwegian University of Science and Technology.

Sartori, I., Santamouris, M., Cipriano, X., & Lopez, J. S. (2014). *How to promote "nearly" Zero-Energy Buildings for new and retrofitting buildings in the municipal context*. World SB14 Barcelona, Spain.

Sartori, I., & Hestnes, A. G. (2007, March). Energy use in the life cycle of conventional and low-energy buildings: A review artcile. *Energy and Building*, *39*(3), 249–257. doi:10.1016/j.enbuild.2006.07.001

Schmidt M. (2015). *Making the city compact more livable and resilient thanks to the greening*. Presentation 19th May 2015 on Hotel de ville de Paris. Retrieved from https://api-site.paris.fr/images/72760

Schultz, P. W., Shriver, C., Tabanico, J. J., & Khazian, A. M. (2004). Implicit connections with nature. *Journal of Environmental Psychology*, *24*(1), 31–42. doi:10.1016/S0272-4944(03)00022-7

Sherratt, A. F. C. (1987). *Applications of heat pumps to buildings*. London: Hutchinson.

Short, W., Packey, D. J., & Holt, T. (1995). *A Manual for the Economic Evaluation of Energy Efficiency and Renewable Energy Technologies*. Retrieved from https://www.nrel.gov/docs/legosti/old/5173.pdf

Singh, B., Roy, P., Spiess, T., & Venkatesh, B. (2015). *Sustainable Integrated Urban & Energy Planning, the Evolving Electrical Grid and Urban Energy Transition. The Centre for Urban Energy*. Ryerson University.

Slovak. (2013). *National plan for increasing the number of nearly zero-energy buildings.* Academic Press.

Slovenia. (2013). *National plan for increasing the number of nearly zero-energy buildings in accordance with Article 9 of the Recast EPBD.* Republika Slovenija, Ministrstvo za Infrastrukturo in Prostor.

Sobel, D. (1996). *Beyond Ecophobia: Reclaiming the Heart of Nature Education.* Great Barrington, MA: The Orion Society.

SOLANOVA. (2003-2006). *Project Solar-supported, integrated eco-efficient renovation of large residential buildings and heat-supply-systems, "Eco-buildings" project of the European Commission in Eastern Europe dealing with a "major renovation" of a large existing building.* Retrieved from www.solanova.eu

Solarek, K., Rynska, D., & Mirecka, M. (2016). *Urbanistyka i architektura w zintegrowanym gospodarowaniu wodami.* Warszawa: Oficyna Wydawnicza Politechniki Warszawskiej.

Solon, J. (2008). Koncepcja "Ecosystem Services" i jej zastosowania w badaniach ekologiczno-krajobrazowych. *Problemy Ekologii Krajobrazu, 21*, 25–44.

Sørensen, A. L., Andresen, I., Walnum, H. T., Justo-Alonso, M., Fufa, S. M., Jenssen, B., & Fjeldheim, H. et al. (2017). *Pilot Building Powerhouse Kjørbo. As Built Report.* The Research Centre on Zero Emission Buildings. NTNU.

Sowa, J. (Ed.). (2017). Budynki o niemal zerowym zużyciu energii. Oficyna Wydawnicza PW.

Sowa, J. (Ed.). (2017). Nearly zero energy buildings [Budynki o niemal zerowym zużyciu energii]. Oficyna Wydawnicza PW.

Sowa, J., Noga-Zygmunt, J., & Ugorowska, J. (2017). *Assessment of indoor air parameters in a university building before the expected modernization to the nearly zero energy standard* [Ocena parametrów powietrza wewnętrznego w budynku wyższej uczelni przed przewidywaną modernizacją do standardu niemal zero energetycznego]. *Building Physics in Theory and Practice.*

Sowa, J. (Ed.). (2017). *Budynki o niemal zerowym zużyciu energii.* Oficyna Wydawnicza, PW.

Sowa, J. (Ed.). (2017). *Budynki o niemal zerowym zuzyciu energii.* Warszawa: Oficyna Wydawnicza PW.

Staab Architekten. (2007-2011). *Darmstadt University of Applied Science. Overhaul and Extension of the C10 Highrise.* Retrieved February 3, 2016 from http://www.staab-architekten.com/

Steel, J., Drogemuller, R., & Toth, B. (2012). Model interoperability in building information modelling. *Software & Systems Modeling, 11*(1), 99–109. doi:10.1007/s10270-010-0178-4

Succar, B. (2009). Building information modelling framework: A research and delivery foundation for industry stakeholders. *Automation in Construction, 18*(3), 357–375. doi:10.1016/j.autcon.2008.10.003

Sullivan, W., & Kuo, E. (1996). Do trees strengthen urban communities reduce domestic violence? *Arborist News, 5*, 33–34.

Sweden. (2012). *Towards nearly zero-energy buildings.* Communication 2011/12:131. Ministry of Industry, Employment and Communications.

Szajda-Binfeld, E., Pływaczyk, A., & Skarżyński, D. (2012). *Zielone dachy. Zrownowazona gospodarka wodna na terenach zurbanizowanych.* Wroclaw: Uniwersytetu Przyrodniczego we Wroclawiu.

Szczepanowska, H. B. (2007). Ekologiczne, spoleczne i ekonomiczne korzysci z drzew na terenach zurbanizowanych. *Czlowiek i Srodowisko, 31*(3-4), 5–26. Retrieved from https://www.igpim.pl/publikacje/str07_3-4/Szczepanowska.pdf

Szczepanowska, H. B. (2015). Drzewa w miescie - zielony kapitał wartosci i usług ekosystemowych. *Czlowiek i Srodowisko, 39*(2), 5–28. Retrieved from https://czlowiekisrodowisko.igpim.pl/wp-content/uploads/2016/06/01-Szczepanowska.pdf

Szczepanowska, H. B., Sitarski, M., & Suchocka, M. (2014). *Funkcjonowanie drzew i krzewow w warunkach oddzialywania infrastruktury technicznej miasta.* Artykul powstal na podstawie referatu, wygloszonego podczas konferencji: MIASTO IDEALNE – MIASTO ZROWNOWAZONE Planowanie przestrzenne terenow zurbanizowanych i jego wpływ na ograniczenie skutkow zmian klimatu Warszawa, 24 pazdziernika 2014 r. Retrieved from http://ucbs.uw.edu.pl/wp-content/uploads/SitarskiSzczepanowska-Funkcjonowanie-POPRAW.pdf

Szczepanowska, H. B. (1984). *Wspoldzialanie czynnikow siedliskowych i ich wpływ na stan i rozwoj roslinnosci na terenach zieleni, Wpływ zieleni na ksztaltowanie srodowiska miejskiego red. Szczepanowskiej H.B.* Warszawa: PWN.

Szczepanowska, H. B. (2001). *Drzewa w miescie.* Warszawa: Hortpress.

Szczurek, M., & Zych, M. (2012). *Dzielo - Dzialka.* Krakow: muzeum Etnograficzne w Krakowie.

Szumacher, I. (2005). Funkcje ekologiczne parkow miejskich. *Prace i Studia Geograficzne, 46.*

Szumacher, I. (2011). Funkcje terenow zieleni miejskiej a swiadczenia ekosystemow. *Prace i Studia Geograficzne, 49.*

Tadeusiewicz, R. (2012). Computer Decision Support. *Proceedings of XVI Conference of Automation Rytro (CD).*

Taylor, J., & Levitt, R. (2004). Understanding and managing systemic innovation in project-based industries. In D. Slevin, D. Cleland, & J. Pinto (Eds.), *Innovations: Project Management Research* (pp. 83–99). Newton Square, PA: Project Management Institute.

The National Centre for Emissions Management (KOBiZE). (2016). *Fuel Values (WO) and CO_2 Emissions (EC) in 2014 for reporting under the Emission Trading Scheme for 2017.* Author. (in Polish)

The National Centre for Emissions Management (KOBiZE). (2017). *CO_2, SO_2, NO_x, CO and TSP emission factors for electricity based on information in the National Greenhouse Gas and Gaseous Emissions Database for 2015.* Author. (in Polish)

The Research Centre on Zero Emission Buildings. (2017). Retrieved from www.zeb.no

Throndsen, W., Berker, T., & Knoll, E. B. (2015). *Powerhouse Kjørbo. Evaluation of construction process and early use phase.* SINTEF Academic Press and Norwegian University of Science and Technology.

Tiba, S., & Omri, A. (2017). Literature survey on the relationships between energy, environment and economic growth. *Renewable & Sustainable Energy Reviews, 69*, 1129–1146. doi:10.1016/j.rser.2016.09.113

Todd, J. A., & Gail, L. (2013). Planning and Conducting Integrated Design (ID) Charrettes. Whole Building Design Guide. National Institute of Building Sciences.

Ungers, O. M. (1981). *A Vertical Garden. Design for the Facade of a Department Store* (Vol. 31). Lotus.

Urban Gardening – Back to the Future. (2013). Cieszyn: Zamek Cieszyn. Retrieved from http://www.zamekcieszyn.pl/baza_w_files/1368089604-59katalog_urban-net.pdf

Uribe, O., Santos, M., Garcia-Alegre, M., & Guinea, D. (2015). A context-awareness architecture for managing thermal energy in an nZEB building. *2015 IEEE 1st International Smart Cities Conference.* Institute of Electrical and Electronics Engineers Inc.

Valuing London's Urban Forest Results of the London i-Tree Eco Project. (2015). Retrieved from https://www.forestry.gov.uk/pdf/LONDONI-TREEECOREPORT151202.pdf/$FILE/LONDONI-TREEECOREPORT151202.pdf

Vilches, A., Garcia-Martinez, A., & Sanchez-Montañes, B. (2017). Life cycle assessment (LCA) of building refurbishment: A literature review. *Energy and Building, 135*, 286–301. doi:10.1016/j.enbuild.2016.11.042

Wagner, A. (2015). *Architektura Politechniki Warszawskiej.* Warszawa: Oficyna Wydawnicza Politechniki Warszawskiej.

Walawender, J. P. (2015). *Wplyw dachow zielonych na warunki klimatyczne w miescie.* Portal Zielona Architektura. Retrieved from http://zielonainfrastruktura.pl/wplyw-dachow-zielonych-na-warunki-klimatyczne-w-miescie/

Warszawa, Centralny Osrodek Informacji Budownictwa. (n.d.). Retrieved from http://www.yego.pl/x_C_I__P_63511184__PZTA_01.html

WDD. (n.d.). *The Worlds of David Darlig.* Retrieved September 3, 2017, from http://www.daviddarling.info/encyclopedia/S/AE_solar_collector.html

Wells, N., & Evans, G. (2003). Nearby Nature: A Buffer of Life Stress Among Rural Children. *Environment and Behavior, 35*(3), 311–330. doi:10.1177/0013916503035003001

Wikipedia. (2017a). *Concentrated Solar Power.* Retrieved September 3, 2017, from https://en.wikipedia.org/wiki/Concentrated_solar_power

Wikipedia. (2017b). *Multi-Junction Solar Cell.* Retrieved September 3, 2017, from https://en.wikipedia.org/wiki/Multi-junction_solar_cell

Wikipedia. (2017c). *List of Solar Thermal Power Stations.* Retrieved September 3, 2017, from https://en.wikipedia.org/wiki/List_of_solar_thermal_power_stations

Wikipedia. (2017d). *List of Cities by Sunshine Duration.* Retrieved September 3, 2017, from https://en.wikipedia.org/wiki/List_of_cities_by_sunshine_duration

Wikipedia. (2017e). *List of Photovoltaic Power Stations.* Retrieved September 3, 2017, from https://en.wikipedia.org/wiki/List_of_photovoltaic_power_stations

Wikipedia. (2017f). *List of Onshore Wind Farms.* Retrieved September 3, 2017, from https://en.wikipedia.org/wiki/List_of_onshore_wind_farms

Wikipedia. (2017g) *List of offshore wind farms.* Retrieved September 3, 2017, from https://en.wikipedia.org/wiki/List_of_offshore_wind_farms

Wikipedia. (n.d.). *Köppen climate classification.* Retrieved September 2, 2017, from https://en.wikipedia.org/wiki/K%C3%B6ppen_climate_classification

WindPower. (n.d.). *Wind turbine power output variation with steady wind speed.* Retrieved September 3, 2017, from http://www.wind-power-program.com/turbine_characteristics.htm

Wines, W. (2016). Green Architecture. *Encyclopaedia Britannica online.* Retrieved from http://www.britannica.com/art/green-architecture

Włoch, M., & Brzeziński, P. (2010). *Warsaw heating network system, and selected European systems.* Academic Press. (in Polish)

Wohlleben, P. (2016). *Sekretne rzycie drzew, wyd.* Kraków: Otwarte.

Wojcicka, I. (1971). *Uciazliwosci klimatu miast i mozliwosci jego poprawy za pomoca roslinnosci.* Warszawa: Inst. Urb. I Arch.

Wolski, P. (2013). Znaczenie okien hygrologicznych. *Problemy Ekologii Krajobrazu, 36,* 129-144. Retrieved from http://agro.icm.edu.pl/agro/element/bwmeta1.element.agro-aee7a572-f4b5-4206-a2a8-2a5153ef6114

WSUEEP. (2009). *Solar Electric System Design, Operation and Installation. An Overview for Builders in the Pacific Northwest.* Retrieved September 3, 2017, from www.energy.wsu.edu

Wytyczne do projektowania, wykonania i pielegnacji dachow zielonych – wytyczne dla dachow zielonych. (2015). Opole: DAFA.

Yeang, K. (2008). *Eco Design. A Manual for Ecological Design.* Chichester, UK: John Wiley & Sons.

Yoshie, R., Mochida, A., Tominga, Y., Kataoka, H., Harimoto, K., Nozu, T., & Shirosawa, T. (2007). Cooperative project for CFD prediction of pedestrian wind environment in the Architectural Institute of Japan. *Journal of Wind Engineering and Industrial Aerodynamics, 95*(9-11), 1551–1578. doi:10.1016/j.jweia.2007.02.023

Yudelson, J. (2009). *Green Building through integrated design. A GreenSource Book.* New York: The McGraw-Hill Companies, Inc.

Zaborowski, M. (2013). *Definition of deep energy retrofits, Working Draft No. 2.* Institute of Environmental Economics. (in Polish)

ZEBRA 2020. (2015). *Nearly Zero-Energy Building Strategy 2020.* Author.

ZEBRA 2020. (2016). *Nearly zero-energy building strategy 2020 - Strategies for a nearly Zero-Energy Building market transition in the European Union.* Co-funded by the Intelligent Energy Europe Programme of the European Union. Retrieved from www.zebra2020.eu

Zeiler, W. (2010). Integral design of NZEB a perquisite for 0 impact buildings. *Proceedings SB10 Euregion: Towards 0-impact buildings and environments.* Academic Press.

Zielonko-Jung, K. (2013). *Kształtowanie przestrzenne architektury ekologicznej w strukturze miasta.* Academic Press.

Zielonko-Jung, K. (2014). *Kompas Lewis & Clark Authentic Models.* Retrieved from http://www.yego.pl/x_C_I__P_63511184__PZTA_01.html

Zielonko-Jung, K. (2013). *Ksztaltowanie przestrzenne architektury ekologicznej w strukturze miasta.* Warszawa: Oficyna Wydawnicza Politechniki Warszawskiej.

Ziembicki, P., & Węglarz, A. (2013). *Expert system algorithm for choosing an energy source for a building.* Ciepłownictwo, Ciepłownictwo, Ogrzewanie, Ciepłownictwo, Ogrzewanie, Wentylacja, 5/2013.

Zimmerman, A. (2006). *Integrated design process guide.* Ottawa, Canada: Canada Mortgage And Housing Corporation.

Zimmermann, A. (2011). *Constructing landscape: materials, techniques, structural components.* Boston: Birkhauser.

Zimny, H. (2005). *Ekologia miasta.* Warszawa: Agencja Reklamowo-Wydawnicza Arkadiusz Grzegorczyk.

Zimny, H., & Wysocki, Cz. (1984). *Produktywnosc trawnikow na terenach zieleni miejskiej Warszawy. In Wplyw zieleni na ksztaltowanie srodowiska miejskiego* (pp. 163–180). Warszawa: PWN.

Zinowiec-Cieplik, K., Guranowska-Gruszecka, K., & Dankiewicz, B. (2015). *Polish Green Wall -research program.* Innotech-K1/I1/40/159571/NCBR/12 - poster for 1st European Urban Green Infrastructure Conference 2015, Vienna, Austria.

Żurawski, J. (2012). *Zero- or almost zero-energy construction in Polish conditions.* Conference "Insulation". (in Polish)

Żurawski, J. (n.d.). Energy efficiency of buildings in Poland present state and development perspectives. [in Polish]. *Lower Silesia Agency for Energy and Environment.*

About the Contributors

Elzbieta D. Ryńska, Prof. MSc. PhD. Arch - since - 2010 position of a Professor, Faculty of Architecture, Warsaw University of Technology, Poland, TEACHING ACTIVITIES 2013-15 - lecturer at a course Civil Engineering in Sustainable Development, Faculty of Civil Engineering, Warsaw University of Technology, design studio – 1st semester Master level, lectures: New trends in Architecture - Harmonious Architecture, Economy of Ecologic Projects. 2010 – one of the initiators and coordinator of a 2 semester interdisciplinary course Architecture and Urban Planning in Sustainable Development, Master Level studies, at Faculty of Architecture Warsaw University of Technology, lectures: New trends in Architecture - Harmonious Architecture, New trends in Architecture – Synergic Design, Economy of Ecologic Projects, design studios on 2nd and 3rd semester MSc level. Diplomas on BSc and MSc levels, Warsaw University of Technology 2008 - lecturer in following subjects: Economy of Building Investment Process, Management in Building Projects, EU and Polish legislation in Building Projects, design studios: 4th,5th, 6th and 7th Semester BSc level, 1st, 2nd, 3rd semester MSc level. Diplomas on BSc and MSc levels, Faculty of Architecture, Warsaw University of Technology 2004-2008 lecturer in following subjects: Economy of Building Investment Process, Management in Building Projects, design studios: 4th,5th, 6th and 7th Semester BSc level, Faculty of Architecture, Warsaw University of Technology. SUPERVISION OF GRADUATE STUDENTS AND POSTDOCTORAL FELLOWS 2006 – 2014 2 Postdocs – FA WUT, supervision of 9 PhD dissertations at FA WUT, 14 Master Diplomas of Architecture at Faculty of Architecture WUT, Poland 2012-2014 Supervision of 1 PhD dissertation as a joint diploma with NTNU, Trondheim 2012-2014 Supervision of 1 PhD dissertation as a joint diploma with Warsaw University of Life Sciences (SGGW), Faculty of Construction and Environment 2000-2014 6 Bachelor Engineer Diplomas specialty: Energy Efficient Buildings at University of Ecology and Management in Warsaw, Poland 2000-2014 30 Bachelor Engineer Diplomas in Architecture, 8 Master Students of Architecture - at University of Ecology and Management in Warsaw, Poland. INTERNATIONAL GRANTS Participation in several international grant, the latter - KodnZeB (2015-2017) – unpublished grant outcomes and analysis from KodnZEB project, supported by Finance Mechanism EOG 2009-2014, within the framework of cooperation Fund Operation Program PL04 Efficient Energy and promotion of alternative energy sources. Project realized with the cooperation of NTNU Trondheim. BOOKS and SCIENTIFIC JOURNALS: Author and co-author of 9 books and ca 40 reports on Sustainability in Architecture in Urban Planning, these are available in English and Polish.

Urszula Koźmińska, PhD.Eng. Arch, is an architect from Warsaw. She studied architecture and design in Warsaw and Madrid and she has been working in architectural offices in Warsaw and Amsterdam. Currently, she has achieved her PhD degree in Architecture in July 2017 at the Warsaw University of Technology with a specialization in sustainable design and the reuse of building materials in architecture and urban planning.

Kinga Zinowiec-Cieplik, PhD in agriculture sciences, Eng. landscape architect, paysagiste CESP (ENSP- Versailles). For many years associated with Departement of Landscape Art on University of Life Sciences - SGGW. Since 2012 - Assistant Professor in the Department of Architecture and Urban Design - on Faculty of Architecture of Warsaw University of Technology. In addition to teaching, conducts professional practice, author of many projects of Warsaw public spaces f.ex. co-author (with Małgorzata Milecka) Park Znicza, (with Piotr Hardecki) Park at Balaton area, (with Katarzyna Jakubowska) Park Henrykowski.

Joanna Rucińska, PhD., Eng. - is currently working as a lecturer at the Faculty of Building Services, Hydro and Environmental Engineering at Warsaw University of Technology (WUT). She in cooperates with the National Energy Conservation Agency (NAPE SA) and Energy Conservation Foundation (FPE). She graduated master's and doctoral studies at the Faculty of Environmental Engineering of Warsaw University of Technology. Both during her basic and postgraduate studies she was involved in matters related to energy, energy efficiency of buildings, energy systems and sources. Education and completed professional training provide a solid basis for performing simulations and analyses in the field of energy, pollution emissions to the environment and economic analyses. In her work, Joanna Rucińska is also involved in the use of renewable energy sources in buildings and their impact on the environment.

***

Jacek Borowski, PhD. assistant professor - Lecturer in dendrology at Faculty of Horticulture, Biotechnology and Landscape Architecture, Warsaw University of Life Sciences - SGGW. In his scientific activity he deals with widely understood dendrology. Author and co-author of numerous scientific and popular publications, including those concerning urban greenery and climbing plants. He developed with co-workers "Photographic method of tree increment evaluation" as well as "Standards of the Greening of Warsaw". President of the Polish Dendrological Society.

Vanda Dimitriou (F) (MEng, MSc, PhD) is a scientist working in the field of energy performance of buildings. Areas of interest include low carbon building design and modelling and data analytics. Developed methodologies and tools to enable the data transfer process from BIM to Building Energy modelling and support energy retrofit decision making for various stakeholders. Participated and led tasks in projects DESIGN4ENERGY and REFIT.

Szymon Firląg, PhD., Eng. was working on building efficiency-related projects for Polish companies and foundations, such as: the National Energy Conservation Agency, the Polish National Energy Conservation Agency and the Energy Conservation Foundation. He was involved in the process of designing, constructing and certification of the first passive buildings in Poland. Currently, he is a lecturer at the Faculty of Civil Engineering, Warsaw University of Technology (WUT) in the team of sustainable development. Currently He is representing BPIE in Poland. Szymon Firląg graduated from WUT with the PhD title. His thesis was on the topic of energy demand and thermal comfort in passive buildings. He conducted his research at the Passivhaus Institut, Darmstadt where he had the chance to work as the Nowicki Foundation and as a DBU fellow. Thanks to a scholarship from the Center of Advanced Studies WUT he conducted research on dynamic windows control at the Lawrence Berkeley National Laboratory. He authored numerous papers and books and took part in campaigns prompting low energy buildings in Poland. Szymon Firląg is interested in various aspects of energy efficiency in buildings, including both nZEB and deep renovation.

Mari Hukkalainen (F) (M. Sc. (Tech), is a Senior Scientist working in the field of Smart cities and Eco efficient districts. She has been developing concepts and innovative solutions for energy systems of smart and sustainable urban areas, including energy efficient buildings, renewable energy production and distribution solutions. She has been analysing and comparing sustainable urban energy solutions and developing their impact assessment methods, and how they can be integrated and utilised in sustainable urban planning to support efficient decision making and end-users' involvement and engagement. She has participated to and led many national and international research and development projects, such as DESIGN4ENERGY, E-HUB, IREEN, and READY4SmartCities. She is also doing a PhD study about low carbon planning of energy systems of urban areas.

Hanna Jędrzejuk, D.Sc., Ph.D., M.Sc., Eng., since 2011 is employed at Warsaw University of Technology, Faculty of Power and Aeronautical Engineering, Institute of Heat Engineering, Division of Refrigeration and Energy in Buildings. She is responsible for the courses: Engineering Physics, Ventilation and Heating, Building Construction, Energy performance of buildings. The main areas of interest are the following: building physics, HVAC systems, control systems in buildings, renewable energy and application of multi-criteria optimisation in engineering. She is a member of Scientific Committees: Conference of Building Physics in Theory and Practice (Łódź, Poland), Architecture - Building - Engineering and Environmental Protection (Solina, Poland). She is an author or co-author of more the 10 scientific articles during the last five years (more information at: http://www.itc.pw.edu.pl/Pracownicy/Naukowo-dydaktyczni/Jedrzejuk-Hanna).

Joanna Klimowicz, PhD. Eng. Arch., education at the Faculty of Architecture Warsaw University of Technology. Doctors degree received in 1999, the topic of doctor's thesis „The influence of new real estate building organization for creation the new human habitation". Since 2000 have been working as a teacher at the Faculty of Architecture WUT. In 2013 published the book "Cultural and social determinants of development in different climate zones. On selected examples of Syrian cities in the context of Polish conditions. ". Published over then 20 different articles, participated in many different scientific conferences (for example ECAADE 2002, „Young scientists towards the challenges of modern technology"). Participated in over 10 scientific exchanges (University of Detroit Mercy, Teschreen University, Private University of Science and Arts in Aleppo). Promoter of the different designs of bachelor degree at the Faculty of Architecture WUT.

Krzysztof Klobut has a Master degree, M.Sc.(Tech.) 1981, from Technical University of Wroclaw (TUW), Poland, and Licentiate, Lic.Sc.(Tech.) 1994, first doctoral degree from Helsinki University of Technology (HUT), Finland. He holds a Senior Scientist position at VTT Technical Research Centre of

Finland Ltd. Klobut has 30+ years of scientific experience in the field of indoor environment quality, heating and ventilation systems, energy consumption in buildings, RES, and district heating. Activities in theoretical modeling, computer simulation, experimental research and energy monitoring. Investigated thermal performance of heating boilers, ovens, heat exchangers, fuel cells and district heating substations. Author or co-author in 100+ publications. Leading and contributing in domestic and international projects aiming at improving energy efficiency in buildings, districts and on a building stock scale.

Michał Pierzchalski, MSc. Eng. Arch. - Director of Energy-Efficient Building Design Laboratory at the Polish National Energy Conservation Agency (KAPE). PhD student at the Faculty of Architecture, Warsaw University of Technology. Co-author and chief editor of the Architektura.info website. Member of the Association of Polish Architects (SARP), The Chamber of Architects and the Polish Green Building Council (PLGBC).

Marta Poćwierz graduated from Faculty of Power and Aeronautical Engineering at Warsaw University of Technology in 1993 (MSc. Eng.) and 2005 (PhD). She works at Faculty of Power and Aeronautical Engineering as a senior lecturer. She participated in projects related to the spread of dust and gas in the turbulent atmosphere. For several years she had been dealing with the vortex methods for three dimensional flows. In 2008 she attended the Lecture series on “Atmospheric Boundary Layer Flows in Air Pollution Modeling” which took place at the Von Karman Institute in Brussels. From 2011 she has cooperated with faculty of Architecture (WUT) and conducts research in the field of airflow around buildings and in particular issues of ventilation space built. She directs many students’ projects from these topics.

Arkadiusz Węglarz, PhD. Eng., is Chief Advisor for Low Emission Economy in The Polish National Energy Conservation Agency, lecturer in The Department of Civil Engineering at Warsaw University of Technology, vice-president of board of Association of Polish Auditors. Co-author of several scientific publications, technical elaborations and press articles about building subjects and energy Author of several expert reports for: Ministry of Economy, Ministry of Finance, Ministry of Environmental Protection, Chief Inspectorate for Environmental Protection, Polish Energy Grids (PSE), Ministry of Transport and Construction, National Fund for Environmental Protection and Water Management. Project manager and team member in projects realized in the framework of programs: Phare, Save, 5th Framework Program, 6th Framework Program, Intelligent Energy Europe and bilateral programs. Awarded with the Gold Cross of Merit by the President of the Republic of Poland for achievements in the field of energy efficiency in building sector.

Katarzyna Zielonko-Jung, an architect, received her MSc in Architecture and Urbanism from Gdańsk University of Technology in 1996 and PhD in Architecture from Warsaw University of Technology in 2003. In 2016 she was recommended by a fully qualified faculty board and granted by the Central Commission for Academic Titles and Degrees. She works as a research and teaching assistant professor in the Chair of Structural Design, Construction and Technical Infrastructure at the Faculty of Architecture of Warsaw University of Technology. She writes and presents widely on issues of ecological architecture and sustainable urban environment. From academic year 2017/18 – she has moved back to Gdańsk University of Technology, where is following her sustainable interests.

Index

G

H

I

K

L

M

N

P

R

S

T

U

V

W

Z

www.ingramcontent.com/pod-product-compliance
Lightning Source LLC
LaVergne TN
LVHW060302011026
849161LV00011B/2758
9781522541059